Advances in Tangible and Embodied Interaction for Virtual and Augmented Reality

Advances in Tangible and Embodied Interaction for Virtual and Augmented Reality

Editors

Jorge C. S. Cardoso
André Perrotta
Paula Alexandra Silva
Pedro Martins

MDPI • Basel • Beijing • Wuhan • Barcelona • Belgrade • Manchester • Tokyo • Cluj • Tianjin

Editors

Jorge C. S. Cardoso
Department of Informatics
Engineering
University of Coimbra
Coimbra
Portugal

André Perrotta
Department of Informatics
Engineering
University of Coimbra
Coimbra
Portugal

Paula Alexandra Silva
Department of Informatics
Engineering
University of Coimbra
Coimbra
Portugal

Pedro Martins
Department of Architecture
University of Coimbra
Coimbra
Portugal

Editorial Office
MDPI
St. Alban-Anlage 66
4052 Basel, Switzerland

This is a reprint of articles from the Special Issue published online in the open access journal *Electronics* (ISSN 2079-9292) (available at: www.mdpi.com/journal/electronics/special_issues/VAR_electronics).

For citation purposes, cite each article independently as indicated on the article page online and as indicated below:

LastName, A.A.; LastName, B.B.; LastName, C.C. Article Title. *Journal Name* **Year**, *Volume Number*, Page Range.

ISBN 978-3-0365-7755-5 (Hbk)
ISBN 978-3-0365-7754-8 (PDF)

Contents

electronics

Editorial

Advances in Tangible and Embodied Interaction for Virtual and Augmented Reality

Jorge C. S. Cardoso [1,*,†], André Perrotta [1,†], Paula Alexandra Silva [1,†] and Pedro Martins [2,†]

1 Department of Informatics Engineering, Centre for Informatics and Systems of the University of Coimbra, University of Coimbra, 3030-790 Coimbra, Portugal; avperrotta@dei.uc.pt (A.P.); paulasilva@dei.uc.pt (P.A.S.)
2 Department of Architecture, Faculty of Sciences and Technology, University of Coimbra, 3030-790 Coimbra, Portugal; pmcarvalho@uc.pt
* Correspondence: jorgecardoso@dei.uc.pt
† These authors contributed equally to this work.

Virtual Reality (VR) and Augmented Reality (AR) technologies have the potential to revolutionise the way we interact with digital content. VR and AR technologies have seen tremendous progress in recent years, enabling novel and exciting ways to interact within virtual environments. An interesting approach to interaction in VR and AR is the use of tangible user interfaces, which leverage our innate understanding of the physical world, how our bodies move and interact with it, and our learned capabilities to manipulate physical objects. This Special Issue explores recent advances in the field of tangible and embodied interactions for virtual and augmented reality, as showcased by the following papers.

- "Design and Implementation of Two Immersive Audio and Video Communication Systems Based on Virtual Reality" [1]: This paper presents two immersive communication systems that combine VR with audio and video to create a more natural and engaging communication experience. The researchers describe the technical details of the systems, including the hardware and software used, and provide a thorough analysis of the results of user testing. They also discuss the potential applications of the systems, such as in remote conferencing and virtual collaboration.

- "An Interactive Augmented Reality Graph Visualization for Chinese Painters" [2]: This paper describes an interactive AR system that allows Chinese painters to explore and visualise complex graphs in an intuitive and immersive way. The authors discuss the design considerations that went into creating the system, such as the use of colour and motion to convey information, and provide detailed examples of how the system can be used in practice.

- "Situating Learning in AR Fantasy, Design Considerations for AR Game-Based Learning for Children" [3]: This paper discusses the use of AR in educational games for children and presents design considerations for creating effective and engaging learning experiences. The authors provide an overview of the current state of the field, including the benefits and challenges of using AR in education, and present a set of best practices for designing AR games for children. They also describe several case studies of AR games that have been successfully implemented in real-world educational settings.

- "Development of a Virtual Object Weight Recognition Algorithm Based on Pseudo-Haptics and the Development of Immersion Evaluation Technology" [4]: This paper presents a virtual weight recognition algorithm that uses pseudo-haptics to create a more realistic and immersive experience. The authors describe the technical details of the algorithm, including the software and hardware used, and provide a thorough analysis of the results of user testing.

- "A 3D Image Registration Method for Laparoscopic Liver Surgery Navigation" [5]: This paper proposes a new method for 3D image registration in laparoscopic liver surgery navigation. The authors introduce a hybrid registration method that combines feature-based and intensity-based registration to improve the accuracy and robustness of the registration process. The method was tested on real patient data and showed promising results. Overall, the paper presents a new approach to improve the precision of laparoscopic AR navigation in minimally invasive abdominal surgery.

- "Gaze-Based Interaction Intention Recognition in Virtual Reality" [6]: This paper explores the use of gaze-based interaction in VR and how it can be used to recognise user intentions. The paper discusses the potential of this technology to unlock intuitive new interaction schemes and proposes a classification model for recognising user intentions based on gaze data. The authors conducted experiments to test the accuracy of their model and found promising results. They also discuss potential future research directions for this technology.
- "Personalized Virtual Reality Environments for Intervention with People with Disability" [7]: This paper discusses the use of personalised VR environments to provide rehabilitation and intervention for people with disabilities. The authors provide an overview of the current state of the field, including the benefits and challenges of using VR in rehabilitation, and describe several case studies of successful interventions using VR. They also discuss the potential implications of the technology for the field of disability services.
- "Preoperative Virtual Reality Surgical Rehearsal of Renal Access during Percutaneous Nephrolithotomy: A Pilot Study" [8]: This paper proposes a preliminary study of PCNL surgical rehearsal using the Marion Surgical PCNL simulator, where preoperative CT scans of a patient are used to create a 3D model of the renal system. An experienced surgeon planned and practised the procedure in the simulator before performing the surgery in the operating room. Preliminary results suggest that surgical rehearsal using a combination of VR and haptic feedback strongly affects decision making during the procedure.
- "Digital Taste in Mulsemedia Augmented Reality: Perspective on Developments and Challenges" [9]: This article reviews how AR can be used to stimulate and modulate the sensation of taste in humans using low-amplitude electrical signals. The article explores techniques from prominent research pools and proposes extensions to the already established technological architecture for taste stimulation and modulation. The goal is to integrate gustatory augmentation into the commercial market and create a viable multichannel medium for the transfer of sensory information. The article highlights benefits and limitations and proposes the use of modern technological extensions, including the Internet of Things, artificial intelligence, and machine learning.
- "Visual Positioning System Based on 6D Object Pose Estimation Using Mobile Web" [10]: The article presents a new method of detecting 3D objects from a single image taken by a smartphone camera in indoor spaces and calculating the location of the smartphone to find users in those spaces. The proposed indoor visual positioning system for mobile devices is inexpensive, as it integrates deep learning and computer vision algorithms and does not require additional infrastructure. The method uses convolutional neural networks (CNNs) and real-time pose estimation to handle the entire 6D pose estimate and determine the location and direction of the camera. The estimated position is addressed to a voxel to determine a stable user position. The proposed voxel-addressed optimisation approach with camera 6D position estimation using RGB images outperforms current state-of-the-art methods using RGB depth or point cloud, and provides users with indoor information in a 3D AR model.
- "Effects of Using Vibrotactile Feedback on Sound Localization by Deaf and Hard-of-Hearing People in Virtual Environments" [11]: This paper proposes a haptic VR suit that helps Deaf and Hard-of-Hearing (DHH) individuals complete sound-related VR tasks efficiently. The VR suit receives sound information wirelessly and indicates the direction of the sound source using vibrotactile feedback. The study suggests that using different setups of the VR suit can significantly improve VR task completion times. The results of mounting haptic devices on different positions on users' bodies

indicate that DHH users can complete a VR task significantly faster when two vibro-motors are mounted on their arms and ears compared to their thighs. In an additional study, it was found that there was no significant difference in task completion time when using four vibro-motors with the VR suit compared to using only two vibro-motors in users' ears without the VR suit.

- "Virtual/Augmented Reality for Rehabilitation Applications Using Electromyography as Control/Biofeedback: Systematic Literature Review" [12]: The article is a systematic literature review that explores whether there is a standardised protocol towards therapeutic applications of surface electromyography (sEMG) signals in VR and AR interfaces. The review found 40 relevant articles that focused on applications, such as neurological motor rehabilitation and prosthesis training, and processing algorithms such as artificial intelligence and direct control. The hardware used includes Myo Armband, Delsys, and proprietary equipment, and the VR/AR interfaces are training scene models, video games, and first-person views. The review concludes that there is no consensus regarding signal processing or classification criteria and proposes that future work should aim to standardise these technologies for adoption in clinical practice.

- "Virtual Reality for Safe Testing and Development in Collaborative Robotics: Challenges and Perspectives" [13]: This paper explores the use of extended reality (XR), specifically VR, to test and develop collaboration between humans and collaborative robots (cobots). The use of XR simulations allows for evaluating collaboration without putting humans at risk, making it useful for dangerous scenarios. XR also enables combining human behavioural data, subjective self-reports, and biosignals to measure human comfort, stress, and cognitive load during collaboration. The paper suggests that XR has the potential to change the way cobots are designed, tested, and trained in a range of applications, from industry to healthcare and space operations.

These papers demonstrate the diverse range of applications and approaches to tangible and embodied interactions in virtual and augmented reality and highlight the potential of these technologies to create more natural and engaging user experiences. As the field continues to evolve, we can expect to see even more innovative and exciting developments in this area.

Funding: This work is funded by national funds through the FCT-Foundation for Science and Technology, I.P., within the scope of the project CISUC-UID/CEC/00326/2020 and by European Social Fund, through the Regional Operational Program Centro 2020.

Acknowledgments: The editors of this Special Issue would like to thank all the authors for their contributions and all the reviewers for their careful and timely reviews that helped improve the quality of this Special Issue.

Conflicts of Interest: The authors declare no conflict of interest.

References

1. Zhang, H.; Wang, J.; Li, Z.; Li, J. Design and Implementation of Two Immersive Audio and Video Communication Systems Based on Virtual Reality. *Electronics* **2023**, *12*, 1134. [CrossRef]
2. Li, J.; Wang, Z. An Interactive Augmented Reality Graph Visualization for Chinese Painters. *Electronics* **2022**, *11*, 2367. [CrossRef]
3. Zuo, T.; Jiang, J.; der Spek, E.V.; Birk, M.; Hu, J. Situating Learning in AR Fantasy, Design Considerations for AR Game-Based Learning for Children. *Electronics* **2022**, *11*, 2331. [CrossRef]
4. Son, E.; Song, H.; Nam, S.; Kim, Y. Development of a Virtual Object Weight Recognition Algorithm Based on Pseudo-Haptics and the Development of Immersion Evaluation Technology. *Electronics* **2022**, *11*, 2274. [CrossRef]
5. Li, D.; Wang, M. A 3D Image Registration Method for Laparoscopic Liver Surgery Navigation. *Electronics* **2022**, *11*, 1670. [CrossRef]
6. Chen, X.L.; Hou, W.J. Gaze-Based Interaction Intention Recognition in Virtual Reality. *Electronics* **2022**, *11*, 1647. [CrossRef]
7. Lagos Rodríguez, M.; García, Á.G.; Loureiro, J.P.; García, T.P. Personalized Virtual Reality Environments for Intervention with People with Disability. *Electronics* **2022**, *11*, 1586. [CrossRef]
8. Sainsbury, B.; Wilz, O.; Ren, J.; Green, M.; Fergie, M.; Rossa, C. Preoperative Virtual Reality Surgical Rehearsal of Renal Access during Percutaneous Nephrolithotomy: A Pilot Study. *Electronics* **2022**, *11*, 1562. [CrossRef]

9. Duggal, A.S.; Singh, R.; Gehlot, A.; Rashid, M.; Alshamrani, S.S.; AlGhamdi, A.S. Digital Taste in Mulsemedia Augmented Reality: Perspective on Developments and Challenges. *Electronics* **2022**, *11*, 1315. [CrossRef]
10. Kim, J.Y.; Kim, I.S.; Yun, D.Y.; Jung, T.W.; Kwon, S.C.; Jung, K.D. Visual Positioning System Based on 6D Object Pose Estimation Using Mobile Web. *Electronics* **2022**, *11*, 865. [CrossRef]
11. Mirzaei, M.; Kán, P.; Kaufmann, H. Effects of Using Vibrotactile Feedback on Sound Localization by Deaf and Hard-of-Hearing People in Virtual Environments. *Electronics* **2021**, *10*, 2794. [CrossRef]
12. Toledo-Peral, C.L.; Vega-Martínez, G.; Mercado-Gutiérrez, J.A.; Rodríguez-Reyes, G.; Vera-Hernández, A.; Leija-Salas, L.; Gutiérrez-Martínez, J. Virtual/Augmented Reality for Rehabilitation Applications Using Electromyography as Control/Biofeedback: Systematic Literature Review. *Electronics* **2022**, *11*, 2271. [CrossRef]
13. i Badia, S.B.; Silva, P.A.; Branco, D.; Pinto, A.; Carvalho, C.; Menezes, P.; Almeida, J.; Pilacinski, A. Virtual Reality for Safe Testing and Development in Collaborative Robotics: Challenges and Perspectives. *Electronics* **2022**, *11*, 1726. [CrossRef]

MDPI

Article

Design and Implementation of Two Immersive Audio and Video Communication Systems Based on Virtual Reality

Hanqi Zhang [1], Jing Wang [1,*], Zhuoran Li [2,*] and Jingxin Li [3]

1   School of Information and Electronics, Beijing Institute of Technology, Beijing 100811, China
2   Aerospace Information Research Institute, Chinese Academy of Sciences, Beijing 100045, China
3   China Electronics Standardization Institute, Beijing 101102, China
*   Correspondence: wangjing@bit.edu.cn (J.W.); lizhuoran@aircas.ac.cn (Z.L.)

Abstract: Due to the impact of the COVID-19 pandemic in recent years, remote communication has become increasingly common, which has also spawned many online solutions. Compared with an in-person scenario, the feeling of immersion and participation is lacking in these solutions, and the effect is thus not ideal. In this study, we focus on two typical virtual reality (VR) application scenarios with immersive audio and video experience: VR conferencing and panoramic live broadcast. We begin by introducing the core principles of traditional video conferencing, followed by the existing research results of VR conferencing along with the similarities, differences, pros, and cons of each solution. Then, we outline our view about what elements a virtual conferencing room should have. After that, a simple implementation scheme for VR conferencing is provided. Regarding panoramic video, we introduce the steps to produce and transmit a panoramic live broadcast and analyze several current mainstream encoding optimization schemes. By comparing traditional video streams, the various development bottlenecks of panoramic live broadcast are identified and summarized. A simple implementation of a panoramic live broadcast is presented in this paper. To conclude, the main points are illustrated along with the possible future directions of the two systems. The simple implementation of two immersive systems provides a research and application reference for VR audio and video transmission, which can guide subsequent relevant research studies.

Keywords: virtual reality; immersive communication; VR conference room; panoramic live broadcast; spatial audio

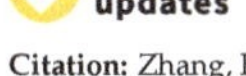
check for updates

Citation: Zhang, H.; Wang, J.; Li, Z.; Li, J. Design and Implementation of Two Immersive Audio and Video Communication Systems Based on Virtual Reality. *Electronics* **2023**, *12*, 1134. https://doi.org/10.3390/electronics12051134

Academic Editors: Jorge C. S. Cardoso, André Perrotta, Paula Alexandra Silva and Pedro Martins

Received: 14 January 2023
Revised: 23 February 2023
Accepted: 24 February 2023
Published: 26 February 2023

1. Introduction

Virtual reality (VR) technology has gone through numerous periods since it was first proposed [1,2]. Currently, applications of virtual reality can be seen in many fields, such as entertainment, education, medical care [3], real estate, etc. [4–6]. All of these benefit from its ability to provide a sense of immersion. The promotion of the concept "Metaverse" has ushered in another peak. The future application scenarios of VR include an end-to-end immersive communication experience combining virtual reality software and hardware, computer graphics and images, immersive communication, audio and video multimedia, and other technologies to ensure that users are able to experience being "immersed" [7]. Long-distance travel usually takes much time and vigor [8]. Many activities are dependent on people attending in person, such as conferences, large-scale competitions, lectures, etc. Today, there are many complete video conference applications or live broadcast applications (e.g., Skype and FaceTime) through which users can participate in remote activities via computers or mobile phones. These existing technologies or solutions make it possible for remote access to information from the perspective of results, but they are far from ideal from the perspective of the immersive experience effect due to a large gap with a real scenario.

Many online conferencing solutions already existed prior to the COVID-19 pandemic, such as multiperson video chat software, but they were far less popular than today. At that time, people were still willing to spend hours traveling to attend meetings that were sometimes less than an hour. The public is also willing to spend days buying tickets to performances or large-scale sports competitions. The reason is that a screen cannot provide the real feeling of immersion. Traditional online video conferences are widespread, and their enabling mechanisms are mostly based on traditional audio and video streams, which allow the main functions of the conference to be realized: video communication, voice communication, and presentation (such as slides). However, the gap between a video conference and an in-person conference is always the fact that is most criticized. This is because communication between people is not merely the simple exchange of information, and there are many other deeper behavioral exchanges that are important information in communication, such as body language, eye contact, gaze perception, etc. [9]. To achieve this sense of reality and immersion, virtual reality technology and equipment are needed. In this paper, we investigate the virtual reality conference room solutions, introduce the research state of the art, and propose a simple virtual conference room prototype that we have implemented.

Compared with traditional video, panoramic video can enhance immersion and bring an unprecedented visual experience. At present, the usage of panoramic video in various fields has been studied [10–12], and it is likely to be widely and deeply applied in the foreseeable future [13]. The panoramic video stream brings new video presentation forms to the on-demand and live broadcast fields. Scenes can be experienced more immersively, with a free choice of the viewing angle. At present, panoramic live broadcast technology has been applied to large-scale performances and event scenes. Thus, the audience can not only feel the atmosphere of the scene but also experience the perspective of the athletes in first person. At present, research on panoramic video streaming is focused on the video [14,15], audio [16,17], and quality evaluation [15]. The main focus is on the video, which involves acquiring the video, projection splicing, coding transmission, and equipment presentation. In this paper, we outline the current state of research in addition to presenting a simple panoramic live broadcast system designed and implemented by us.

1.1. Methodology and Contributions

In this article, a literature review investigation on the current research on VR conferencing and panoramic live broadcast is presented. By analyzing the related works, we determine the features and obstacles of VR conferencing. The procedure and key technologies of panoramic live broadcasting are also presented as well. With the use of available tools, the simple implementation of two immersive systems is given, aiming to provide a research and application reference for VR audio and video transmission, which can guide subsequent relevant research studies.

1.2. Article Structure

The paper is organized as follows:

- Section 2: overview on background of VR conferencing and panoramic live broadcast.
- Section 3: features and obstacles of virtual conference designing.
- Section 4: procedure and key technologies of panoramic live broadcasting.
- Section 5: essential role of 3D audio in the immersive experience.
- Section 6: methods of realizing the two typical immersive scenarios.
- Section 7: the main points of the article are summarized.

2. Related Work

Typical scene applications of virtual reality audio and video communication technology include VR conferencing and panoramic live broadcast. Panoramic live broadcast was developed as early as 2016, which is also considered the "first year of VR". Now, 5G, VR, and 8K have come into the public eye. With the rapid development of VR software and hardware technology, panoramic live broadcast has become increasingly widely used in various industries. In recent years, due to the impact of COVID-19, researchers and the industry have begun to pay a greater attention to the development, design, promotion, and application of VR conferencing to improve the experience of remote collaboration. A VR conferencing room is one such typical application scenario. In addition, the international standards organizations Moving Picture Experts Group (MPEG) and 3rd Generation Partnership Project (3GPP), as well as the Audio and Video Coding Standards Workgroup of China (AVS), have been carrying out standardization research and development work related to immersive media, involving transmission protocols, audio and video coding and decoding, immersive communication systems, and other aspects, especially in the VR field, where the goal is to achieve a six-dof (degree of freedom) immersive experience. The research state of the art of VR conferencing and panoramic live broadcast in the academic research field is discussed next.

2.1. VR Conferencing

The purpose of video conference is to facilitate the possibility of instant communication between participants who are not physically together, through using network and multimedia technology, for participants to appear to be physically together. The traditional approach is to transmit the video and audio data [18,19] of each participant in their different spaces to all other participants through a multipoint control unit (MCU). At present, the technology for video conference has become increasingly flawless and has been widely used in the current COVID-19 pandemic. However, as video conference cannot provide a sense of presence, an immersive experience cannot be generated, opening the way for the introduction of virtual reality conference [20]. So far, the research and application of VR conference rooms is still in the start-up stage of development and exploration, with no widely accepted systematic theory.

In [21], a virtual reality conference room based on the traditional MCU architecture was proposed, in which participants could be placed in a virtual space for communication. A three-dimensional image in a virtual environment was rendered by processing the two-dimensional video information of the participants. This method could display the real image of people in a virtual space, as well as the body information in real time. However, the participants in the images still wore a head display, which may reduce the effectiveness of communication. A novel network video conference system based on MR display devices and AI segmentation technology was proposed in [22]. This was a robust, real-time video conference software application that made up for the simple interaction and lack of immersion and realism of traditional video conference, allowing users to interact in a conference in a new way. However, similar to [21] mentioned above, the participants in the images wore a head display. The presentation of real-time images of human subjects lacked realism if the observer's position was much different from the camera.

In [23], a more complex virtual reality conference system was proposed, which involved the generation of avatars and animation, especially facial expression animation. The authors provided three feasible means of multiperson cooperation: VR painting, slide display, and model import. There was also the general design of the network module. To finish, the experience of wearing the head display device was evaluated, and the conclusion was that the experience of wearing the head display device was better. A novel video transmission scheme of a virtual space conference was proposed in [24], which was based on perceptual control. In the system, the perception of each participant was calculated, with the construction of a perceptual space matrix based on this, wherein the matrix parameters were used to control the codec, effectively encoding the video object and thereby reducing the total bandwidth. The same approach was applied to audio objects.

In [25], the methods for evaluating the authenticity of immersive virtual reality, augmented reality (AR), and hybrid reality were summarized. The conclusion was drawn following the analysis of a large number of relevant articles. Most methods consisted of a combination of objective and subjective measures. The most commonly used assessment tools were questionnaires, many of which were customized and unverified. The existence questionnaire was the most commonly used one , which was usually used to evaluate the authenticity and participation of one's existence and perception.

The research on VR conferencing is still in its early stages. Ref. [26] discussed the features and obstacles found within virtual conference solutions (both 2D and 3D) through a systematic literature review investigation. As a result, 67 key features and 74 obstacles users experience when interacting with virtual conferencing technologies were identified. The current VR conferencing solutions mainly focus on the visual aspect, and researchers focus on the representation of avatars in the virtual space. Under the limited equipment conditions, it is difficult to achieve a realistic avatar that makes people feel good. The exploration of sound and interaction is relatively small. Based on the above papers, the current research on VR conference can be divided into the following aspects: the presentation of characters, the interaction form in the virtual conference room, transmission optimization, and the evaluation of the conference experience effect.

2.2. Panoramic Live Broadcast

Panoramic video is also called 360-degree video, and it can facilitate more immersive feelings in people than traditional video. Audiences can observe from different directions at the same position in the video scene and feel all the information around the scene. Since the video picture contains all the information around the scene, in order to ensure that the resolution of pictures from all angles is not lower than that of traditional video, there must be more data for panoramic video, and the demand for bandwidth is therefore larger [27].

The acquisition and production of panorama requires multiple fisheye lenses for picture acquisition, and the pictures of each lens are spliced to form a frame picture. The panoramic video image acquired synchronously is then presented on a spherical surface, but its data form is not suitable for storage, transmission, compression, or other forms of processing by classical methods. Thus, projection is required [14,28]. The projection of panoramic video is the process of mapping the three-dimensional spherical information to a two-dimensional plane. Several projection methods used in the development of the Joint Video Exploration Team (JVET) coding standards are illustrated in [14]. As the main carrier of panoramic video, the projected image contains all the contents of the captured picture, and the distortion should be reduced as much as possible during splicing to improve the quality of picture playback. At the same time, [14] studied the impact of different projection methods on coding efficiency according to existing experiments.

The resolution of a panoramic video is very high. A larger network bandwidth is therefore needed while also using traditional video compression methods. In order to reduce the pressure of the network bandwidth, it is necessary to optimize the coding method for projected pictures. The panoramic live broadcast system can be divided into three types according to different encoding optimization methods [15]: full-view video stream, viewport-based video stream, and tile-based video stream. The full-view video stream transmits the whole frame, using the same encoding transmission mode as the traditional video stream. This approach requires a high processor performance and network bandwidth. Viewport-based streaming provides a high-quality transmission for the part viewed by the user based on the position of the user's current viewpoint, while the transmission quality of the rest is low. On the client side, the streaming endpoint device detects the user's head movement and receives only the specific required video frame area, dynamically selecting the video stream region of the viewport and adjusting the viewport quality. In this way, the bitrate of the video stream can be reduced. The server stores multiple adaptive sets related to a user's direction and performs matching and viewport position prediction according to the network state. Ref. [29] proposed two dynamic viewport selection approaches, which adapted the streamed regions based on content complexity variations and positional information to ensure viewport availability and smooth visual angles for VR users. Tile-based coding technology is used to divide each frame in the video stream into multiple blocks. According to the location of the viewport, the resolution between different blocks will be different. For example, ref. [28] proposed a video streaming system that used the divide-and-conquer method to separate the video space into multiple blocks and encapsulate them during encoding. Ref. [30] proposed a sight-guidance scheme based on tile-based coding aiming at minimizing the weighted sum of the average traffic load and users' watching preference derivation. MPEG-DASH's (Dynamic Adaptive Streaming over HTTP) SRD (spatial relationship description) [31] is used to describe the relationship between blocks in the 360-degree space and is tiled in the field of view (FoV). Ref. [32] designed a feasible panoramic stereo video live-broadcast framework based on the current situation, which is similar to our system below.

At present, research on panoramic video is in a mature stage. Now and in the future, researchers will still explore better approaches to reduce transmission bandwidth and improve image quality. However, the research on the 3D sound effect of panoramic video needs to be expanded. Meanwhile, more research on applications on panoramic video is required as a supplement.

3. VR Conferencing

3.1. What a Virtual Reality Conferencing Application Needs to Succeed

As mentioned above, a virtual reality conference room is proposed with the aim of obtaining a feeling of immersion for users that is lacking in traditional video conference rooms. The design of the system focuses on this.

- Virtual space: Immersion means the users' experience of the scene. A virtual scene is the main part of the visual information and is an important source of immersion. The layout of the scene should conform to the appearance of the conference room, with tables and chairs for virtual avatars to move and interact. So far, there are many modeling tools as well as models made by others that can be obtained, which can be easily imported into the game engine for use.
- Avatar: Facial presentation is the most important and complicated factor in this section. One way to achieve this is to update the avatar's facial animation in real time by capturing the participants' facial information in real time, such as the eye rotation, lip movement, etc. When a user wears a helmet-mounted display, it becomes difficult to capture the face data, and the action of the face needs to be controlled by the content of the participants' speech [23]. As multiple participants join the conference room, each participant can see the virtual avatar of all other participants. When a user's position status or body movements change, the new status should be synchronized with other users, for which a network module that synchronizes the status of all players is required.
- Audio sense of space: Vision and hearing are presently critical aspects of bringing immersion to virtual reality devices. In order to obtain the same real experience as in the real conference room, auditory perception is as important as vision. To realize realistic auditory perception, spatial audio technology should be applied to virtual reality conferences. When in the same virtual conference room, each participant should be able to correctly sense the orientation of the other speakers as well as the reverberation of sound after multiple reflections from the room walls. While novel interactions are designed in the conference scenario, users can perceive the sound effect generated by the interaction when some interactions occur.
- Interaction: Communication in the meeting room should be supplemented by content presentation, such as slides, pictures, or videos. A virtual screen can be set up as the display area in the scene. In addition to basic interactions, there are many possibilities for interaction. In [23], VR painting and model import were mentioned. There are also new attempts to interact among participants.

3.2. Current Challenges

Nowadays, the technology of building virtual reality conference rooms is still incomplete and faces many challenges [26].

The challenge of network delay still exists. In contrast to a traditional video conference, there is no need to transmit the video stream of each participant, so the total bandwidth is reduced. However, to obtain the above spatial audio effect, it is necessary to transmit the audio stream of other participants to each participant while the system simultaneously updates the position, posture, and other information of all other participants for all users. The bandwidth demand is therefore still very large. Thus, a feasible scheme [24] is needed to reduce the bandwidth of audio transmission.

Another challenge is the lack of avatar authenticity. The user's impression of an avatar's posture and animation is still quite different from the real situation. The methods for avatar generation include avatars based on the user's real image and pure virtual avatars. An avatar based on the real image has a more realistic visual effect, but expensive equipment is required to achieve the ideal effect. Although completely virtual avatars cannot show the real appearance of participants, they tend to be more integrated with the virtual environment, and the body animation is more natural [33].

There is no available solution for interaction design. Traditional video conference is the mainstream form at present and will likely remain so for a long time. One reason is that its interaction design is mature, and we can experience similar interaction modes even using different video conferencing software, thus it could save us time in getting familiar with new software. In such circumstances, virtual reality conferencing has a long way to go [34]. Moreover, unlike familiar operations on computers, participants usually need more time to get familiar with their interaction in a virtual reality environment.

How to attract more people to virtual reality conference rooms is also a critical challenge. In [8], it was mentioned that the virtual conference itself reduces the opportunities for interaction between people. Another viewpoint is that important meetings should be face to face, and people can better prepare for the meeting by taking advantage of the long-distance travel time [35].

4. Panoramic Live Broadcast

4.1. Steps to Implement Panoramic Live Broadcast

In October 2015, the standardization of the panoramic video packaging format was launched by MPEG, called omnidirectional media format (OMAF) [36], which was jointly developed by experts from high-tech enterprises, research institutions, and universities around the world.

It is stipulated in the OMAF standard that a panoramic video spliced on the server side can be streamed using a DASH or MPEG media transport protocol after content preprocessing, encoding, and encapsulation. On the one hand, an OMAF player can receive, parse, and present relevant media content; on the other hand, it needs to track a user's head movements, eye movements, and other interactive operations to feed back window information in real time. Figure 1 shows the panorama video processing flow specified in the OMAF standard.

Shooting and splicing: A panoramic camera is required to obtain panoramic videos, whose information is sourced from a composite of multiple cameras. In cases where a higher video resolution is required, more cameras of higher resolution are needed. Stitching technology is used to place the pictures of all cameras around the observer to generate the panoramic picture. If there are six ideal identical cameras, they should be placed on a bracket with a strict position and rotation to take pictures in four horizontal directions and in the up and down directions; thus, pictures that are taken should be spliced together to form a square. In that case, the results can be used directly without optimization. However, it can be difficult and even impossible to accurately meet the above standards. Therefore, some researchers use methods [37] to reserve image redundancy and correctly identify and process the areas where the images of two cameras overlap and then splice them.

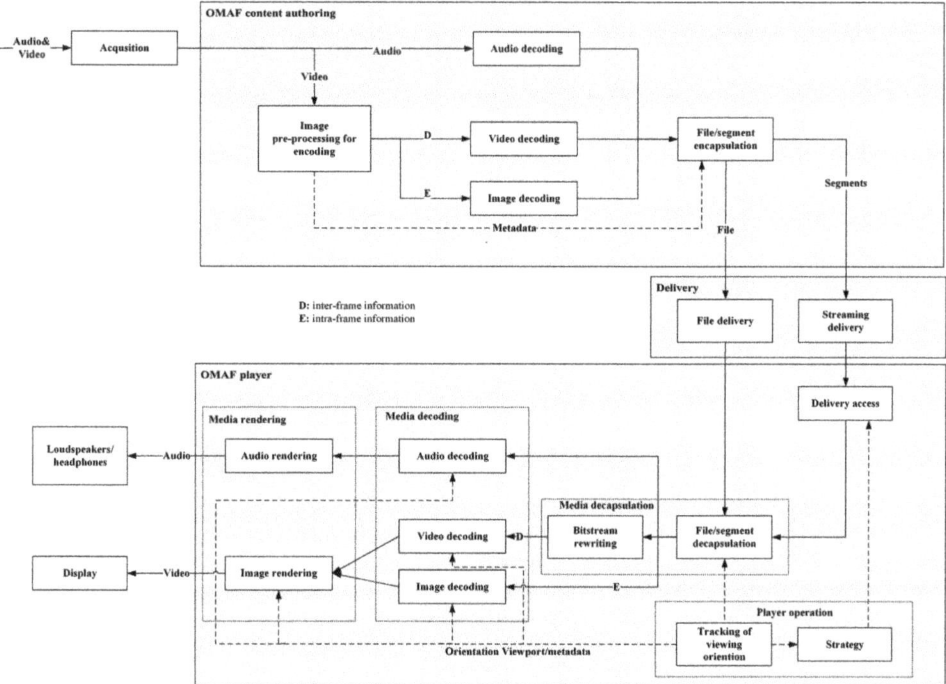

Figure 1. The OMAF architecture for panorama video processing flow. Three main steps are involved: content authoring, delivery, and playing. Content authoring is completed by a server, the delivery is over the Internet, and the content obtained is generally displayed on a personal computer [36].

Projection and back projection: Each frame of a panoramic video is presented in the form of a sphere. However, the original video coding standards are applied to the transmission of panoramic video. That means there should be preprocessing before transmission. The process of mapping the spliced picture to a rectangular picture is called projection. At present, the OMAF standard only supports longitude and latitude maps, equirectangular projection (ERP), and cube mapping (CMP) [36]. ERP is similar to the generation of the world map. In ERP, people look from the center of the sphere outward to the surface of the sphere inward, while for the world map, they look at the sphere from the outside to the inside. In CMP, the complete sphere is divided into six regions, which are projected onto each of the six faces of the cube. The bottom, back, and top faces need to be arranged together with the other three faces to form a rectangular frame through a specific rotation operation. In order to improve the coding efficiency, the operation principle of the three rotating surfaces is to maintain the consistency of the media content at the junction of the arrangement time surface and the surface [36].

Encoding and decoding: Traditional video encoding and decoding schemes may meet the codec requirements of panoramic video streams. However, because panoramic video itself is different from traditional video, many problems arise when traditional schemes are applied to panoramic video, such as geometric distortion, discontinuous pictures, etc. As a result, various optimization schemes for panoramic video codecs have been put into practice [38,39].

Streaming: It was described in the second part of the related work that there were three mainstream panoramic live broadcast schemes: full-view video streaming, viewport-based video streaming, and tile-based video streaming. Determining how to make a better choice between improving video quality and reducing transmission bandwidth is still the main direction of current and future research.

4.2. Current Challenges for Panoramic Live Broadcast

Panoramic video provides an immersive volume of video that is not available in traditional 2D. A panoramic video is spherical in nature, which brings about many challenges regarding its acquisition, storage, coding, transmission, and display.

There is a variety of distortions from capture to display. To solve the distortion problem in 360-degree video streams, efficient splicing, projection, and encoding methods with better exploration results and a lower bandwidth should be explored. With the popularity of virtual reality technology and the formulation of the next-generation video coding standard, 360-degree video has been receiving increasing attention from academia and various fields in industry. Innovating video projection methods and efficient compression coding to meet bandwidth and quality requirements represent the current direction of mainstream research and exploration [14].

Sphere-to-plane projection is a common procedure for panoramic video before encoding. There are, so far, many projection formats. Considering that different plane projection formats may be adapted to different applications, we often need to convert from one projection format to another, for which the interpolation algorithm is thus critical [40].

It is necessary to focus on designing quality evaluation methods and indicators for panoramic video. This is a complex challenge, for traditional video QoE (quality of experience) models are not suitable for 360-degree content. Although most studies have carried out various subjective and panoramic video objective evaluations, most evaluation methods still follow the traditional video evaluation standards. Due to the lack of unified and standardized factors affecting 360-degree video, the standard evaluation method has not been finalized. This represents a complex and challenging problem [15].

5. The Importance of VR Audio

Also known as virtual audio, spatial audio, and immersive audio [41], 3D audio includes two forms of playback based on binaural audio and speaker playback. The former is widely used in VR devices, usually called binaural audio. The basic principle of binaural audio is to simulate the sound field generated by a sound source at a certain point in space in two ears such that the listener can have a sense of where the source was emitted from. This technology is also known as binaural acoustic technology [42,43]. Because of the inherent characteristics of the human auditory system, there are often certain differences between the sound that people subjectively feel and natural sound. The effect of the human auricle and other structures on sound waves can be seen as a filter, called head-related transfer function (HRTF). HRTF simulates the human ear's perception of sound direction and distance in space, which plays an important role in the generation of binaural virtual audio.

With the development of virtual reality devices, more and more immersive applications are emerging. Although most research is focused on how to improve the resolution and frame rate of frames, hearing plays as equally an important role as vision in providing immersion [44,45]. Audio is one of the important outputs of electronic games [44]. In first-person shooting games, players need to identify the position of the enemy according to the direction of the sound for an effective reaction. In addition, game audio can play an important technical role in providing basic user feedback by providing player behavior confirmation or warning of in-game activities. Research in [46] showed that in a shared space, with the inclusion of spatial audio and video, users can identify speakers better, retain more information, and have an increased comprehension from video conference meetings.

As is indicated in [47], sound may have a greater influence on immersion than immersion has on images. Sound is one of the most important sources of human cognition. VR/AR content matching of audio effects has a great impact on enhancing users' memory of content. Dale, an American audiovisual educator, put forward the theory of a "tower of experience" in his book *Audio-Visual Methods in Teaching*. This theory considers how human experience is derived. Figure 2 shows the top three layers in Dale's tower of experience (the main theoretical basis of audiovisual education). The third row from the top represents audio/recording/photos, and the layer above is visual signals and language symbols, indicating that the combination of listening and audition is more conducive to memory and understanding of knowledge.

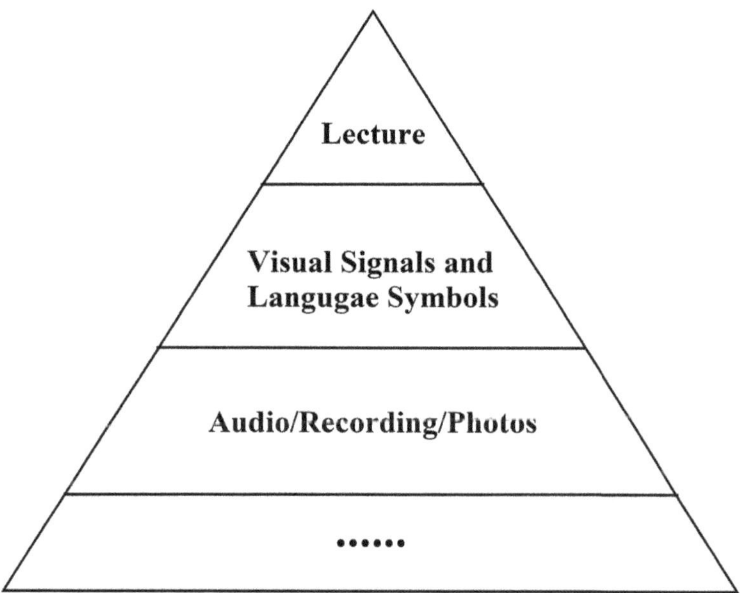

Figure 2. The top three layers in Dale's tower of experience [48].

The demand for VR will increase along with the improvement of hardware, and the demand for immersive audio is also growing. It is pointed out in [49] that we need better authoring tools to support the creation of high-quality immersive audio works, regardless of whether the creators understand the underlying principles of audio, just as many video creators do not understand the underlying codec of a video.

6. Methodology of Sample Implementation

Two simple implementations for the above two scenarios are presented in this section, while the implementation of each module or step is introduced. The results are given at the end of each part.

6.1. Implementation of VR Conferencing

Now, we introduce a simple virtual reality conference room, which includes a scene with audio and video experience created by a Unity engine and simple network communication.

6.1.1. Architecture

The overall architecture of the system is shown in Figure 3.

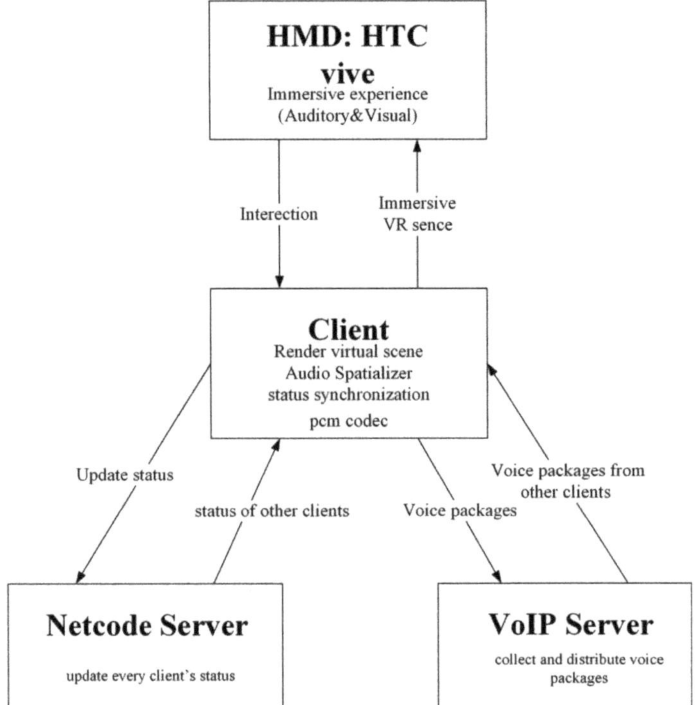

Figure 3. Architecture of the virtual reality conference room. The two servers are for Netcode and VoIP, respectively. The spatial audio effect is realized on the client computer. The immersive feeling is generated by a HMD.

The whole system is comprised of mainly three parts: server, client, and a HMD.

- Server: We use two servers, a Netcode server and a VoIP server. Netcode is a concept in game development. The Netcode server in our system is to synchronize each user's status, allow the user to see the precise and fluid representation of the room state, and influence the scene state shared in common. The VoIP server controls the transmission of audio packages on network. It receives voice packages from each client and distributes them to each other clients.
- Client: The client, usually a high-performance computer, is responsible for collecting the microphone input from the HMD, obtaining the pulse code modulation (PCM), compressing the PCM packets, and sending it to the VoIP server. Meanwhile, it receives the PCM packets from other clients sent by the server, decompresses them, and processes them with the spatial audio algorithm [50] to play the audio with room reverberation and orientation. The client synchronizes the position to the Netcode server and updates the position of other clients obtained from the server.
- HMD: The virtual scene is rendered in real time and could be experienced in an HMD, which meanwhile transmits a user's physical state and interaction data to the client. In our system, we used HTC Vive as the HMD shown in Figure 4.

Figure 4. HTC Vive.

6.1.2. Scenes, Avatars, and Interactions

To simplify the workload, the modeling of the scene [51] and avatar [52] were from free resources on the Internet. Figure 5 shows how our VR conferencing room looks. Figure 6 is the avatar of the user. After a participant enters the virtual conference room, they are assigned to a random seat. At this time, they can change the direction of view and observe the surrounding environment by moving the mouse or rotating the HMD (head-mounted display). When another participant also enters the conference room, the two participants can see each other in the scene and talk to each other by voice. The view of a user is shown in Figure 7. The sound effect produces a real sense of space and orientation. When there are multiple participants in the conference room, everyone can feel the orientation of other participants.

Figure 5. Virtual conference scene.

Figure 6. Avatar.

Figure 7. View of a user.

6.1.3. NetCode for Status Synchronization

In online games, in order to allow multiple players to play games together on different computers, a mechanism is needed to ensure that all computers are synchronized such that players can accurately see the performance of each player smoothly; players' input should influence the game state. We use the NetCode [53] network architecture of Unity to synchronize the status information of roles in multiple terminals, mainly including location and orientation information. The network code has always been one of the most difficult parts in game development. The Unity NetCode software package provides a dedicated server model with client prediction, which can be used to simplify the development of multiplayer games. It is currently in the experimental phase. Listing 1 is some pseudocode for Netcode.

Listing 1. pseudocode of Netcode in unity

```
1   Vector3  Position  =  new  Vector3;
2
3   Move()
4   {
5       if  (this.IsServer)
6       {
7           newPosition  =  GetPosition();
8           Position  =  randomPosition;
9       }
10      else
11      {
12          SubmitPosition();
13      }
14  }
15  [ServerRpc]
16  SubmitPosition()
17  {
18      Position.Value  =  GetPosition();
19  }
20
21  void  Update()
22  {
23      transform.position  =  Position.Value;
24  }
```

6.1.4. Simple VoIP

The traditional multiperson VoIP voice conference refers to a server that obtains the voice code stream of each client, synthesizes it, and sends it to each client. In order to realize the sense of direction of each participant's voice, the voice data stream of each participant should not be merged on the server but, instead, separately distributed to each other. Without lowering the voice quality, this method requires more bandwidth than traditional voice conferencing. Moreover, as the number of participants increases, the pressure on network bandwidth grows rapidly. Both the server and the client have to bear the pressure of the network bandwidth, and the demand on server grows faster. The network pressure can be relieved by limiting the number of participants. As the local area network (LAN) environment is in good condition, we simplified the VoIP process into the following sequence: collect audio signals, encode and package them, serialize them, send them to the server, transmit them to other clients, and the client receives a packet, deserializes it, unpack and decode it, and play the audio. We used the User Datagram Protocol (UDP) for network transmission and the Speex open source library [54] for the encoding and decoding.

Listing 2 is some pseudocode for the network transmission.

Listing 2. pseudocode of VoIP

```
1   // voip server
2   sendData(data)
3   {
4       //send audio package to every client
5       for (int i = 0; i < client_count; i++){
6           if (hasClient[i]){
7               mySocket.SendTo(..., Remotes[i]);
8           }
9       }
10  }
11
12  //voip client
13  startSocket()                          //start udp socket
14  {
15      mySocket = new Socket(UDP);
16      IPEndPoint sender = new IPEndPoint();
17      Remote = (EndPoint)sender;
18      int recv = mySocket.ReceiveFrom(data, Remote);
19      while (true){
20          //receive audio package from server
21          mySocket.Receive(data, Remote);
22          process(data);                 //process audio data
23      }
24  }
```

6.1.5. Result

In the VR conference room solution given above, the basic functions of the conference were achieved. Users could switch perspectives horizontally, make speeches, and communicate with other participants. In terms of listening feeling, users heard the voices of different participants from different directions and experienced the effect of room reverberation. The NetCode latency and VoIP latency in the LAN were acceptable. Character animation and scenes could be optimized by referring to 3D games, and more applicable interactive functions could be further explored.

6.2. Implementation of Panoramic Video Broadcast

6.2.1. Architecture

Figure 8 shows the architecture of our panoramic video broadcast system.The following are the main procedures for making panoramic live broadcasts: audio and video capture (Figures 9 and 10), projection (Figure 11), compression, encapsulation, and data stream transmission. The receiver obtains the code stream and then decodes it. After backprojection transformation, the panorama video can be displayed. Next, is the process of implementing a simple panoramic live broadcast.

Video capture:

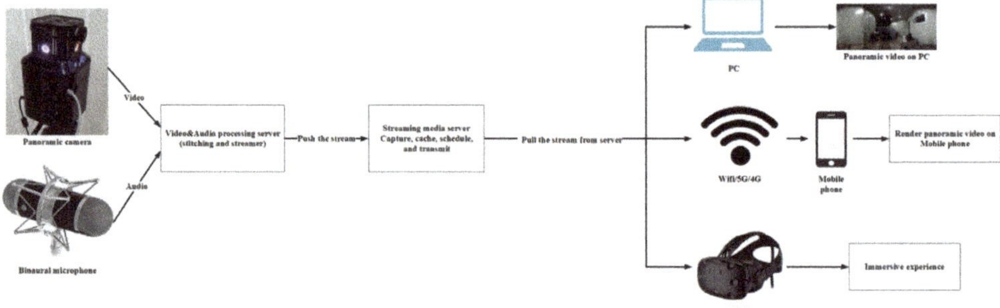

Figure 8. Architecture of panoramic video broadcast system. This is a simplified implementation of Figure 1. The cameras and microphone are used to capture the original input. Through the simple processing of the server, the content can be distributed to various terminal devices.

Figure 9. Four fisheye camera, integrated with projection transformation, compression coding, and network transmission modules.

Audio capture:

Figure 10. A binaural microphone that records a directional dual-channel audio stream. The sound direction is determined by its orientation.

Projection:

Figure 11. Four pictures of one frame spliced into one plane for coding and compression.

6.2.2. Network Transmission

For the network transmission protocol part, RTMP (Real-Time Message Transmission Protocol) was adopted. The 2D images after projection splicing and audio coding were encapsulated and then pushed to the network. On the client side, the corresponding RTMP stream was pulled. The content in Figure 11 is also the real-time frame of RTMP stream obtained by the client through testing. There was a sense of direction of the sound, but it was still difficult to achieve a three-dof sound direction change through the perspective transformation.

6.2.3. End-to-End System Delay Analysis

Based on the panoramic video live broadcasting system, we conducted an end-to-end delay analysis, and the system set the duration of each frame at 30 ms.

(1) The camera captured and cached one frame, resulting in a 30 ms delay;

(2) The splicing box could collect up to three cached frames and generated a maximum delay of 90 ms;

(3) The maximum splicing buffer was three frames, resulting in a maximum delay of 90 ms;

(4) The maximum encoding buffer was two frames, resulting in a maximum delay of 60 ms;

Therefore, the maximum delay was 270 ms at the VR panoramic video content production end.

(5) The delay was determined by the bandwidth and the bit rate of the transmitted video. We use a wired local area network in the laboratory according to the previous implementation. In the case of the network test, when the overall minimum delay was 500 ms–1 s, the minimum delay of this part was within 500 ms, although wireless transmission would be higher;

(6) + (7) + (8) The delay at the playback end was less than 20 ms for mainstream display devices. Network transmission delay was still the main factor causing end-to-end delay. After the above analysis, the following system (Figure 12) delay analysis composition diagram was obtained.

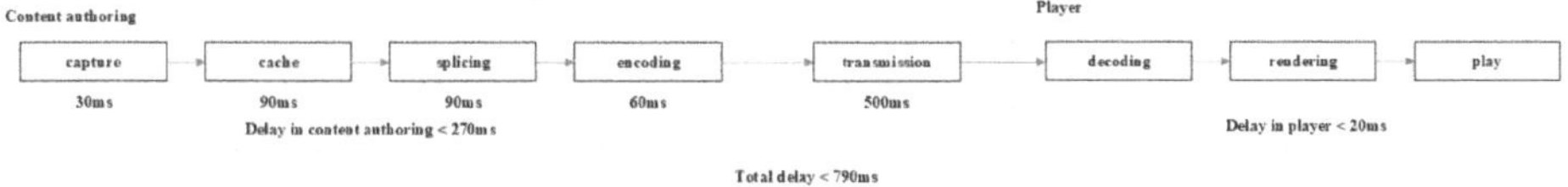

Figure 12. System delay analysis. The delay mainly comes from the content authoring and transmission, and the delay proportion of the player is very small.

7. Conclusions and Future Work

As virtual reality technology and equipment develop, the demand for immersive experience will continue to grow. We presented two main patterns of immersive experience. One was the pure virtual environment represented by games, in which users can interact. Most of these scenes are created by game engines such as Unity or Unreal. The other was the panoramic video obtained by the processing of real scenes captured by cameras. In this article, we discussed these two kinds of immersive audio and video scenes and provided sample implementation approaches for these two systems.

We discussed the design and implementation of virtual reality conference environments, outlined the current research status in this field, and analyzed the many challenges. Then, we designed and implemented a simple virtual reality conference room, for which the virtual scene design, audio transmission, and realization of the spatial orientation were described. In the future, research will increasingly focus on how to generate more realistic avatars as well as facial expression optimization. There will be more exploration of the interaction patterns in conference rooms. With the development of virtual reality hardware equipment, new forms of interactions in virtual reality conference rooms may arise. In future work, discussions about the advantages and disadvantages of VR conferencing [8,35,55] will continue.

In this article, we built a panoramic live broadcast system. By designing the system architecture and completing each module, we implemented a panoramic live broadcast system. The delay analysis was presented. In the future, we will do more research based on this system. Today, panoramic video live broadcast is widely used, especially in some live events. How to use a lower network bandwidth to achieve a better playback effect will be the focus of future studies. Since panoramic video is presented as a sphere, the client playing the video needs to project the picture onto the sphere, which imposes certain requirements regarding the rendering speed of the computer. Future research should consider how to improve the rendering speed and reduce the consumption of performance on the premise of ensuring that the picture is not distorted. In addition to vision, the source of immersion of panoramic video should be the sense of the audio space. It is worth paying attention to facilitating users' experience in sensing the audio space in the panoramic video. Specifically, the content that is suitable for a traditional video presentation may not be suitable for panoramic video and vice versa. We should explore more appropriate content for panoramic video, which may also improve demand for panoramic video.

Author Contributions: Conceptualization, H.Z. and J.W.; methodology, H.Z. and J.W.; software, H.Z.; validation, H.Z., J.W. and Z.L. ; formal analysis, H.Z.; investigation, H.Z. and J.W.; resources, J.L.; data curation, H.Z. J.W. and Z.L.; writing—original draft preparation, H.Z.; writing—review and editing, H.Z.; visualization, J.L.; supervision, J.L.; project administration, H.Z.; funding acquisition, J.L. All authors have read and agreed to the published version of the manuscript.

Funding: National Natural Science Foundation of China (grant no. 62071039) and Beijing Natural Science Foundation (grant no. L223033).

Data Availability Statement: Not applicable.

Acknowledgments: The authors wish to thank all participants who volunteered for this study. Informed consent was obtained from all subjects involved in the study.

Conflicts of Interest: The authors declare no conflict of interest.

References

1. Mel, S.; Sanchez-Vives, M.V. Enhancing Our Lives with Immersive Virtual Reality. *Front. Robot. AI* **2016**, *3*, 74.
2. Santana, A.; Lira, A.; Lara, G.; Pena, A. Evolution of Virtual Reality's Interaction Devices. In Proceedings of the 2018 7th International Conference On Software Process Improvement (CIMPS), Guadalajara, Mexico, 17–19 October 2018.
3. Hameed, B.M.Z.; Somani, S.; Keller, E.X.; Balamanigandan, R.; Mahapatra, S.; Pietropaolo, A.; Tonyali, S.; Juliebo-Jones, P.; Naik, N.; Mishra, D.; et al. Application of Virtual Reality, Augmented Reality, and Mixed Reality in Endourology and Urolithiasis: An Update by YAU Endourology and Urolithiasis Working Group. *Front. Surg.* **2022**, *9*, 866946. [CrossRef] [PubMed]

4. Zhang, T.; tian, F.; Lü, W.; Wang, Y.; Huang, C. Survey of VR applications in interactive films and games. *J. Shanghai Univ.* **2017**, *23*, 342.
5. Tan, Q.P.; Huang, L.; Xu, D.; Cen, Y.; Cao, Q. Serious Game for VR Road Crossing in Special Needs Education. *Electronics* **2022**, *11*, 2568. [CrossRef]
6. Maskeliūnas, R.; Damaševičius, R.; Blažauskas, T.; Canbulut, C.; Adomavičienė, A.; Griškevičius, J. BiomacVR: A Virtual Reality-Based System for Precise Human Posture and Motion Analysis in Rehabilitation Exercises Using Depth Sensors. *Electronics* **2023**, *12*, 339. [CrossRef]
7. Cho, Y.; Hong, S.; Kim, M.; Kim, J. DAVE: Deep Learning-Based Asymmetric Virtual Environment for Immersive Experiential Metaverse Content. *Electronics* **2022**, *11*, 2604. [CrossRef]
8. Roos, G.; Oláh, J.; Ingle, R.; Kobayashi, R.; Feldt, M. Online conferences—Towards a new (virtual) reality. *Comput. Theor. Chem.* **2020**, *1189*, 112975. [CrossRef]
9. Ishii, H.; Kobayashi, M.; Grudin, J. Integration of interpersonal space and shared workspace. *ACM Trans. Inf. Syst. (TOIS)* **1993**, *11*, 349–375. [CrossRef]
10. Blair, C.; Walsh, C.; Best, P. Immersive 360° videos in health and social care education: A scoping review. *BMC Med. Educ.* **2021**, *21*, 590. [CrossRef]
11. Pirker, J.; Dengel, A. The Potential of 360-Degree Virtual Reality Videos and Real VR for Education—A Literature Review. *IEEE Eng. Med. Biol. Mag.* **2021**, *41*, 76–89.
12. Ouglov, A.; Hjelsvold, R. Panoramic video in video mediated education. In Proceedings of the Storage Retrieval Methods and Applications for Multimedia, San Jose, CA, USA, 18–19 January 2005; pp. 326–336.
13. Li, W.; Qian, L.; Feng, Q.; Luo, H. Panoramic video in education: A systematic literature review from 2011 to 2021. *J. Comput. Assist. Learn.* **2022**, *39*, 1–19. [CrossRef]
14. Da, A.; Jiujun, D.; Nan, L.; Ying, L. Advance of 360-degree video coding for virtual reality: A survey. *Appl. Res. Comput.* **2018**, *35*, 1606–1612.
15. Ruan, J.; Xie, D. A Survey on QoE-Oriented VR Video Streaming: Some Research Issues and Challenges. *Electronics* **2021**, *10*, 2155. [CrossRef]
16. Shimamura, R.; Feng, Q.; Koyama, Y.; Nakatsuka, T.; Morishima, S. Audio–visual object removal in 360-degree videos. *Vis. Comput.* **2020**, *36*, 2117–2128. [CrossRef]
17. Li, D.; Langlois, T.R.; Zheng, C. Scene-aware audio for 360 videos. *ACM Trans. Graph. (TOG)* **2018**, *37*, 1–12. [CrossRef]
18. Sabri, S.; Prasada, B. Video conferencing systems. *Proc. IEEE* **2005**, *73*, 671–688. [CrossRef]
19. Watanabe, K.; Murakami, S.; Ishikawa, H.; Kamae, T. Audio and visually augmented teleconferencing. *Proc. IEEE* **2005**, *73*, 656–670. [CrossRef]
20. Valenti, A.; Fortuna, G.; Barillari, C.; Cannone, E.; Boccuni, V.; Iavicoli, S. The future of scientific conferences in the era of the COVID-19 pandemic: Critical analysis and future perspectives. *Ind. Health* **2021**, *59*, 334–339. [CrossRef]
21. Dijkstra-Soudarissanane, S.; Assal, K.E.; Gunkel, S.; Haar, F.T.; Hindriks, R.; Kleinrouweler, J.W.; Niamut, O.A. Multi-sensor capture and network processing for virtual reality conferencing. In Proceedings of the the 10th ACM Multimedia Systems Conference, Amherst, MA, USA, 18–21 June 2019.
22. Sun, B.; Gao, X.; Chen, W.; Sun, Q.; Cui, X.; Guo, H.; Kevin, C.R.; Liu, S.; Liu, Z. Video Conference System in Mixed Reality Using a Hololens. *CMES-Comput. Model. Eng. Sci.* **2023**, *134*, 383–403. [CrossRef]
23. Pazour, P.D.; Janecek, A.; Hlavacs, H. Virtual Reality Conferencing. In Proceedings of the 2018 IEEE International Conference on Artificial Intelligence and Virtual Reality (AIVR), Taichung, Taiwan, 10–12 December 2018; pp. 84–91.
24. Ling, L.I.; Tian, S.Z.; Sun, L.F.; Zhong, Y.Z. The Video Transmission Scenario Based on Awareness in the Virtual Space Teleconference. *Appl. Res. Comput.* **2004**, *21*, 209–211.
25. Goncralves, G.; Monteiro, P.; Coelho, H.; Melo, M.; Bessa, M. Systematic Review on Realism Research Methodologies on Immersive Virtual, Augmented and Mixed Realities. *IEEE Access* **2021**, *9*, 89150–89161. [CrossRef]
26. Hurst, W.; Withington, A.; Kolivand, H. Virtual conference design: Features and obstacles. *Multimed. Tools Appl.* **2022**, *1*, 16901–16919. [CrossRef] [PubMed]
27. Li, J.; Sun, R.; Wang, G.; Fan, M. Panoramic video live broadcasting system based on global distribution. In Proceedings of the Chinese Automation Congress (CAC), Hangzhou, China, 22–24 November 2019; pp. 63–67.
28. Hosseini, M.; Swaminathan, V. Adaptive 360 VR Video Streaming: Divide and Conquer. In Proceedings of the IEEE International Symposium on Multimedia (ISM), San Jose, CA, USA, 11–13 December 2016; pp. 107–110.
29. Yaqoob, A.; Togou, M.A.; Muntean, G.M. Dynamic Viewport Selection-Based Prioritized to Adaptation for Tile-Based 360 degrees Video Streaming. *IEEE Access* **2022**, *10*, 29377–29392. [CrossRef]
30. Dai, Y.; Han, S. Sight Guidance Enhanced VR Video Transmission. In Proceedings of the 2022 16th IEEE International Conference on Signal Processing (ICSP), Beijing, China, 21–24 October 2022; pp. 305–309.
31. Hosseini, M.; Swaminathan, V. Adaptive 360 VR Video Streaming based on MPEG-DASH SRD. In Proceedings of the 2016 IEEE International Symposium on Multimedia (ISM), San Jose, CA, USA, 11–13 December 2016; pp. 407–408.
32. Zheng, M.; Tie, Y.; Zhu, F.; Qi, L.; Gao, Y. Research on Panoramic Stereo Live Streaming Based on the Virtual Reality. In Proceedings of the 2021 IEEE International Symposium On Circuits And Systems (Iscas), Daegu, Republic of Korea, 22–28 May 2021; pp. 1–5.

33. Lugrin, J.L.; Latt, J.; Latoschik, M.E. Avatar Anthropomorphism and Illusion of Body Ownership in VR. In Proceedings of the 2015 IEEE Virtual Reality Conference (VR), Arles, France, 23–27 March 2015; pp. 229–230.

34. Qi, W.; Sun, S.; Niu, T.; Zhao, D. Research and prospects of virtual reality systems applying exoskeleton technology. In *Universal Access in the Information Society*; Springer: Berlin/Heidelberg, Germany, 2022.

35. Palmer, C. Are Virtual Conferences Here to Stay? *Engineering* **2021**, *7*, 280–281. [CrossRef]

36. Hannuksela, M.M.; Wang, Y.K. An Overview of Omnidirectional MediA Format (OMAF). *Proc. IEEE* **2021**, *109*, 1590–1606. [CrossRef]

37. Ren, H.; Ren, Q. Real-time Panoramic Video Mosaic system Based on Mapping Table and GPU Acceleration. In Proceedings of the 33RD Chinese Control and Decision Conference (CCDC), Kunming, China, 22–24 May 2021; pp. 812–816.

38. Liu, Y.; Mai, X.; Chen, L.; Li, S.; Wang, Z. A novel rate control scheme for panoramic video coding. In Proceedings of the 2017 IEEE International Conference on Multimedia and Expo (ICME), Hong Kong, China, 10–14 July 2017; pp. 691–696.

39. Liu, Y.; Yang, L.; Xu, M.; Wang, Z. Rate Control Schemes for Panoramic Video Coding. *J. Vis. Commun. Image Represent.* **2018**, *53*, 76–85. [CrossRef]

40. Zhang, S.; Yang, F.; Wan, S.; Di, P. Spherical Lanczos Interpolation in Planar Projection or Format Conversions of Panoramic Videos. *IEEE Access* **2020**, *8*, 9655–9667. [CrossRef]

41. Sun, X. Immersive audio, capture, transport, and rendering: A review. *Apsipa Trans. Signal Inf. Process.* **2021**, *10*, e13. [CrossRef]

42. Begault, D.R. *3-D Sound for Virtual Reality and Multimedia*; CD-ROM Professional:462 DANBURY RD, WILTON, CT 06897-2126 USA, 1994; pp. 127–128.

43. Fu, Y.; Wang, L.; Chen, Z. Review on 3D audio technology. *J. Commun.* **2011**, *32*, 130–138.

44. Rogers, K.; Ribeiro, G.; Wehbe, R.R.; Weber, M.; Nacke, L.E. Vanishing Importance: Studying Immersive Effects of Game Audio Perception on Player Experiences in Virtual Reality. In Proceedings of the 2018 CHI Conference on Human Factors in Computing Systems, Montreal, QC, Canada, 21–26 April 2018.

45. Çamcı, A.; Hamilton, R. Audio-first VR: New perspectives on musical experiences in virtual environments. *J. New Music. Res.* **2020**, *49*, 1–7. [CrossRef]

46. Wong, M.; Duraiswami, R. SharedSpace: Spatial Audio and Video Layouts for Videoconferencing in a Virtual Room. In Proceedings of the 2021 Immersive And 3d Audio: From Architecture To Automotive (I3da), Online, 8–10 September 2021.

47. Cummings, J.J.; Bailenson, J.N. How Immersive Is Enough? A Meta-Analysis of the Effect of Immersive Technology on User Presence. *Media Psychol.* **2016**, *19*, 272–309. [CrossRef]

48. Witt, P.W.F. Audio-Visual Methods in Teaching. *Audiov. Commun. Rev.* **1954**, *2*, 291—298. [CrossRef]

49. Camci, A.Some Considerations on Creativity Support for VR Audio. In Proceedings of the 2019 IEEE Conference on Virtual Reality and 3D User Interfaces (VR), Osaka, Japan, 23–27 March 2019; pp. 1500–1502.

50. MicroSoft Spatial Sound In Unity. Available online: https://docs.microsoft.com/zh-cn/windows/mixed-reality/develop/unity/spatial-sound-in-unity (accessed on 4 February 2022).

51. Unity Assetstore Furnished Cabin. Available online: https://assetstore.unity.com/packages/3d/environments/urban/furnished-cabin-71426 (accessed on 4 February 2022).

52. Adobe Mixiaomo. Available online: www.mixamo.com (accessed on 4 February 2022).

53. U. Technologies. Unity Netcode 0.6.0-Preview.7. Available online: https://docs.unity3d.com/Packages/com.unity.netcode@0.6/manual/index.html (accessed on 4 February 2022).

54. Speex: A Free Codec For Free Speech. Available online: https://www.speex.org (accessed on 4 February 2022).

55. Salomon, D.; Feldman, M.F. The future of conferences, today: Are virtual conferences a viable supplement to "live" conferences? *EMBO Rep.* **2020**, *21*, e50883. [CrossRef]

Article

An Interactive Augmented Reality Graph Visualization for Chinese Painters

Jingya Li and Zheng Wang *

School of Architecture and Design, Beijing Jiaotong University, Beijing 100044, China; lijy1@bjtu.edu.cn
* Correspondence: wangzheng@bjtu.edu.cn

Abstract: Recent research in the area of immersive analytics demonstrated the utility of augmented reality for data analysis. However, there is a lack of research on how to facilitate engaging, embodied, and interactive AR graph visualization. In this paper, we explored the design space for combining the capabilities of AR with node-link diagrams to create immersive data visualization. We first systematically described the design rationale and the design process of the mobile based AR graph including the layout, interactions, and aesthetics. Then, we validated the AR concept by conducting a user study with 36 participants to examine users' behaviors with an AR graph and a 2D graph. The results of our study showed the feasibility of using an AR graph to present data relations and also introduced interaction challenges in terms of the effectiveness and usability with mobile devices. Third, we iterated the AR graph by implementing embodied interactions with hand gestures and addressing the connection between the physical objects and the digital graph. This study is the first step in our research, aiming to guide the design of the application of immersive AR data visualization in the future.

Keywords: augmented reality; embodied interaction; data visualization; node-link diagram

Citation: Li, J.; Wang, Z. An Interactive Augmented Reality Graph Visualization for Chinese Painters. *Electronics* **2022**, *11*, 2367. https://doi.org/10.3390/electronics11152367

Academic Editors: Jorge C. S. Cardoso, André Perrotta, Paula Alexandra Silva and Pedro Martins

Received: 30 June 2022
Accepted: 24 July 2022
Published: 28 July 2022

Publisher's Note: MDPI stays neutral with regard to jurisdictional claims in published maps and institutional affiliations.

1. Introduction

Augmented Reality (AR) provides immersive experience in three dimensions (3D), offering new possibilities for engaging, embodied, and interactive data visualization [1,2]. Recent research in the area of Immersive Analytics [3,4] has indicated that the stereoscopic views and visual scalability of AR could improve the performance in data analysis tasks [5,6] and navigation processes [7]. In addition, its ability to connect the digital and physical world enables collaboration [4] and supports the perception of network structures [8].

A node-link diagram, often contains a set of nodes and a set of edges and can facilitate users understanding the overall structure of a graph and identifying existing relationships between two nodes [9,10]. Typical use cases include presenting structure and relations in social networks (e.g., [11–13]), road networks (e.g., [14]), as well as software networks (e.g., [15]). Additionally, a node-link diagram can represent knowledge and deliver information in a clear, novel and effective way, making it also an important method for educational activities [16].

2D visualization of a node-link diagram has been widely studied while challenging problems for 2D techniques in the design and layout of the graph, such as the edge crossings, the scalability of the data in terms of size and complexity, and the limitation of screen displays for visualization, interaction, as well as navigation in complex graph, have been identified [9,17,18].

Consequently, 3D visualization of node-link diagram has been proposed to eliminate issues of 2D techniques and offer a more effective way to identify links between certain nodes [19]. However, 3D layouts on 2D displays might introduce additional stereo cues and extra viewpoint navigation, such that the graph will appear differently based on the rotation, zoom, and angle adjustment on the screen, making the graph become more complicated to understand and interact with [9].

Utilizing the capability of AR to visualize node-link diagram on top of the real world has the potential to overcome the limitations. Combining AR and a node-link diagram can provide an immersive and ubiquitous experience that engages users to explore, embed richer information and content that extends the boundaries of the graph, and deal with data sets that have a strong connection to the physical objects or space. Belcher et al. [20] also identified three potential benefits of an AR node-link diagram, including increased comprehension, larger display, and enhanced spatial recall of the graph.

Despite the great potential of an AR node-link diagram, there is still a lack of research on how to design the AR graph to enable engaging, interactive, and effective data visualization [1]. With the increasing importance of AR visualization and the growing size of data, how to design an AR graph is of utmost importance, together with user studies to practically evaluate the visualization [8].

Our motivation for this study is twofold: first, we aim to explore opportunities of combining AR and a node-link diagram to create immersive visualization. Second, we also aim to investigate the design space of an AR graph to better utilize its embodied interactions and the ability to connect the physical objects and the digital graph.

In this paper, we first systematically describe the design process of the AR graph. We use the social network data of early Chinese painters in our study based on the database we have been built since 2014. We build an AR graph to visualize the teacher-student relationship of these painters. Then, we validate our AR concept by implementing a prototype and conduct a user study with 36 participants to examine users' behaviors in the AR graph and the 2D graph. Third, we report on feedback collected from the user study and present the AR graph again with iterated features and interactions. With this exploratory work, we provide inspirations and an informed basis to guide the development of an AR graph in the future and identifying opportunities of the application of immersive AR data visualization.

2. Related Work

2.1. Augmented Reality Data Visualization

AR data visualization has attracted considerable attention in recent research area [2,21,22]. Research has concentrated on examining the advantages of AR visualization in comparison to traditional means. For example, Bach et al. [23] examined the effectiveness of three visualization environments for common 3D scatter-plots exploration, including an AR headset, a desktop setting, and a tablet-based AR. They found that each of the immersive AR environments was more effective for highly interactive tasks that require detailed manipulation. Similarly, Kirshenbaum et al. [24] compared the effectiveness and user engagement of geographical data projected on a 2D surface and a physical 3D terrain for geo-visualization tasks. According to their study, the 3D visualization supported the tasks better due to its ability to provide physical shapes.

Some research has investigated the integration of AR technology with traditional displays. Reipschläger et al. [25] proposed the combination of AR with large interactive displays for information visualization in order to enhance the data exploration and analysis. Wang et al. [26] conducted an observational study to understand how an immersive AR headset could be used as an extension to the traditional PC-based data analysis tools to analyze particle physics simulations. The results of their qualitative study showed that the AR HMD effectively improved experts' understanding of particle collision events. Langner et al. [17] proposed MARVIS, which was a conceptual framework for immersive visualization with tablet devices and AR headsets. Their case studies demonstrated the benefit of combining multiple mobile devices with AR headsets for data visualization and data analysis. Hubenschmid et al. [21] also presented a spatially-aware tablets combined with AR headsets for immersive data visualization and revealed that the novel interaction concept was appreciated by users. They also provided various design insights to foster the development of spatially-aware touch devices in AR settings.

Beyond extending traditional visualization techniques, some research has also addressed the capability of AR to combine the virtual content with physical objects. Mahmood et al. [27] presented a way to create multiple coordinated spaces for data analysis in a physical environment and found the great flexibility of combining visualization on 2D displays and AR. Chen et al. [28] designed and implemented an AR environment for static visualization that combined the digital world with physical objects, such as books, posters, projections, and wall-sized visualization. They reported a user study that showed high user satisfaction when using the proposed AR system and confirmed that the system was quick and easy to operate.

Several studies have also carried out practical toolkit on how to transfer 2D data into 3D representation to create immersive data visualization in AR, such as DXR [22], MARVisT [28], and IATK [29].

Overall, previous works demonstrated the value and the great potential of AR data visualization in general. Meanwhile, previous research on AR data visualization involved diverse graph types, such as scatter-plots, bar-plots, line-charts, etc., but did not focus on node-link diagram specifically. A node-link diagram could lead to less unified user experience due to the lack of predefined axes, dimensions and directions, making it more complicated to carry out design decisions and solutions [8]. Thus, we delve deeper into AR visualization for node-link diagram specifically in the next section.

2.2. Augmented Reality Node-Link Diagram Visualization

A node-link diagram, a traditional statistical visualization strategy, is the most commonly used graph type to visualize the overall network structure and specific relational links [9,30]. With the ever-increasing size of the data and and complexity in data analysis, researchers have advocated for innovative and interactive features to visualize a node-link diagram [31,32].

With the advantages of the novel and immersive technology, there has been an increasing interest in making AR available for node-link diagram visualization. For example, Drochtert et al. [33] explored the design and implementation of an AR visualization prototype that consisted of several mobile devices, which were used as tracking targets in AR. The prototype also allowed simultaneously interactions among multiple users. However, the study lacked a design rationale and did not develop further evaluation in respect to the effectiveness of the interaction and visualization in AR.

Other research has addressed the evaluation of the usefulness and effectiveness of an AR node-link diagram. Ware and Franck [18] examined the effect between a 2D and 3D node-link diagram on monitor screens, using path-finding tasks with different stereo cues to evaluate users' performance. Their study results showed that a 3D graph enhanced accuracy and spatial comprehension for abstract data analysis with the depth and motion cues it provided. Belcher et al. [20] compared the effect of using AR for graph link analysis to a desktop interface. According to their study, a tangible AR interface was better suited in the graph exploration than the desktop interface, while the stereo-graphic viewing had little effect on comprehension and task performance.

In previous studies, a Head-Mounted Display (HMD) has also been utilized as a data visualization tool, but many only focused on virtual reality and did not consider AR settings. Cordeil et al. [34] compared a CAVE-style environment and a VR HMD in the analysis of network connectivity and presented an in-depth analysis of the difference between a CAVE environment and a VR HMD for collaborating data analysis tasks. According to their study, a VR HMD could lead to faster collaboration, making it a suitable alternative to the more expensive CAVE devices, which is more accessible to reach a larger audience. Kwon et al. [35] also studied a VR HMD for graph visualization, comparing its usage for the 2D representation and the 3D representation. Bacim et al. [36] conducted a study, in which the authors studied how display fidelity affected graph analysis performance in a

VR HMD and found evidence that more immersive displays could offer significantly better overall task performance with higher display fidelity.

The detailed design and interactions of an AR node-link diagram, which can be especially challenging in immersive environments, have also been addressed in previous studies. Büschel et al. [37] examined how to visualize edges in a 3D node-link diagram to support efficient analysis in AR. They implemented eight edge variants and compared these different variants with eight participants. They reported on the findings of an initial study in which they compared these different variants and found that most variants achieved similar results. Participants slightly preferred to have colored edges in the AR graph, while blinking edges were rated low. Büschel et al. [1] conducted another study focusing on the design space for edge styles through comparing six different variants, including straight edge, curved edge, dashed edge, animated edge, glyph edge, and tapered edge, and reported on the results of a quantitative user study with 18 participants. The tasks applied in the study were typical graph exploration tasks, where participants were asked to find paths between two highlighted nodes. The results of the study showed that all participants were able to solve the tasks and there was no significant difference in terms of task completion time or accuracy rates. Notably, the tasks with longer paths were perceived as more difficult. The authors did not give a single clear recommendation for the edges to apply and developers were suggested to freely choose from the variants when designing for their use cases in particular.

Sun et al. [38] specifically investigated on how users would manage their spatial relationship with an AR node-link diagram with different graph scales, namely the room-scale and the table-scale. Through a user study with 16 participants solving two logical reasoning tasks by interacting with the AR graph, the study revealed three types of spatial arrangements and studied different user preferences under different scale conditions. However, the study did not examine the factors that would influence users' performance and behaviors in completing tasks. Moreover, the study demonstrated the graph with seven edges only. Schwajda et al. [39] transformed 2D graph data visualization for planar displays into AR spatial space with pull and drag gestures and identified a variety of factors influencing users' perception with an AR HMD. The authors believed that with proper design choices, the data visualization transitions from 2D to AR can leverage the efficiency and productivity of data analysis tasks. However, the study required further empirical study.

Our review of the related work shows that although previous studies have shown the potential of using AR for node-link diagram visualization, the research area is still in the early stage of development and the structured research into the design of an AR graph is largely underrepresented. Existing studies focused on the comparison of an AR graph to other methods, but did not pay much attention to the aspects such as interactions. Some studies lacked empirical evaluation. The research that investigated the embodied interactions with mobile devices was rare too. There is a rich space worthy of extensive and systematical exploration. As an important step toward a better understanding of this design space and to address existing gaps, we present and study a mobile based AR graph.

3. Design

3.1. Design Process

We are interested in exploring the design space of an AR graph for relationship visualization. To that end, we built an AR graph based on the database of Chinese painters. The database includes painters of all dynasties in China, using a relational database management system with information from academic works, papers, historical documents, etc., as the basic source of data. The database not only contains basic information of painters such as their names, places of origin and works, but also the social network of their family relationship, teacher-student relationship, and friendship relationship. As of 2021, a total of 29,893 painters have been included in the database. In this study, we first included painters involved in the path of a teacher–student relationship between two famous Chinese painters, resulting in 31 nodes (painters) and 108 edges (relations) in the graph.

We developed a prototype that could process JSON data in Unity3D, to generate nodes with text on top representing the names of painters and edges to link them in an AR environment. Figure 1 shows the application of the AR graph in an art gallery.

Figure 1. The application of the AR graph with data of Chinese painters in an art gallery (the text on top of the nodes are the Chinese names of the painters).

3.1.1. Layout Strategy

The typical visualization methods for a 2D node-link diagram are constraint-based and force-directed in 2D depending on whether it is based on a mechanical model [40]. The force-directed layout is a popular layout technology, especially for the network diagram visualization [41,42], which can display the overall structure of the network and the connections between nodes [16]. The algorithm of the force-directed approach described in [41] is achieved by utilizing attractive and repulsive forces between nodes: the attractive force will connect nodes with links, and the repulsive forces will push each other away if they are getting too close to each other.

In our study, we utilized the physics engine in Unity to simulate the two types of forces. To be more specific, we used the game component *Spring Joint* in Unity as the attractive force, which can connect two nodes to each other so that if one node moves the other one will also move together. With this feature, the two nodes that are linked to each other can be connected. What's more, we used the game component *Collider* in Unity to represent the repulsive force. The *Collider* defines the boundary of a node so that it can remain a distance set by the size of the *Collider* between nodes.

3.1.2. Interaction Paradigms

In this paper, we followed an approach that was solely based on mobile devices. We deliberately considered no other display since that mobile devices are ubiquitous and widely used in everyday life [17]. Users do not need to spend extra time on learning new types of interactions and will focus on the interaction with the AR graph. We first designed the interactions with the touch capability of mobile devices since it could suit precision requirements for engaging with detailed visualization tasks [17]. We will examine its effectiveness by observing users' behaviors with three basic interactions:

Click. By performing a single touch on a node, it will be highlighted, together with the nodes that are connected to it. See Figure 2A.

Drag. By performing a long touch on a certain node, the user can move it around and place the node in different positions. All the nodes that are related to the dragged node will be moved together. See Figure 2B.

Move. By moving physically around the room, the user can manipulate the viewpoint. The user can obtain an overview by stepping back, or explore certain nodes in details by moving close to the nodes. It should be noted that the text showing the names of the painters will always face to users no matter what angles they rotate. See Figure 2C.

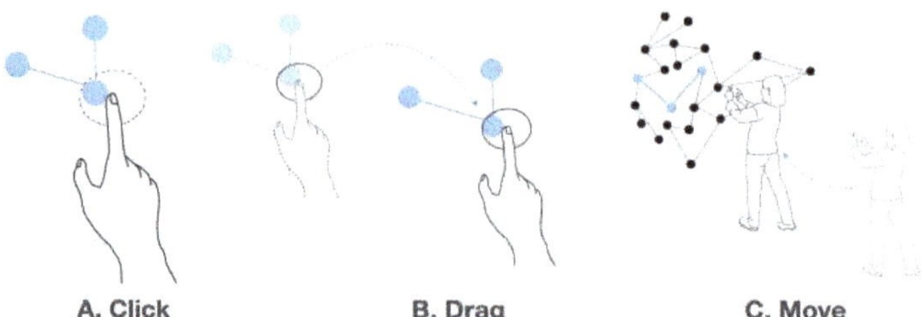

Figure 2. Interaction paradigms.

3.1.3. Colors and Scale

In the study of Büschel et al. [37], the authors mapped different colors to the edges and found that participants felt that they could discern different colors. However, there was no conclusion for the preferred colors in the graph. In their later study [1], they used colors of gray and blue in the graph, the sphere to represent the nodes, and six variants for the edges. Regarding to the scale and size of the AR graph, Sun et al. [38] presented an initial study to compare a room-scale graph and a table-scale graph. According to their study, the room scale brought benefits of a clearer view and wider interaction space, as well as the convenience to obtain different perspectives of the visualization. Overall, there has been a lack of research regarding the aesthetics of the graph, leaving unanswered to the basic question of how to design the nodes and edges in AR settings in terms of their colors, shapes, and size [8]. In our current study, we used the colors the same as in our database, which are also similar to the colors used in the work of Büschel et al. [1], including gray and blue. The shape of the nodes was spherical, and for the edge it was a line, as most 2D graphs use circle and line. The graph was in a room-scale with size of 50 (w) $\times$ 120 (d) $\times$ 100 (h) in cm referring to the size of the laboratory room.

4. User Study

The purpose of this study is to evaluate users' experience with the AR graph and explore the design space for further development. We also developed a 2D graph so that participants could compare the two methods for their pros and cons and make improvement suggestions to the AR graph. The nodes in 2D graph were randomly distributed, applying the same interaction paradigms and appearances as the AR graph. It should be noted that we were not aiming to compare 2D and the AR graph. Instead, we shifted the evaluation focus towards the interaction and design of the AR graph by comparing a novel visualization with a traditional visualization, and further identifying envisioned improvements. To narrow the scope of our study, we chose tasks that do not require pre-knowledge of visual data analysis. We collected data from each participant to gain further insights. See Figure 3 below.

Figure 3a shows the laboratory room. The room has no windows and the color of the light is white. The color of the wall is light gray. The experiment settings were the same for users to experience with the AR graph and the 2D graph. Figure 3b,c depict the AR graph. With the AR graph, the user saw the graph through the camera of the mobile device with the nodes distributed in the space. The user had to move around the room by walking and turning around to see the nodes. Figure 3d,e show the 2D graph, where the user interacted with the nodes via a 2D screen.

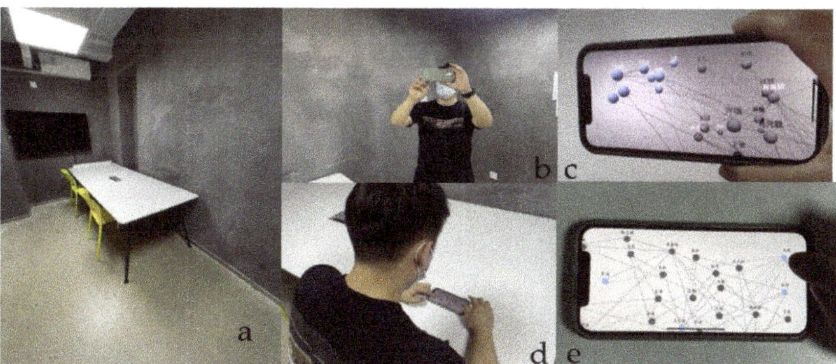

Figure 3. Experiment settings: (**a**). laboratory room; (**b**). the user experiences the AR graph; (**c**). AR graph; (**d**). the user experiences the 2D graph; (**e**). 2D graph (the text on top of the nodes are the Chinese names of the painters).

4.1. Experiment Design

We designed a within-subjects study with a counter-balanced order across participants. Participants performed the 2D graph and the AR graph with the data of 31 nodes and 108 edges. Lee et al. [43] defined the most common analysis tasks of graph data, including topology-based tasks, browsing tasks, overview tasks, and attribute-based tasks. Our study focused on the low-level topology-based tasks to assess the potential use of the AR graph.

The first type of task was to count all nodes that are related directly to the given node. Based on the number of the linked nodes, we divided the tasks into five levels. To be more specific, in task 1, we asked about the painter who had teacher-student relationships with two other painters (*2D: Who have the direct teacher-student relationship with Huang Shen? AR: Who have the direct teacher-student relationship with Chen Yuansu?*). In task 2, we asked about the painter who had teacher-student relationships with three other painters (*2D: Who have the direct teacher-student relationship with Gao Xiang? AR: Who have the direct teacher-student relationship with Lu Zhi?*). In task 3, we asked about the painter who had teacher-student relationships with four other painters (*2D: Who have the direct teacher-student relationship with Jiang Yanxi? AR: Who have the direct teacher-student relationship with Li Shan?*). In task 4, we asked about the painter who had teacher-student relationships with five other painters (*2D: Who have the direct teacher-student relationship with Chen Chun? AR: Who have the direct teacher-student relationship with Zhua Da?*). In task 5, we asked about the painter who had teacher-student relationships with seven other painters (*2D: Who have the direct teacher-student relationship with Ni Zan? AR: Who have the direct teacher-student relationship with Wen Zhengming?*). For the same task level, the questions involved the same number of nodes for the 2D graph and the AR graph, but for different painters. Task 6 involved second-degree teacher-student relationship, meaning that participants would need to click on the given node first, and then after the related nodes (first degree) showed up, they would need to click on the nodes to find out the nodes with second-degree relationship (*2D: Who have the second-degree teacher-student relationship with Hong Ren? AR: Who have the second-degree teacher-student relationship with Wen Ding?*).

Figure 4 shows the example of each task. The given node (given painter) in the diagram was highlighted in red circle. The related nodes (first degree) were highlighted in orange.

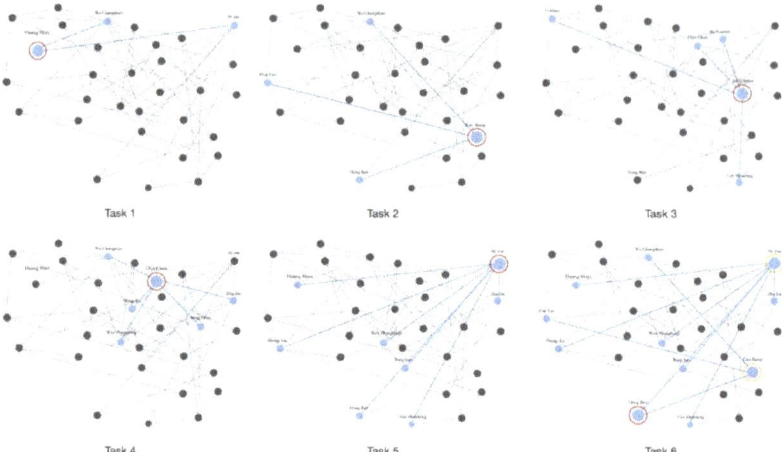

Figure 4. Examples of task 1 to task 6.

4.2. Measurements

The following measures were captured over the course of the experiment. We recorded the completion time to finish each task in seconds and the accuracy rate as scores in completing the tasks (score 1 for obtaining the correct answer, score 0 for the wrong answer). In addition, the user experience questionnaire (UEQ) [44] was applied as the evaluation of the exploring experience with a five-point Likert-scale. UEQ contains 6 scales and 24 items, including attractiveness, perspicuity, efficiency, dependability, stimulation, and novelty. The pre-test and post-test regarding the knowledge of Chinese painters were filled in to evaluate and compare the learning performance. The pre-test worked as the baseline of participants with six questions related to the student-teacher relationships among the 31 painters. The pre-test included the same questions chosen randomly from questions asked in task 1–6. Another two post-tests were conducted after using the AR graph and the 2D graph with the questions the same as the tasks they just finished and the average scores were calculated. The experimental process was also observed and recorded with the consent of the participants. Figure 5 depicts the overall procedure of the experiment.

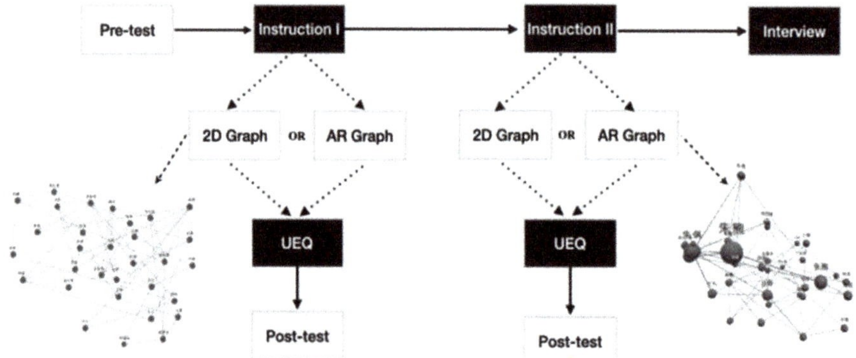

Figure 5. Experiment procedure (the text on top of the nodes are the Chinese names of the painters).

4.3. Participants and Procedures

In total, 36 undergraduate and postgraduate students (18 men and 18 women) volunteered for this experiment. They ranged in ages from 18 to 27. We asked participants how familiar they were with Chinese painters. Only five participants indicated that they knew about Chinese painters a little bit. The rest of the participants stated that they did not have any prior knowledge.

The experiments were conducted in a laboratory room with two desks and two chairs. Experimental stimuli were presented on the iPhone 12.

Participants first filled in a pre-test with six questions to find out their prior knowledge of Chinese painters. Then the instructions of how to use the graph as a tool to complete the tasks were presented to the participants. In the experimental process, participants used a 2D and AR graph to complete the six tasks respectively and filled in the questionnaire for each graph. The starting point for the AR graph was marked on the ground and the virtual scene stationary was placed in the middle of the room.

After experiencing each graph, they also finished a post-test with the same six questions to test their learning performance. At the end of the experiment, participants were interviewed to express more opinions regarding to both graphs.

The entire experiment, including pre-test, instructions, formal experiment, and post-test, took around 30–40 min to complete.

5. Results

In the following section, we present the results of the experiment in details. For each dependent measure of completion time, accuracy rate, learning performance, and user experience, we used the repeated measures ANOVA in SPSS, analyzing the within-subjects' factors of different types of graphs and different tasks.

5.1. Completion Time

The ANOVA showed significant main effects on completion time for different types of graphs ($F(1,35) = 20.11$, $p = 0.000$, $\eta^2 = 0.365$). It took significantly more time to complete all tasks with the AR graph (M = 49.93 s, SD = 15.73 s) than with the 2D graph (M = 38.05 s, SD = 10.71 s). See Figure 6.

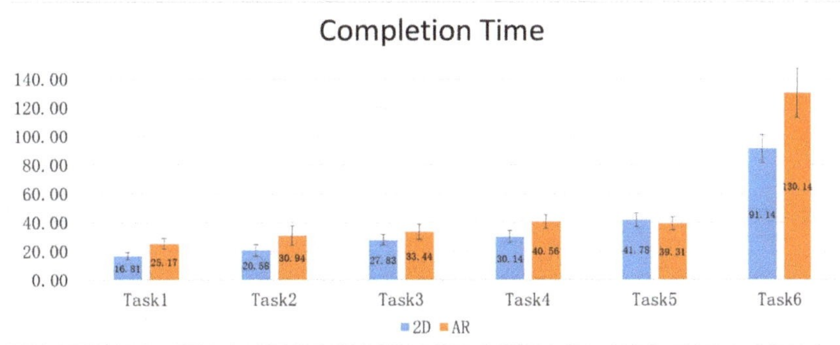

Figure 6. Completion time (seconds) for the six tasks with the 2D graph and the AR graph.

There were also significant main effects on completion time for different tasks ($F(5,175) = 109.04$, $p = 0.000$, $\eta^2 = 0.757$). The more nodes and edges involved, the more time was spent on completing the tasks: task 1 (M = 20.99 s, SD = 1.45 s), task 2 (M = 25.76 s, SD = 2.27 s), task 3 (M = 30.64 s, SD = 2.26 s), task 4 (M = 35.35 s, SD = 2.23 s), task 5 (M = 40.54 s, SD = 2.21 s), and task 6 (M = 110.64 s, SD = 6.96 s).

There was a significant interaction effect of graph types and different tasks ($F(5,175) = 8.618$, $p = 0.000$, $\eta^2 = 0.198$). Post-hoc tests showed that when visualized in a 2D graph, the more complicated tasks that involved more nodes and edges took significantly longer time to complete: there was no significant difference in completion time for the first four tasks, while task 5 took significantly more time than task 1 ($p = 0.000$), task 2 ($p = 0.000$), and task 3 ($p = 0.041$). Task 6 took significantly more time than all the other five tasks ($p = 0.000$).

When visualized in the AR graph, the numbers of nodes and edges involved in the tasks seemed to not influence the completion time: there was no significant difference in the first five tasks, while task 6, which was more complex with indirect relations, took significantly more time than task 1 to 5 ($p = 0.000$).

5.2. Accuracy Rate

Regarding the accuracy rate, ANOVA showed significant main effects for graph types ($F(1,35) = 12.80$, $p = 0.001$, $\eta^2 = 0.268$). Completing tasks with the 2D graph (M = 0.77, SD = 0.20) was more accurate than completing with the AR graph (M = 0.62, SD = 0.22).

There were also significant main effects on the accuracy rate for different tasks ($F(5,175) = 12.28$, $p = 0.000$, $\eta^2 = 0.260$). The accuracy rate and the complexity of the tasks were negatively correlated: task 1 (M = 0.83, SD = 0.04), task 2 (M = 0.81, SD = 0.05), task 3 (M = 0.72, SD = 0.05), task 4 (M = 0.74, SD = 0.04), task 5 (M = 0.68, SD = 0.05), and task 6 (M = 0.40, SD = 0.06). See Figure 7.

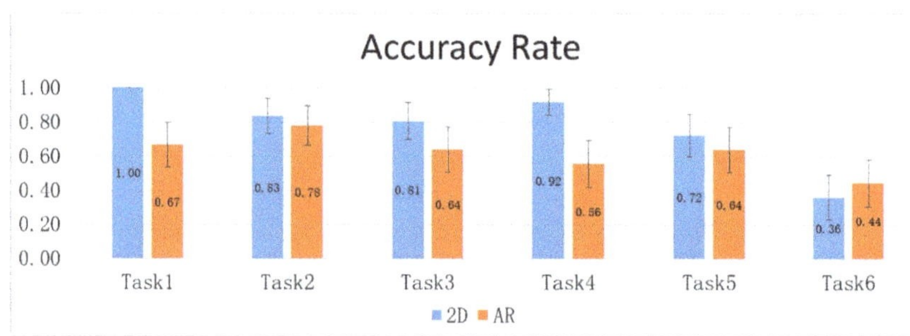

Figure 7. Accuracy rate for the six tasks with the 2D graph and the AR graph.

There was a significant interaction effect of graph types and different tasks ($F(5,175) = 2.67$, $p = 0.023$, $\eta^2 = 0.070$). Post-hoc tests showed that for the 2D graph, there was no significant difference in the accuracy rate for the first five tasks except that task 1 was answered significantly more accurate than task 5 ($p = 0.026$). Tasks 1 to 5 were completed significantly more accurately than task 6 ($p \leq 0.000$).

When completed with the AR graph, there was no significant difference in all the tasks for their accuracy rate. In addition, the 2D graph was significantly more accurate than the AR graph when the task was relatively easy (e.g., task 1, $p = 0.000$). For more complicated tasks such as task 5 and task 6, there was no significant difference.

5.3. Learning Performance

There were significant main effects in the post-test for different graph types ($F(1,35) = 12.87$, $p = 0.001$, $\eta^2 = 0.269$). The AR graph (M = 0.28, SD = 0.17) resulted in a better learning effect than the 2D graph (M = 0.18, SD = 0.13). When compared on the pre-test and the post-test, the learning performance for the AR graph was significant for the 2D graph, the learning performance was not significant. See Figure 8.

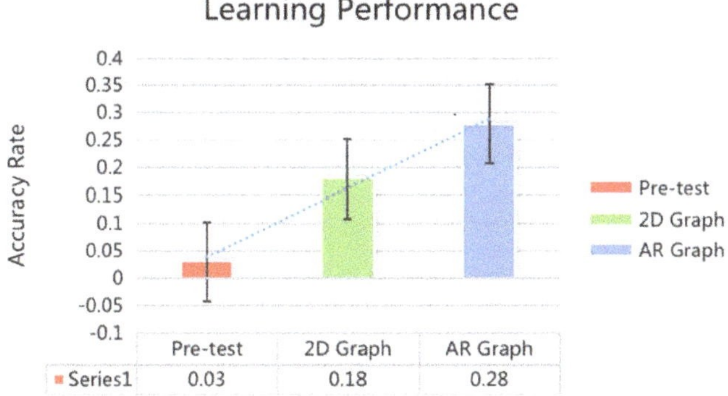

Figure 8. Learning performance between the pre-test and the post-test (2D graph and AR graph) (the dot line shows the growing trend among the learning performance).

5.4. User Experience

All categories in UEQ had significant differences for different graph types. The 2D graph was rated higher in terms of perspicuity (F(1,35) = 71.73, p = 0.000, η^2 = 0.672), efficiency (F(1,35) = 132.00, p = 0.000, η^2 = 0.790), and dependency (F(1,35) = 6.36, p = 0.016, η^2 = 0.154). To be more specific, the 2D graph was perceived as easier to become familiar with than the AR (perspicuity: 2D (M = 4.78, SD = 0.34) vs. AR (M = 3.78, SD = 0.57)); participants could solve tasks with the 2D more efficiently without unnecessary effort than with the AR (efficiency: 2D (M = 4.45, SD = 0.35) vs. AR (M = 3.49, SD = 0.50)); and they felt in control of the interaction with the 2D graph more than with the AR (dependency: 2D (M = 4.29, SD = 0.58) vs. AR (M = 4.04, SD = 0.30)).

On the other hand, the AR was rated higher in terms of attractiveness (F(1,35) = 4.28, p = 0.046, η^2 = 0.109), stimulation (F = 61.89, p = 0.000, η^2 = 0.639), and novelty (F = 240.06, p = 0.000, η^2 = 0.873). Participants liked the AR graph more than the 2D graph in general (attractiveness: 2D (M = 4.08, SD = 0.37) vs. AR (M = 4.23, SD = 0.31)); and it was perceived as more exciting and motivating to use the AR graph than the 2D (stimulation: 2D (M = 3.82, SD = 0.49) vs. AR (M = 4.51, SD = 0.39)). The AR graph was also considered as more innovative and creative than the 2D (novelty: 2D (M = 3.33, SD = 0.55) vs. AR (M = 4.79, SD = 0.21)). See Figure 9.

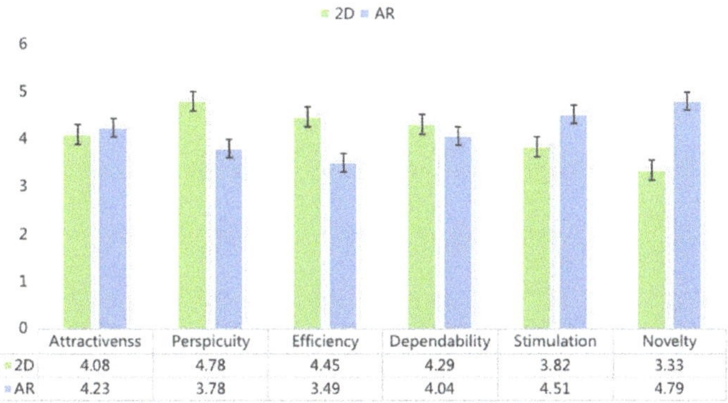

Figure 9. User experience with the 2D and the AR graph.

5.5. Interview and Observation

During the interview session, we asked participants to express their feelings on the two graphs. The answers from participants were aligned with the results of UEQ. Participants thought the 2D graph was more direct and efficient for completing tasks: *"it is more direct and clearer, we can use it to find the relationship easily"* (p2); *"we can use it to do assignments and get the information quickly"* (p11); *"I can move the graph by simply touching the screen, it is more direct"* (p21). The AR graph was preferred by participants with the reasons of fun and immersive experience, and perceiving the sense of space: *"it was more fun with the AR graph, like a game, while the 2D graph could be useful for homework"* (p32); *"everything was happening in the real world, I could see different things from different positions"* (p34).

Although we told participants that they could use as much time as they needed, most of them still tried to complete tasks as fast as they could without taking time to remember the relationship among the painters. When asked how to improve the learning performance with the AR graph, they would like to see richer information about a certain painter to help them discover more knowledge: *"more elements can be filled into the AR graph, such as the descriptions of the artists and their paintings"* (p5).

Meanwhile, we observed that when interacting with the AR graph, participants frequently touched the wrong nodes over the one they expected via the touch screen. Another common behavior in participants was trying to scale the AR graph with their fingers pinching on the screen rather than physically moving in the room, especially in the beginning of the experiment, *"when I was touching and moving the nodes, I always forgot that I could move around to adjust my view point"*(p17); *"it was interesting to explore in the room, but it could be better if I could adjust the angles and scales through the screen as well"* (p12).

5.6. Discussion of the Results

The results of our study showed that the overall completion time with the 2D graph was shorter than with the AR graph. This could be caused by the time spent walking around the room physically. Meanwhile, our results revealed that in the 2D graph, it took a longer time to complete tasks with more nodes and edges, while it will not affect the performance in the AR graph, indicating the potential to use the AR graph to deal with larger data set.

For the accuracy rate, the 2D graph outperformed the AR graph in more simple tasks with less nodes and edges or one-degree. When it was becoming more complicated, the accuracy rate in the 2D dropped, but had no effect in the AR. This also supports the capabilities of AR to visualize more complicated graphs. The lower accuracy rate in AR might be caused by ignoring the nodes behind when participants did not move or change angles in the physical world, or by mis-touching on the wrong nodes in AR.

The scores of the pre-test were extremely low due to the lack of knowledge of early Chinese painters among students. Experiencing with the AR graph, the learning performance was better improved than with the 2D. One possible reason for this is that participants spent more time in AR. Another possible explanation is that the spatial layouts and embodied movements could enhance the recall and comprehension of the graph, which was also mentioned in previous works (e.g., [20]).

The results of UEQ were divided into two parts. On one hand, participants perceived the AR graph as more exciting and motivating, innovative and creative, and attractive over the 2D graph. On the other hand, the 2D graph was easier to become familiar with and could help them solve their tasks without unnecessary effort. They felt in control of the interaction more with the AR graph.

6. Iterations

Design

The study results demonstrated that AR could be utilized as an engaging way for data visualization especially with graphs with more nodes and edges. However, the results also revealed that current interactions in the AR graph with a touch-screen and embodied movements might introduce interaction challenges in terms of effectiveness and usability. Based on these insights, we iterated the current AR graph to implement more:

- Embodied interactions with hand gestures

- Connection between the physical objects and the digital graph

Figure 10 shows the iterated interaction process: (0) when scanning a painting in the real world with the mobile device, the node of the painter will show up (the blue sphere). (1) If the user use their hand to ***touch*** a certain node, the node will be highlighted in blue, and the other nodes that are directly linked to the touched-node will show up. (2) The user can ***hold*** a certain node to obtain further information, e.g., the painting of the painter being held will show up. (3) The user can then ***grab*** the virtual painting and move it next to the real painting or to anywhere in the space. The interaction paradigms are applicable for all nodes, applying the *AR Foundation SDK* and *ManoMotion SDK* on iPhone 12.

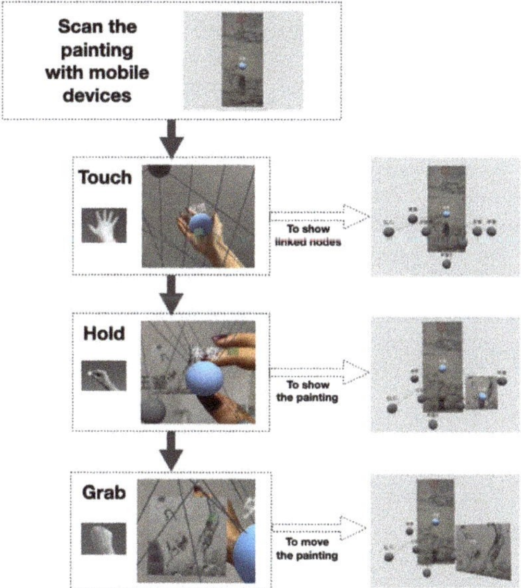

Figure 10. The interaction process with the iterated AR graph (the text on top of the nodes are the Chinese names of the painters).

We asked six participants to try the iterated AR graph and collected initial feedback from them. According to the participants, it was easy to interact with the virtual objects with their hands, *"the interaction was easy to understand, I can touch or move the virtual objects just like in the real world"* (p1); *"it is convenient to interact with hands"* (p5). With the hand-gestures, participants did not touch the screen but moved physically in the room in a more natural way.

Participants also appreciated the feature to obtain the painting in the real world connected to the graph network, *"I would use this feature when I am in a museum"* (p2). *"I get to see more paintings in AR, and I can put them next to the real painting on the wall, it is pretty cool"* (p4).

7. Conclusions and Future Work

Overall, our study showed the feasibility of using the AR graph to present data relations, validated its possibilities for education and opportunities to engage a wider audience, and demonstrated its potentials for visualizing a complex graph with more nodes and edges. However, we identified challenges of applying AR interactions with touch-screen on mobile devices. To address the challenges, we further iterated the AR graph with more embodied interactions. Moreover, to better utilize the ability of AR to overcome the limitation of the screen size and put more information into the graph, we addressed the connection between the physical objects and the digital graph in more depth.

For example, we started with a real painting and displayed the painter's information and social networks within the AR environment.

Although the application of the AR graph is promising, there are still unanswered questions in the design and many aspects that require further research. This study is the first step in our research. In the future, we will conduct studies on the AR embodied interactions proposed in this study with different analysis tasks. The methods of transferring a data set into an AR graph can be extended to other domains to create immersive experiences. We can investigate how the AR interactions and features can be applied in immersive data visualization for general context. We can also take an in-depth look into the details of how to design the nodes and edges with different shapes, sizes, colors, etc. We can examine how to couple the AR graph to the physical world, such as the locations and the objects. The freedom to define the axes, dimensions, and directions in AR also leaves a space for more findings. It is also important to determine the performance of the AR graph with more complex and larger data set with different layout strategies. Last but not least, the research on the real-time collaboration in the AR graph is also a future direction.

Author Contributions: Conceptualization, J.L. and Z.W.; methodology, J.L. and Z.W.; software, J.L. and Z.W.; validation, J.L. and Z.W.; formal analysis, J.L.; investigation, J.L. and Z.W.; resources, J.L. and Z.W.; data curation, J.L. and Z.W.; writing—original draft preparation, J.L.; writing—review and editing, J.L. and Z.W.; visualization, J.L. All authors have read and agreed to the published version of the manuscript.

Funding: This research received no external funding.

Acknowledgments: The authors wish to thank all participants who volunteered for this study. Informed consent was obtained from all subjects involved in the study.

Conflicts of Interest: The authors declare no conflict of interest.

References

1. Büschel, W.; Vogt, S.; Dachselt, R. Augmented reality graph visualizations. *IEEE Comput. Graph. Appl.* **2019**, *39*, 29–40. [CrossRef] [PubMed]
2. Satriadi, K.A.; Smiley, J.; Ens, B.; Cordeil, M.; Czauderna, T.; Lee, B.; Yang, Y.; Dwyer, T.; Jenny, B. Tangible globes for data visualisation in augmented reality. In Proceedings of the CHI Conference on Human Factors in Computing Systems, New Orleans, LA, USA, 29 April–5 May 2022; pp. 1–16.
3. Chandler, T.; Cordeil, M.; Czauderna, T.; Dwyer, T.; Glowacki, J.; Goncu, C.; Klapperstueck, M.; Klein, K.; Marriott, K.; Schreiber, F.; et al. Immersive analytics. In Proceedings of the IEEE International Symposium on Big Data Visual Analytics, Hobart, TAS, Australia, 22–25 September 2015; pp. 1–8.
4. Marriott, K.; Chen, J.; Hlawatsch, M.; Itoh, T.; Nacenta, M.A.; Reina, G.; Stuerzlinger, W. Immersive analytics: Time to reconsider the value of 3d for information visualisation. In *Immersive Analytics*; Springer: Cham, Switzerland, 2018; pp. 25–55.
5. ElSayed, N.; Thomas, B.; Marriott, K.; Piantadosi, J.; Smith, R. Situated analytics. In Proceedings of the 2015 Big Data Visual Analytics (BDVA), Hobart, TAS, Australia, 22–25 September 2015.
6. Ens, B.; Irani, P. Spatial analytic interfaces: Spatial user interfaces for in situ visual analytics. *IEEE Comput. Graph. Appl.* **2016**, *37*, 66–79. [CrossRef] [PubMed]
7. Büschel, W.; Reipschläger, P.; Langner, R.; Dachselt, R. Investigating the use of spatial interaction for 3D data visualization on mobile devices. In Proceedings of the 2017 ACM International Conference on Interactive Surfaces and Spaces, Brighton, UK, 17–20 October 2017; pp. 62–71.
8. Kraus, M.; Fuchs, J.; Sommer, B.; Klein, K.; Engelke, U.; Keim, D.; Schreiber, F. Immersive analytics with abstract 3D visualizations: A survey. *Comput. Graph. Forum* **2022**, *41*, 201–229. [CrossRef]
9. Alper, B.; Hollerer, T.; Kuchera-Morin, J.; Forbes, A. Stereoscopic highlighting: 2d graph visualization on stereo displays. *IEEE Trans. Vis. Comput. Graph.* **2011**, *17*, 2325–2333. [CrossRef] [PubMed]
10. Gibson, H.; Faith, J.; Vickers, P. A survey of two-dimensional graph layout techniques for information visualisation. *Inf. Vis.* **2013**, *12*, 324–357. [CrossRef]
11. Henry, N.; Fekete, J.D. Matlink: Enhanced matrix visualization for analyzing social networks. In *IFIP Conference on Human-Computer Interaction*; Springer: Berlin/Heidelberg, Germany, 2007; pp. 288–302.
12. Henry, N.; Fekete, J.D.; McGuffin, M.J. NodeTrix: A hybrid visualization of social networks. *IEEE Trans. Vis. Comput. Graph.* **2007**, *13*, 1302–1309. [CrossRef] [PubMed]
13. Pinaud, B.; Vallet, J.; Melançon, G. On visualization techniques comparison for large social networks overview: A user experiment. *Vis. Inform.* **2020**, *4*, 23–34. [CrossRef]

14. Greilich, M.; Burch, M.; Diehl, S. Visualizing the evolution of compound digraphs with TimeArcTrees. *Comput. Graph. Forum* **2009**, *28*, 975–982. [CrossRef]
15. Burch, M.; Beck, F.; Weiskopf, D. Radial Edge Splatting for Visualizing Dynamic Directed Graphs. In Proceedings of the International Conference on Computer Graphics Theory and Applications and International Conference on Information Visualization Theory and Applications-IVAPP, (VISIGRAPP 2012), Rome, Italy, 24–26 February 2012; pp. 603–612. [CrossRef]
16. Sun, K.; Liu, Y.; Guo, Z.; Wang, C. Visualization for knowledge graph based on education data. *Int. J. Softw. Inform.* **2016**, *10*, 1–13. [CrossRef]
17. Langner, R.; Satkowski, M.; Büschel, W.; Dachselt, R. Marvis: Combining mobile devices and augmented reality for visual data analysis. In Proceedings of the 2021 CHI Conference on Human Factors in Computing Systems, Yokohama, Japan, 8–13 May 2021; pp. 1–17.
18. Ware, C.; Franck, G. Evaluating stereo and motion cues for visualizing information nets in three dimensions. *ACM Trans. Graph.* **1996**, *15*, 121–140. [CrossRef]
19. Ware, C.; Mitchell, P. Visualizing graphs in three dimensions. *ACM Trans. Appl. Percept.* **2008**, *5*, 1–15. [CrossRef]
20. Belcher, D.; Billinghurst, M.; Hayes, S.E.; Stiles, R. Using augmented reality for visualizing complex graphs in three dimensions. In Proceedings of the Second IEEE and ACM International Symposium on Mixed and Augmented Reality, Tokyo, Japan, 10 October 2003; pp. 84–93.
21. Hubenschmid, S.; Zagermann, J.; Butscher, S.; Reiterer, H. Stream: Exploring the combination of spatially-aware tablets with augmented reality head-mounted displays for immersive analytics. In Proceedings of the 2021 CHI Conference on Human Factors in Computing Systems, Yokohama, Japan, 8–13 May 2021; pp. 1–14.
22. Sicat, R.; Li, J.; Choi, J.; Cordeil, M.; Jeong, W.K.; Bach, B.; Pfister, H. DXR: A toolkit for building immersive data visualizations. *IEEE Trans. Vis. Comput. Graph.* **2018**, *25*, 715–725. [CrossRef]
23. Bach, B.; Sicat, R.; Beyer, J.; Cordeil, M.; Pfister, H. The hologram in my hand: How effective is interactive exploration of 3D visualizations in immersive tangible augmented reality? *IEEE Trans. Vis. Comput. Graph.* **2017**, *24*, 457–467. [CrossRef]
24. Kirshenbaum, N.; Hutchison, J.; Theriot, R.; Kobayashi, D.; Leigh, J. Data in context: Engaging audiences with 3D physical geo-visualization. In Proceedings of the Extended Abstracts of the 2020 CHI Conference on Human Factors in Computing Systems, Honolulu, HI, USA, 25–30 April 2020; pp. 1–9.
25. Reipschlager, P.; Flemisch, T.; Dachselt, R. Personal augmented reality for information visualization on large interactive displays. *IEEE Trans. Vis. Comput. Graph.* **2020**, *27*, 1182–1192. [CrossRef]
26. Wang, X.; Besançon, L.; Rousseau, D.; Sereno, M.; Ammi, M.; Isenberg, T. Towards an understanding of augmented reality extensions for existing 3D data analysis tools. In Proceedings of the 2020 CHI Conference on Human Factors in Computing Systems, Honolulu, HI, USA, 25–30 April 2020; pp. 1–13.
27. Mahmood, T.; Butler, E.; Davis, N.; Huang, J.; Lu, A. Building multiple coordinated spaces for effective immersive analytics through distributed cognition. In Proceedings of the 2018 International Symposium on Big Data Visual and Immersive Analytics (BDVA), Konstanz, Germany, 17–19 October 2018; pp. 1–11.
28. Chen, Z.T.; Su, Y.J.; Wang, Y.F.; Wang, Q.; Qu, Q.; Wu, Y. MARVisT: Authoring glyph-based visualization in mobile augmented reality. *IEEE Trans. Vis. Comput. Graph.* **2020**, *26*, 2645–2658. [CrossRef]
29. Cordeil, M.; Cunningham, A.; Bach, B.; Hurter, C.; Thomas, B.H.; Marriott, K.; Dwyer, T. IATK: An immersive analytics toolkit. In Proceedings of the 2019 IEEE Conference on Virtual Reality and 3D User Interfaces (VR), Osaka, Japan, 23–27 March 2019; pp. 200–209.
30. Vieira, C.; Parsons, P.; Byrd, V. Visual learning analytics of educational data: A systematic literature review and research agenda. *Comput. Educ.* **2018**, *122*, 119–135. [CrossRef]
31. Dyckhoff, A.L.; Zielke, D.; Bültmann, M.; Chatti, M.A.; Schroeder, U. Design and implementation of a learning analytics toolkit for teachers. *J. Educ. Technol. Soc.* **2012**, *15*, 58–76.
32. Husain, S.S.; Kalinin, A.; Truong, A.; Dinov, I.D. SOCR Data dashboard: An integrated big data archive mashing medicare, labor, census and econometric information. *J. Big Data* **2015**, *2*, 1–18. [CrossRef]
33. Drochtert, D.; Geiger, C. Collaborative magic lens graph exploration. In Proceedings of the SIGGRAPH Asia 2015 Mobile Graphics and Interactive Applications, Kobe, Japan, 2–6 November 2015; pp. 1–3.
34. Cordeil, M.; Dwyer, T.; Klein, K.; Laha, B.; Marriott, K.; Thomas, B.H. Immersive collaborative analysis of network connectivity: CAVE-style or head-mounted display? *IEEE Trans. Vis. Comput. Graph.* **2016**, *23*, 441–450. [CrossRef]
35. Kwon, O.H.; Muelder, C.; Lee, K.; Ma, K.L. A study of layout, rendering, and interaction methods for immersive graph visualization. *IEEE Trans. Vis. Comput. Graph.* **2016**, *22*, 1802–1815. [CrossRef]
36. Bacim, F.; Ragan, E.D.; Scerbo, S.; Polys, N.F.; Setareh, M.; Jones, B.D. The effects of display fidelity, visual complexity, and task scope on spatial understanding of 3D graphs. In Proceedings of the Graphics Interface 2013, Regina, SK, Canada, 29–31 May 2013; Volume 2, pp. 25–32.
37. Büschel, W.; Vogt, S.; Dachselt, R. Investigating link attributes of graph visualizations in mobile augmented reality. In Proceedings of the CHI 2018 Workshop on Data Visualization on Mobile Devices, Montreal, QC, Canada, 21–26 April 2018; Volume 18.
38. Sun, Z.; Han, F.; Ma, X. Exploring the Effects of Scale in Augmented Reality-Empowered Visual Analytics. In Proceedings of the Extended Abstracts of the 2018 CHI Conference on Human Factors in Computing Systems, Montreal, QC, Canada, 21–26 April 2018; pp. 1–6.

39. Schwajda, D.; Pointecker, F.; Boess, L.; Anthes, C. Transforming Graph-based Data Visualisations from Planar Displays into Augmented Reality 3D Space. In Proceedings of the ISS'21: Interactive Surfaces and Spaces, Łódź, Poland, 14 November 2021.
40. Chen, Y.; Guan, Z.; Zhang, R.; Du, X.; Wang, Y. A survey on visualization approaches for exploring association relationships in graph data. *J. Vis.* **2019**, *22*, 625–639. [CrossRef]
41. Fruchterman, T.M.; Reingold, E.M. Graph drawing by force-directed placement. *Softw. Pract. Exp.* **1991**, *21*, 1129–1164. [CrossRef]
42. Itoh, T.; Muelder, C.; Ma, K.L.; Sese, J. A hybrid space-filling and force-directed layout method for visualizing multiple-category graphs. In Proceedings of the 2009 IEEE Pacific Visualization Symposium, Beijing, China, 20–23 April 2009; pp. 121–128.
43. Lee, B.; Plaisant, C.; Parr, C.S.; Fekete, J.D.; Henry, N. Task taxonomy for graph visualization. In Proceedings of the 2006 AVI Workshop on BEyond Time and Errors: Novel Evaluation Methods for Information Visualization, Venezia, Italy, 23 May 2006; pp. 1–5.
44. Schrepp, M.; Hinderks, A.; Thomaschewski, J. Applying the user experience questionnaire (UEQ) in different evaluation scenarios. In *International Conference of Design, User Experience, and Usability*; Springer: Cham, Switzerland, 2014; pp. 383–392.

electronics

MDPI

Article

Situating Learning in AR Fantasy, Design Considerations for AR Game-Based Learning for Children

Tengjia Zuo [1,*], Jixiang Jiang [2], Erik Van der Spek [1,*], Max Birk [1] and Jun Hu [1]

[1] Department of Industrial Design, Eindhoven University of Technology, 5612 AZ Eindhoven, The Netherlands; m.v.birk@tue.nl (M.B.); j.hu@tue.nl (J.H.)
[2] Department of Design, Politecnico di Milano, Politecnico di Milano, 20133 Milan, Italy; jixiang.jiang@polimi.it
* Correspondence: t.zuo@tue.nl (T.Z.); e.d.v.d.spek@tue.nl (E.V.d.S.); Tel.: +31-682449194 (T.Z.); +31-622334852 (E.V.d.S.)

Abstract: (1) Background: Augmented reality (AR) game-based learning, has received increased attention in recent years. Fantasy is a vital gaming feature that promotes engagement and immersion experience for children. However, situating learning with AR fantasy to engage learners and fit pedagogical contexts needs structured analysis of educational scenarios for different subjects. (2) Methods: We present a combined study using our own built AR games, MathMythosAR2 for mathematics learning, and FancyBookAR for English as second-language learning. For each game, we created a fantasy and a real-life narrative. We investigated player engagement and teachers' scaffolding through qualitative and quantitative research with 62 participants aged from 7 to 11 years old. (3) Results: We discovered that fantasy narratives engage students in mathematics learning while disengaging them in second-language learning. Participants report a higher imagination with fantasy narratives and a higher analogy with real-life narratives. We found that teachers' scaffolding for MathMythosAR2 focused on complex interactions, for FancyBookAR, focused on story interpretation and knowledge explanation. (4) Conclusions: It is recommended to mix fantasy and real-life settings, and use simple AR interaction and pedagogical agents that enable teachers' scaffolding seamlessly. The design of AR fantasy should evaluate whether the story is intrinsically related to the learning subjects, as well as the requirements of explicit explanation.

Keywords: augmented reality; game-based learning; fantasy; situated learning; serious game

Citation: Zuo, T.; Jiang, J.; Spek, E.V.d.; Birk, M.; Hu, J. Situating Learning in AR Fantasy, Design Considerations for AR Game-Based Learning for Children. *Electronics* **2022**, *11*, 2331. https://doi.org/10.3390/electronics11152331

Academic Editors: Jorge C. S. Cardoso, André Perrotta, Paula Alexandra Silva and Pedro Martins

Received: 8 July 2022
Accepted: 24 July 2022
Published: 27 July 2022

Publisher's Note: MDPI stays neutral with regard to jurisdictional claims in published maps and institutional affiliations.

1. Introduction

Augmented reality (AR), presenting virtual graphics, 3D models, animation, and sound effects into physical materials or spaces, enables users to engage with real-world objects and virtual interactive overlays simultaneously [1,2]. Combining digital game features and physical situations, AR game-based learning allows games to be integrated into the classroom in situated learning in an engaging way. As the core of situated learning is to intrinsically link the knowledge to be learned to physical and cultural contexts where learning occurs [3], AR provides contexts that allow users to sense real-world situations and adapt digital instruction to them [4]. Integrating game features into instruction, game-based learning enables its participants to engage in learning without particularly focusing on the learning content, with a fun and immersive experience [5].

AR game-based learning has been used to playfully augment serious contexts and thereby present instructional goals in an engaging manner and has received increased attention in recent years [6–8]. Research on AR game-based learning incorporates computer science, pedagogy, and experience design [7]. To explore the application of AR game-based learning in 21st-century education, researchers need to simultaneously consider the current situations in learning, technological affordance, and user experience. One approach to improve technological benefits, enriching educational contexts, and player experience, is by incorporating gaming features such as avatar customization [9], fantasy [10], or goal-setting [11] into AR learning. Among them, the game element that can provide motive-specific affective incentives [12,13] is fantasy, i.e., unreal and fictional settings that deviate from everyday life [14,15]. Fantasy can positively affect game-based learning, increasing players' engagement, especially for children who are not self-motivated to study [16–18].

Ways of incorporating fantasy in learning contexts have sparked academic interest in recent years [17,19,20]. To move beyond the entertaining features of fantasy, we need to consider pedagogical requirements when incorporating fantasy elements into game-based learning applications. An appropriate learning environment features authentic contexts that allow for the natural complexity of the real-world [21], coaching and scaffolding during the initial steps [22], and articulation [23] to enable explicit expression of tacit knowledge [24], creating spaces for exploration of participants [23]. How to design a novel game-based learning experience that situates learning in AR fantasy, engages learners, and adheres to pedagogical principles requires disciplinary and context-specific analysis. For example, research to investigate the classroom context in which learning with AR fantasy is employed, exploratory research on the subjects such as language and mathematics to be learned, and more.

To further investigate situating learning in AR fantasy and game-based learning for different learning contexts, we present a combined study using our own designed AR games: MathMythosAR2, i.e., a game for mathematic learning and practising, and FancyBookAR, i.e., a game for English as second-language learning and practising. We prepared two narratives for each game: the fantasy and the real-life version. By comparing the fantasy and real-life versions, we explore the effect of fantasy on engagement and scaffolding and develop design considerations for AR game-based learning with fantasy or real-life contexts. By contrasting the effect of AR fantasy with the different subject matter, we develop strategies for adjusting fantasy structures to the specific qualities of the subject matter in the classroom.

This research benefits the pedagogy domains by offering inspiring practice, and practical solutions for the needs of situating learning in technological-rich contexts with a game-based learning experience. It connects learning situations with immersive technology with a novel, immersive and enjoyable experience that meets the needs of 21st learning. we offer innovative solutions that connect students and teachers with an engaging learning experience and strategies for scaffolding learning with AR fantasy. This research also contributes to the design research society in terms of offering innovative game design and practical design strategies for designing with AR fantasy, bringing the future of immersive technology closer to our lives.

2. Related Work

2.1. Augmented Reality in Game-Based Learning

Augmented Reality (AR) is a technology that enhances real-world environments with an interactive virtual overlay that engages the user's visual, auditory, and haptic senses [25]. Perceptually, AR enhances the actual environment, leading to intensely immersive experiences [26]. AR promises to bring a better play–learn experience by (1) visualizing knowledge and concepts from alternative perspectives, for example by bringing to life 3D invisible and abstract concepts [27], (2) facilitating social interactions with tangible and virtual materials, regardless of geography or time restrictions [28], and (3) by bridging formal and informal learning by eliminating barriers between virtual and physical worlds [29,30]. To incorporate these AR characteristics into game-based learning, designers should first understand how game aspects can be designed with AR and their implications in educational contexts.

2.2. The Magic Circle and Motivational Effect of AR Fantasy

As a term that describes fictional, imaginative, or unreal contexts, fantasy exists in games featuring narratives, interactions, and settings that deviate from the real world [15,31]. Playful experience with fantasy game elements can invite users into an immersive world, which is called the "magic circle [32]," with a temporal and spatial boundary separating users from the real world. Such boundaries are blurred by AR technology, where players enter the virtual world, sensing the physical world simultaneously [33]. AR Fantasy can create a circle Stapleton et al. refer to as a "mixed fantasy continuum" [34], which creates compelling venues and content prototypes to engage the audience's imagination. Certain imaginary mental activities involving creating new realities or relating to existing ones are what Choi et al. (2003) defined as imagination and analogy states of fantasy. These mental states are keys to a convincing mixed fantasy experience that allows participants to step into the magic circle with a "suspension of disbelief" [33]. To identify whether fantasy elements build a convincing magic circle, we need to understand players' fantasy states meanwhile analyze their engagement [35].

Engagement is a process of getting involved and connected with [36]. According to research on indicators that reflect different aspects of learning engagement [37–39], we synthesize a table of indicators for engagement (Table 1). Indicators include emotional, behavioral, and cognitive aspects of engagement, and can be used to identify the effect of integrating AR fantasy on engagement through qualitative methods.

Table 1. Indicators of emotional, behavioral, and cognitive engagement.

Dimension	Positive Indicators	Negative Indicators
Emotional Engagement	Thrilled, Curious, Express Values and Feelings, Focused, Interested, Enthusiastic, Happy.	Anxiety, Bored
Behavioral Engagement	Confident, Preference for Challenges, Extra Effort, Expressing the Value.	Frustration for Failure
Cognitive Engagement	Extra Activity, Comprehension of Knowledge, Attention, Active Participation.	Forced to Play, Not Following Rules, Confused

Players' emotional engagement comes from the pleasure of immersion [36], which fantasy elements can facilitate [40]. Players' cognitive engagement reflects "one's effort to put into self-regulated learning, involving a process of making sure oneself comprehend the game content." [41]. Previous research indicates that players can be more motivated in fantasy contexts but find it challenging to integrate the newly learned information with their prior knowledge [20]. Therefore, it is still an open question if incorporating fantasy in AR game-based learning may result in users' sufficient understanding that leads to positive behavioral and cognitive engagement.

In addition to qualitative methods of measuring engagement, we collect players' self-reported enjoyment parts of intrinsic motivation through intrinsic motivation inventory (IMI) in this study. Malone et al. connect endogenous fantasy with intrinsic motivation, in which learners are driven by qualities such as enjoyment and self-fulfilment. Endogenous fantasy in game-based learning can let players engage with affective incentives [13] without focusing on the learning goals [42].

2.3. Pedagogical Principles for Integrating Fantasy in AR Game-Based Learning

2.3.1. Fantasy and Situated Learning

Situated learning refers to learning practical applications through communication and interaction with environments [3]. Immersive technologies such as AR potentially provide various contexts and alternative perspectives to support situated learning. However, some questions remain about incorporating AR fantasy in game-based learning. Guidelines of situated learning often suggest designers embed contexts for authentic learning, where students explore, discuss and construct concepts of real-world issues they can relate to [43]. Authentic learning refers to activities that involve learning by solving issues, studying cases, and practising situations that are similar to those encountered in complex real-life situations [44]. Some scholars perceive authentic contexts as the opposite of fantasy, suggesting that only real-life contexts can draw learners to engage instead of passively receiving [45]. However, tracing back to the criteria of authentic learning, we found the essential was to learn grounded knowledge in the fields instead of memorizing abstract knowledge [46]. Fantasy contexts can be integrated with knowledge in fields and positively engage learners. The critical question is whether a convincing and immersive magic circle to an accessible learning context is constructed.

2.3.2. Scaffolding and the Role of Teachers

Balancing learners' motivation and comprehension in an immersive game-based learning magic circle requires careful design from game designers and proper guidance from teachers. The role of teachers in situated game-based learning is different from a traditional tutoring context. In the traditional education context, teachers guide students through scaffolding, which means offering temporal support to complete learning targets [47]. Whereas in a game-based learning context, players can also be guided by instructional content with natural progressions from the game's storyline [48]. Students who are fully immersed in the environment might require a less direct explanation from teachers [49]. Certain circumstances, allowing teachers to turn over controls of the learning context to students, encourage autonomous learning by students [50]. However, it also challenges teachers in terms of assisting students and tuning students' attention on the learning focus without interfering with their autonomy [51], i.e., their determination to interact based on their interests and values [52]. AR engages students with the virtual world while also maintaining a portion of their attention in the real-life classroom [53]. Properly designed AR game-based learning allows students to seamlessly move between virtual exploration and real-world interaction, creating spaces to insert teachers' roles.

Incorporating the role of teachers necessitates a detailed assessment of how students perform in different educational contexts. Pivec et al. summarize a criterion for different degrees of problem-solving skills of tasks, based on Vygotsky's work: (1) tasks can be accomplished alone by a student; (2) tasks that can be completed with the assistance of others; and (3) tasks that cannot be performed even with the assistance of others [54,55]. A further in-depth assessment of students' task completion using this criterion, in combination with their engagement analysis, can help situate teachers' roles in scaffolding towards specific contexts.

To construct a convincing and immersive magic circle in an accessible learning context, designing fantasy in game-based learning needs careful consideration based on an analysis of participants' real-world interaction and reaction to subject matter learning contexts. An exploratory study with design intervention is needed to explore the effect of AR fantasy on player engagement and teachers' scaffolding. Through the design study, we would like to understand the following research questions about situating learning in AR fantasy of game-based learning:

- RQ1: Whether fantasy in AR game-based learning for a classroom creates more engaging and immersive mental states for players?
- RQ2: What is teachers' role in scaffolding children in fantasy construction and learning?
- RQ3: How to situate learning in the AR fantasy of game-based learning to improve learners' engagement, experience, and the teachers' scaffolding?

3. Material and Methods

To answer the proposed research questions, we introduce our design, MathMythosAR2, and FancyBookAR, two storybook-based AR games focusing on learning mathematics and English. According to our findings through experiments and data analysis, we answer RQ1 and RQ2 and address design strategies to answer RQ3.

Both FancyBook AR and MathmythosAR 2 were designed with Unity3D, an engine that enables interactive experience through scene building and C#, and Vuforia, an engine that supports the functionality of AR through image recognition. The educational contents of both games are derived from actual educational practices. For the game MathMythos AR2, we combined the form of mathematical word problems [56] with game narrative and interactions. Similarly, in FancyBookAR, game features were integrated with the cloze test, a practice that requires participants to place the missing language items [57]. We invited 2 local English teachers to have the content checked. The fantastical game features used in FancyBookAR are comparable to those found in MathMythosAR2, employing a similar portrayal of innovation [15]—magic, technology, and aesthetics. The same game mechanics were used, such as cards, storybooks, and phone applications (Figure 1). Narrative genres of fantasy were applied in both games. To understand the difference in players' experiences under fantasy and real-life scenarios, we provided two narrative versions for each game: the fantasy and the real-life versions.

(a)

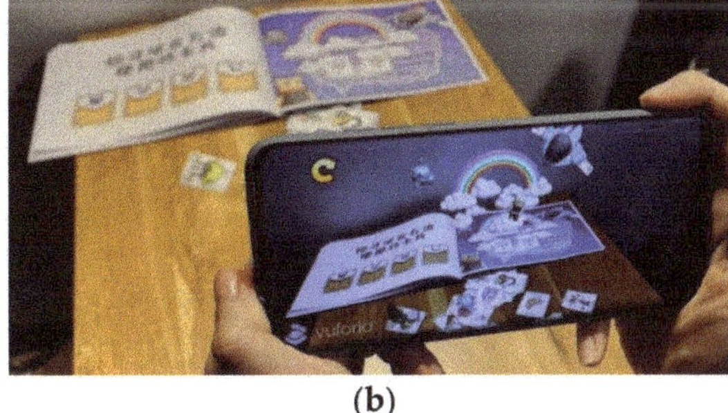

(b)

Figure 1. (**a**) MathmythosAR 2; (**b**) FancyBook AR.

3.1. MathMythosAR2

MathMythosAR2 is an AR game that encourages children to learn the addition techniques of mathematic calculation and practice their learned techniques. There are seven chapters in each game version. Students are first introduced to its narrative by scanning the pattern in the storybook. Then they use a virtual button by covering a specific pattern on the storybook with a hand to react to dialogue between NPC (Non-Player Character). Then players will be introduced to the addition technique. Players need to hold the virtual button and trigger the randomly generated numbers (randomization of correct answers $\pm$ 3), releasing when the right numbers pop up. Players receive cards with values to embark on the final practice, scan the cards with values, and sum up their total values.

In the fantasy version, players play a role of a magician. Players learn math magic to save a village from evil. Players complete the game task by making choices about the magic narratives and completing calculation practice about the math magic. We list a table of key narratives and options available in each section with screenshots in Table 2.

Table 2. The fantasy version of the game MathMythosAR2.

Section Screenshots	Key Narratives	Options
	Wake up! Rubin! Wake up!	N/A
	Hey Rubin, why are you sleeping here? The class is off.	1. Who are you? 2. Hi Mr Steven, I have some questions about math magic.
	There are two kinds of magic power with stones: single-digit and two-digit. (The screenshot shows how the NPC reacts to wrong answers.)	Choices available are between the correct answer ± 3.
	Brunwich was a beautiful town in the east. An invasion of evil magic shattered the peaceful area. Residents are in trouble, and only we can help them.	N/A
	It is a monster! Innocent people are turned into monsters by evil sorcery! To save them, we must use math magic to break the spell! Are you ready?	1. Wait! What was that again? I'm not ready. 2. Yes! I'm ready!
	Every time you sum up the magic power on the two magic stones will break one level of the evil spell. (One time of single-digit addition, two times of two-digit addition.)	Choices available are between the correct answer ± 3.
	Oh, you saved my life! Thank you, young magician!	1. What??? You are a human? 2. I'm glad you are back! Is everything ok?

The "real-life" version (Table 3) of MathMythosAR2 retains the same narrative framework and gameplay mechanics. However, players play a role of a daily life student, Robin. The main tasks are summing up total prices and shopping for the class party. Key narratives and options are presented in Table 3.

Table 3. The real-life version of the game MathMythosAR2.

Section Screenshots	Key Narratives	Options
	Wake up! Robin! Wake up!	N/A
	Hey Robin, why are you sleeping here? The class is off.	1. Who are you? 2. Hi Mr Steven, I have some math questions.
	There are two kinds of fruit prices on the board, the single-digit number and the two-digit number.	Choices available are between the correct answer $\pm$ 3.
	I heard you are organizing the class party for the new year and some materials are missing. I can help you buy them.	N/A
	Let us shop for party snacks. I will give you some coupons. When shopping, please use the addition techniques you just learned. Are you ready?	1. Wait! What was that again? I'm not ready 2. Yes! I'm ready!
	What is the total price of the coupons? (one time of single-digit addition, two times of two-digit addition)	Choices available are between the correct answer $\pm$ 3.
	The cashier machine is broken. I am sorry for the inconvenience. Thank you for your assistance with the calculation.	1. I would not have come to your store if I knew that. 2. No problem, I am willing to help. It is just simple mathematics.

3.2. FancyBookAR

FancyBookAR is a set of AR games that encourage children to learn English words and form sentences using their learned words. There are five sections in the game. In each section, players will first receive vocabulary cards and a storybook with incomplete sentences that need words to fill in the blanks. To complete the sentences, players need to understand the meaning of the words on each card and the possible context wherein they fit. A correctly constructed sentence with cards will trigger the completion of a game scene. As a result of completing tasks on the current scene, players activate the animation of the characters and complete the stories. There are two types of narrative in FancyBook AR, a fantasy and a real-life version.

In the fantasy narrative version, players play a role of a magician, the catperson. Players need to complete the story by choosing the options set in a fantasy genre. We list a table of key narratives and options available in each section with screenshots in Table 2.

The "real-life" version (Table 4) of FancyBook AR retains the same narrative framework and gameplay mechanics. However, the catperson is clothed in a regular outfit and lives in its own normal home this time. The narratives and options are set close to real-life situations (Table 5).

Table 4. The fantasy version of the game FancyBook AR.

Section Screenshots	Key Narratives	Options
	It is too dark. I use the __ to light up the sky.	1. Magic stick 2. Star bottle 3. Open 4. Turn on
	The star is too messy! I use ___ to clean the ground.	1. A robot 2. A magic book 3. A trash-eating monster
	I need to provide food and drinks. I take __ from the sky.	1. A moon croissant 2. Rainbow drink 3. A cotton candy
	I _ a _ to pick up my friend.	1. Ride 2. Drive 3. Dragon 4. Cloud 5. Spaceship
	I play with my friend in a __.	1. Mushroom trampoline park 2. Candy park 3. Magic forest

Table 5. The real-life version of the game FancyBook AR.

Section Screenshots	Key Narratives	Options
	It is too dark! I__ the __.	1. Turn on 2. Open 3. Window 4. Light
	The room is so messy! I use ___ to clean the __.	1. A broom 2. A vacuum 3. A rag 4. Bed 5. Ground 6. Table
	I need to provide food and drink. I buy __ in the supermarket.	1. Chocolate 2. Milk 3. A mango 4. A hamburger
	I __ to pick up my friend.	1. Ride a scooter 2. Take a bus 3. Skateboard
	It is playtime! I _with my friend.	1. Dance 2. Watch TV

Using the two games described above, each having two narrative versions, we experimented with game types (FancybookAR vs MathMythosAR2) as the between-subject factor and the narrative versions (real-life vs fantasy) as the within-subject factor. We compare the mathematic and language learning games that AR fantasy incorporates to explore ways and patterns of using AR fantasy to achieve increased engagement with two distinct learning subjects. We compare the fantasy and real-life versions to see the effect of fantasy on players' engagement and experience. We answered the research questions using the findings of the qualitative and quantitative analyses, investigating the phenomena and causes in light of previous work.

3.3. Participants and Procedure

This research involved 62 (male = 30, female = 32) participants from two different locations in China. Thirty-one participants from Qingdao, Shandong province, with an average age of 9.2, ranging from 7 years old to 11 years old, were invited to play the game MathMythosAR2. Thirty-one participants from Hefei Anhui province, with an average age of 10.6, ranging from 8 years old to 14 years old were invited to play the game FancyBook AR. Participants played two versions of the narrative for both games: fantasy and real-life versions. Both versions were offered in a counterbalanced order to mitigate possible sequence effects. The assistant invited each participant to a preset room with a Huawei Android phone, a Lenovo ThinkPad laptop, and sets of cards together with storybooks. Additionally, we set up a GoPro Hero9 camera to record the procedure.

Before the gameplay started, an assistant introduced the general gameplay to children using a blank version with only text and 3D sphere models, letting participants get used to scanning image targets and holding virtual buttons. The students would then be asked to play one of the versions by the assistant, who also served as an instructor and gave help if children asked. After completing the version, the student would be asked to complete a questionnaire before going to the other version and completing another questionnaire. Besides the experience and engagement-related questions, the questionnaire collected their demographic information, including their self-identified gender and age. We asked them to fill in their nicknames instead of real names to maintain the data anonymously. Informed consent was achieved by participants, the schoolteachers and participants' guardians. The procedure, game contents, and questionnaire were approved by the Eindhoven University of Technology ethical review board with the approval number ERB2020ID165.

3.4. Data Collection

We collected and analyzed both qualitative and quantitative data to answer our research question in this research. To answer RQ1, we collected the children's self-reported immersion through the immersion section of the Player experience of need satisfaction questionnaire (PENS). PENS is a questionnaire that measures the psychology of satisfying experiences of playing through aspects of competence, autonomy, relatedness, and immersion [58,59]. To identify their engagement qualitatively during gameplay, we invited two experts to analyze the gameplay videos and code the children's engagement using the indicators (Table 1) we summarized from a literature review. Experts were also invited to code using words outside of the pool of indicators if they needed to highlight any context-specific phenomenon about engagement. To answer RQ2, we identified students' understanding and the teacher's scaffolding through video analysis of how they completed the tasks. Timelines including tasks that were well understood and completed by students independently were marked as "independent." Tasks that students completed with the teacher's assistance were marked as "with assistance." Tasks that students did not complete even with help from the teachers were marked "uncompleted." Questions students raised when seeking help were also recorded.

To further investigate players' fantasy states during the learning process, we collected their self-reported data of imagination and analogy using the fantasy state scale (FSS). We use the autonomy and enjoyment sections from the PENS questionnaire and the intrinsic motivation inventory (IMI) to collect their self-reported data on autonomy and enjoyment. Since FSS, IMI and PENS were not designed specifically for children, and previous studies suggest some children have difficulty understanding "neutral" and double negative sentences [60], we used a 4-point animated smiley scale developed by Li, Van Der Spek, Hu, and Feijs (2019) (Figure 2) for the PENS and IMI questionnaires [61]. We translated from the original English version and had a language expert check the translation.

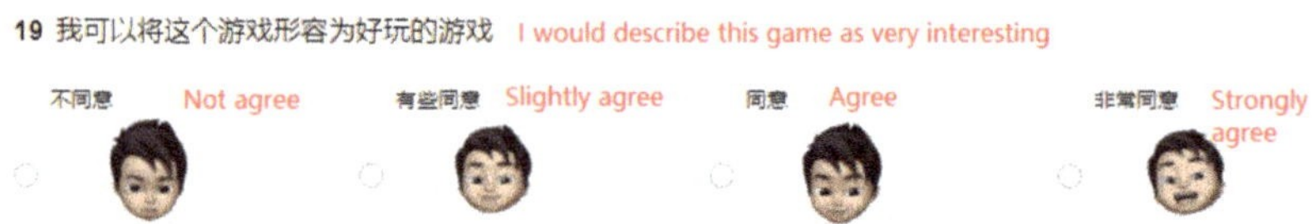

Figure 2. The scale was modified according to Li et al's 4-point animated smiley scale.

3.5. Data Analysis

Two experts coded the students' types of engagement, comprehension, and teacher's assistance using MAXQDA 2020. There were 12 video recordings from 10 participants for each narrative type in each game, for a total of 40 video clips. Experts engaged in closed coding by marking the video's timeline using a pre-defined pool of keywords we determined regarding engagement types (Table 1) and task performance (independent, with assistance, uncompleted). In addition, the experts were allowed to open code salient events themselves. To further understand the qualitative data experts coded, we used an affinity diagram [62] to organize the keywords of user engagement in a structure containing game types, narrative versions, and engagement types. Based on the areas of confusion when users wanted assistance, we drew up representative user journey maps that included students' engagement and teachers' scaffolding.

3.6. Reliability Test

The Cronbach's alpha values for data collected in Hefei using FancybookAR are generally at an acceptable level (0.7~0.8). However, data from Qingdao using MathMythosAR2 has a low Cronbach's alpha for the presence (0.6) and enjoyment section (0.5). After deleting the reversed question, we found an increase in the value, with enjoyment = 0.7 and presence = 0.8. Given that children from Qingdao had a slightly lower average age than those from Anhui, young children around this age (M = 9.2) with their understanding of reversibility still developing [63] may find it difficult to answer reversed questions. The mean scores for enjoyment and presence sections in the Qingdao group are reported without the reversed question.

4. Results

4.1. Players' Self-Reported Experience and Engagement

We analyzed the students' presence, imagination, analogy, enjoyment, and autonomy data using SPSS. We performed a within-between mixed-factor analysis of variance (ANOVA), setting the game type (MathMythosAR2 or Fancybook AR) as the between-subject factor and the game version (fantasy or real-life) as the within-subject factor. Tables 6–9 show the results, with significant outcomes highlighted in bold.

Table 6. Mixed two-way ANOVA with the game type (MathmythosAR2/FancybookAR) as between-subject, the narrative type (fantasy/real-life) as within-subject variable.

DV	Source	SS	F	p
Autonomy	**Game type**	**4.27**	**6.92**	**0.01**
	Narrative type	0.02	0.10	0.76
	Interaction	0.40	1.68	0.20
Presence	**Game type**	**11.65**	**17.56**	**0.00**
	Narrative type	0.00	0.00	1.00
	Interaction	**0.42**	**4.37**	**0.04**
Enjoyment	**Game type**	**2.73**	**5.01**	**0.03**
	Narrative type	0.30	3.10	0.08
	Interaction	0.21	2.16	0.15
Imagination	**Game type**	**5.29**	**7.81**	**0.01**
	Narrative type	**4.19**	**22.25**	**0.00**
	Interaction	0.06	0.34	0.56
Analogy	**Game type**	**5.45**	**6.35**	**0.01**
	Narrative type	**1.16**	**4.04**	**0.05**
	Interaction	0.81	2.80	0.10

DV = dependent variable, SS = sum of squares, F = value on the F distribution, variables of significance: $p \leq 0.05$.

Table 7. The mean and standard deviation values of different sections, presented as M (SD).

		Autonomy	Presence	Enjoyment	Imagination	Analogy
MathmythosAR2	Fantasy	3.66 (0.44)	3.65 (0.37)	3.67 (0.48)	3.48 (0.58)	3.43 (0.61)
	Real-life	3.52 (0.58)	3.54 (0.49)	3.69 (0.45)	3.16 (0.60)	3.46 (0.59)
FancyBookAR	Fantasy	3.17 (0.72)	2.92 (0.79)	3.29 (0.71)	3.12 (0.67)	2.85 (0.98)
	Real-life	3.26 (0.81)	3.04 (0.72)	3.47 (0.59)	2.70 (0.77)	3.20 (0.78)

Table 8. Pairwise comparisons Fantasy vs Real-life, Bonferroni adjusted.

Measure	Game Type	Mean Difference	Std. Error	Sig.b
Autonomy	MathmythosAR2	0.140	0.123	0.260
	FancyBookAR	−0.086	0.123	0.487
Presence	MathmythosAR2	0.117	0.079	0.145
	FancyBookAR	−0.117	0.079	0.145
Enjoyment	MathmythosAR2	−0.016	0.079	0.839
	FancyBookAR	**−0.181 ***	**0.079**	**0.026**
Imagination	**MathmythosAR2**	**0.323 ***	**0.110**	**0.005**
	FancyBookAR	**0.413 ***	**0.110**	**0.000**
Analogy	MathmythosAR2	−0.032	0.136	0.814
	FancyBookAR	**−0.355 ***	**0.136**	**0.012**

* $p < 0.05$

Table 9. Pairwise comparisons MathmythosAR2 VS Fancybook AR, Bonferroni adjusted.

Measure	Game Version	Mean Difference	Std. Error	Sig.b
Autonomy	**Fantasy**	**0.484 ***	**0.152**	**0.002**
	Real-life	0.258	0.179	0.154
Presence	**Fantasy**	**0.730 ***	**0.157**	**0.000**
	Real-life	**0.496 ***	**0.157**	**0.002**
Enjoyment	**Fantasy**	**0.379 ***	**0.154**	**0.016**
	Real-life	0.215	0.133	0.113
Imagination	**Fantasy**	**0.368 ***	**0.159**	**0.024**
	Real-life	**0.458 ***	**0.175**	**0.011**
Analogy	**Fantasy**	**0.581 ***	**0.208**	**0.007**
	Real-life	0.258	0.175	0.146

* $p < 0.05$

4.2. Players' Engagement through Video Analysis by Experts

We created an affinity map of engagement patterns (Figure 3) for each game and narrative version by synthesizing the most frequent types of player engagement. The result suggests both games engage participants more emotionally than the real-life versions. Part of this finding contradicts what we discovered in their self-reported data, in which children reported more enjoyment with the real-life version of FancyBookAR. Such discrepancies show that participants' attitudes towards fantasy in FancyBookAR changed at the beginning when being observed and at the end when being reported. Despite their positive emotional connection with fantasy versions, lots of children showed negative engagement on the cognitive level, showing confusion about the context, while they were more confident about playing the real-life version and showed positive emotional engagement such as "focused."

Regarding specific game types, some children had a negative emotional and behavioral engagement in the real-life narrative of the game MathMythosAR2, including "bored" and "anxious." Those who perceived boredom especially showed reluctance to stay and finish the game tasks showed more positive emotion and active participation with the same game's fantasy narrative.

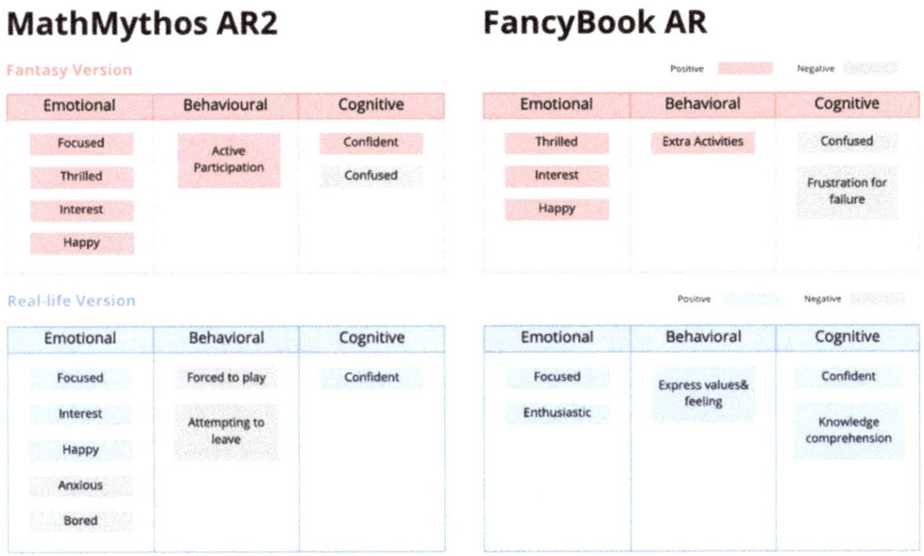

MathMythos AR2

Fantasy Version

Emotional	Behavioural	Cognitive
Focused	Active Participation	Confident
Thrilled		Confused
Interest		
Happy		

Real-life Version

Emotional	Behavioral	Cognitive
Focused	Forced to play	Confident
Interest	Attempting to leave	
Happy		
Anxious		
Bored		

FancyBook AR

Positive   Negative

Emotional	Behavioral	Cognitive
Thrilled	Extra Activities	Confused
Interest		Frustration for failure
Happy		

Positive   Negative

Emotional	Behavioral	Cognitive
Focused	Express values& feeling	Confident
Enthusiastic		Knowledge comprehension

Figure 3. The affinity map of engagement by game types and narrative types (with coloured tags representing positive signs, grey tags representing negative signs).

Children who played both versions of FancyBookAR showed positive emotional engagement, such as focused and enthusiastic engagement in the real-life version, as well as affective emotions when engaging in the fantasy version. When children encountered unfamiliar words and phrases in the fantasy version, they expressed signs of confusion and frustration when repeated attempts failed. In the real-life version, students spoke out more frequently about the meaning of the sentence in their native language when completing the tasks, being more confident about the result.

4.3. Players' Comprehension and the Teacher's Assistance through Video Analysis by Experts

The timelines as a result of the expert analysis are displayed in Figures 4 and 5 in the form of a user journey map, where ten horizontal lines represent ten samples in the video analysis in each game's narrative. Based on the game scenes, we also highlighted the negative and positive engagement moments described in Figure 3. We investigated the teachers' assistance and students' engagement to assess students' comprehension and the role of teachers with different game types and narrative types.

4.4. MathMythosAR2: Differences between Two Narratives Versions

In the fantasy version, when students had difficulty adding two-digit numbers together, the teacher helped them break it down into single-digit addition to recall the addition they had already memorized. It is possible that this help differentially affected the students' fantasy experience of the game, since a school teacher helping students with math questions is part of the real-life narrative of MathMythos AR2, whereas in the fantasy narrative it is a magic teacher, and a school teacher intervening could take students out of the experience. The experiment showed that reported presence was higher in the fantasy condition, so if this effect occurred, it was likely not very strong.

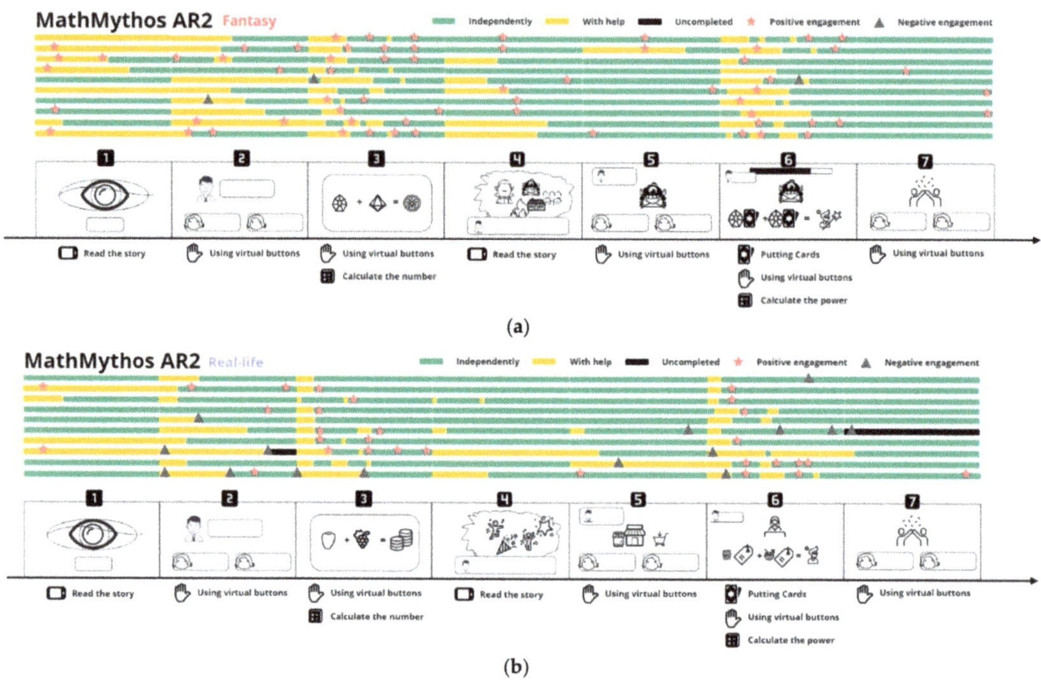

Figure 4. User journey maps of MathMythos AR2 with (**a**): the real-life version and (**b**): the fantasy version.

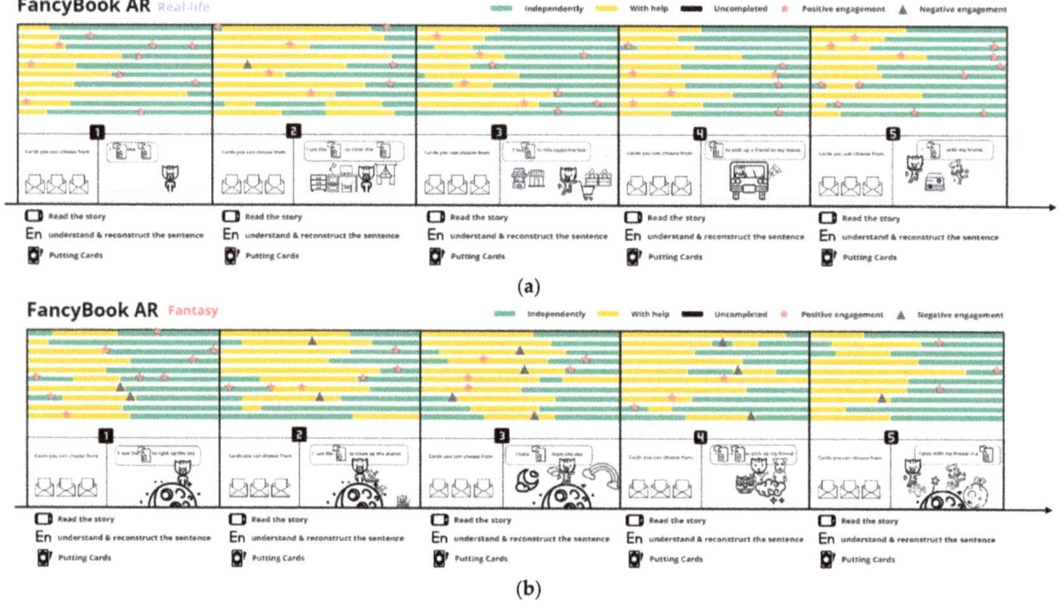

Figure 5. User journey maps of FancyBook AR with (**a**): the real-life version and (**b**): the fantasy version.

In both narratives, we used pink magic (fantasy) and gold coins (real-life) to represent 2-digit numbers, blue magic (fantasy) and silver coins (real-life) to represent 1-digit numbers. When completing tasks independently, participants did not refer to two concepts in their calculations but often recalled in a whispered voice the mathematical operations they had learned before. We also found differences in teachers' role of instruction on story-telling and gameplay. Students listened carefully to the teacher's introduction on the real-life version's interaction first and then tried it out with the teacher's guidance. In the fantasy version, students showed more tendency in trying the game first instead of waiting for instruction (more green lines appear before yellow lines). However, it was still frequent that their initial attempts had failed, requiring scaffolding from teachers. With the fantasy version, the teacher spent more time helping students understand the narrative than with the real-life version.

An additional sign of positive engagement marked by experts was that students in the fantasy condition tended to hold the cards they received in Figure 5, section 3 when most students put cards aside in the real-life version. Only a few students showed signs of disengagement with the fantasy narrative, confused about the fantasy narrative and the virtual buttons' insensitivity at the beginning of the gameplay. Most students showed positive engagement with or without assistance in the fantasy narrative. In contrast, most positive engagements occurred when students finished the task independently in the real-life version. However, more students showed anxiousness at the beginning of the gameplay in the real-life version, whereas others expressed boredom at later stages where most tasks were repetitive practice.

4.5. FancyBook AR: Commonalities between the Two Narrative Versions

In both versions, students needed assistance understanding the meanings of words and sentences. Their understanding of the game narrative and the learning content highly depended on their prior knowledge and the tutoring from teachers. Teachers sometimes played the role of checking whether students properly understood the sentence they constructed, as some children just picked random cards to activate the animation effect without fully understanding the sentence. Therefore, the teacher and students interacted more frequently, but the communication states vary depending on narrative versions.

4.6. FancyBook AR: Differences between Two Narratives Versions

More students read along with the sentence and showed more initiative to tell the teacher the corresponding Chinese translation with the real-life narrative. With this version, participants were more active in asking questions, showing confidence in understanding words related to the context. Whereas with the fantasy version, the meaning and application of words such as "trash-eating monster" and "magic stick" were alien to the students. They showed signs of confusion and prolonged hesitation even after teachers translated these words into Chinese.

5. Discussion

5.1. Whether Fantasy in AR Game-Based Learning for a Classroom Creates More Engaging and Immersive Mental States for Players?

We combined quantitative and qualitative results to address this research question. In general, the fantasy narrative leads to stronger mental states of imagination, while the real-life narrative leads to stronger analogy. Learning English with the game FancyBook AR, students perceived significantly higher enjoyment and immersion in the real-life version than the fantasy version. Combining their self-reported enjoyment (Table 8) with patterns of engagement (Figure 3), we found fantasy in FancyBook led to less enjoyment and negative cognitive engagement for students. Despite children having stronger aroused emotional engagement at the beginning of the gameplay with the fantasy narrative of FancyBook AR, their confusion with the fantasy narrative setting eventually negatively influenced their play experience. This also explains the conflict results: participants perceived more aroused emotional engagement at the beginning with the fantasy FancybookAR, but they self-reported they enjoyed the fantasy version less than the real-life version after the gameplay.

Students learning mathematics with the game MathMythosAR2 reported slightly but statistically insignificant higher enjoyment and presence with fantasy narratives. Their engagement pattern suggests more positive emotional engagement in the fantasy narrative. Although the fantasy narrative caused confusion, it does not influence children's enjoyable and immersive experiences. Children showed signs of negative engagement such as boredom, anxiety, and being forced to play with the real-life narrative. Despite the results, simply interpreting the data as fantasy settings may favourably engage children in math learning while adversely engaging children in second-language learning are inadequate. Instead, we wish to understand these occurrences via the lens of children's experiences and games' fantasy integration within subject matter features.

5.1.1. Contextual Reasons for Differences in Players' Experience and Engagement

The learning of English for pupils is topic-based, taught via subject matter material, which is a key contrast between mathematics learning and second-language acquisition [64]. With FancyBookAR, the subject matter learners simultaneously acquired the narrative content and linguistic knowledge of English, where the unfamiliarity of the content influenced their language comprehension and vice versa. The fantasy narrative of FancyBookAR contains more subject-specific contexts than the real-life narrative, which is regarded as an application of existing language proficiency in new contexts [65]. However, unproficiency with the fantasy-oriented terms hindered the application of knowledge, leading to a poor understanding of the narrative. Such a situation might refer to what O'Malley and Chamot described as foreign language anxiety and content-related anxiety, causing negative engagement such as feelings of frustration [66]. Similarly, pervasive and immersive games often need "reinterpretation of meaning conventions [67]." Players may have failed to achieve a suspension of disbelief if they failed to make the meaning out of the fantasy narrative when learning a second language.

The situation changed in MathMythos AR2 since solving mathematical problems such as calculation can be tackled independently of story comprehension with abstract thinking methods [68]. The phenomenon we found that students murmured the mathematical formulae during the gameplay also partly reflected a natural separation of narrative understanding and mathematical operation when completing mathematical game tasks. The narrative theme in this game is closed to the "mathematical theme" concept, which creates contexts for participants to apply their mathematical knowledge. Such extrinsically tied narrative to the learning content is not a fixed body of knowledge [69]. Therefore, confusion occurred among some children with math learning in the fantasy narrative, but this did not seem to influence their enjoyable, immersive experience and overall engagement. When children are satisfied with meaning interpretation and construction [67], they achieve a suspension of disbelief, feeling more immersed and engaged with their emotional response [70]. Conversely, not understanding the words immediately impacted the understanding of the story in the FancyBookAR.

5.1.2. Analogy and Imagination for Comprehension and Engagement

Our research found that the fantasy narrative triggered higher states of imagination, and the real-life narrative activated stronger analogy states. Analogy helps learners make sense of phenomena, while imagination is often regarded as the prerequisite to engaging in education in most domains [71]. Furthermore, imagination connecting the real world to the virtual world in a mixed reality experience is a driving force in make-believe and creating an immersive experience [40]. Both mental states are fundamental in AR game-based learning for different stages of play. A potential point is that challenges may have an inverse effect on the imagination. Learners need to connect to prior knowledge when making sense of things [20]. If the game is too challenging, participants are more comfortable with things they can relate to their prior experience. Therefore, how much participants are open to fantasy could depend on how confused they are while playing.

5.2. What Is the Teachers' Role in Scaffolding Children with Fantasy Play in Game-Based Learning?

An important way of resolving students' confusion is through teachers' scaffolding. Teachers' scaffolding in game-based learning mainly focused on three areas during the gameplay: interaction, narrative, and knowledge.

5.2.1. Interaction

Children are familiar with digital interactions such as screen touch and physical interactions such as turning pages. The interaction with the augmented physical world, which requires eyes on the digital layer and hands on the physical layer is unfamiliar and therefore sometimes challenging for children, especially at the beginning. Although we designed animated arrows and related instruction to highlight important areas and ways of interaction, we noticed that the teachers' instructions in the real world are still more effective than the virtual instructional content, especially for complex interactions. For example, with the game MathMythosAR2, teachers used verbal instructions and corrective gestures to help some students understand that triggering a virtual button is conducted by covering a pattern with the palm of their hand rather than clicking with their fingers. As the interaction in MathMythosAR2 is more complicated and various than FancyBookAR, we found teachers spent more effort instructing students on how to interact. Despite the necessity of scaffolding at the beginning, too much direct intervention in the middle of the game can interrupt students' immersion and flow experience [72]. This aspect led us to consider the necessity of designing complex interactions such as virtual buttons. A more straightforward and consistent approach to interacting, such as scanning cards in FancyBookAR, is effort-saving for teachers and helps students concentrate on the main task.

5.2.2. Narrative

We found that the teachers' role in facilitating students' understanding of the narrative differed between the two games. Using MathMythosAR2 for mathematics learning, students had more questions regarding the fantasy narrative. However, they also showed more initiative to try things out first regardless of whether the narrative was correctly interpreted or not. The teacher normally assisted students in gaining a better understanding of the narrative by describing the stories in simpler words when students asked related questions. They sometimes inserted themselves into the pedagogical agent, teacher Steven, reading or retelling his lines. This phenomenon might be due to the similarity between the teachers' and the pedagogical agent's roles in guiding students and focusing their attention. To reduce feelings of interruption and provide a smoother play experience for the children, a pedagogical agent can therefore be designed that functions as an easy self-insert for the teacher.

Teachers' scaffolding students' narrative comprehension with the game FancyBookAR for learning English was more complicated because the narrative is intrinsically related to the language learning content. Teachers were more careful when interpreting narratives. Rather than just telling children what the story was, they frequently encouraged children to discuss their understanding first or waited for them to ask questions. Such ways of scaffolding connect to strategies of second-language learning.

5.2.3. Knowledge

Common language learning strategies in real-world contexts are implicit knowledge acquisition and explicit knowledge learning. Implicit knowledge acquisition represents unconscious acquisition through exposure to a second-language context. In contrast, explicit knowledge learning consists of a conscious introduction and instruction of the rules of languages [66]. FancybookAR aims to engage second-language English learners through playful games without necessarily focusing on the learning contents, creating contexts for implicit knowledge acquisition. Research suggests explicit knowledge learning can facilitate implicit knowledge acquisition. The lack of explanation for unfamiliar words and explicit introduction to grammar makes self-directed learning challenging for students with the game FancyBookAR. Teachers, therefore, played a role in introducing the learning contents explicitly and directing students' attention to key learning points. Due to the unfamiliar vocabulary and its unusual application in fictional contexts, learning with a fantasy narrative required even more explicit explanation from teachers. Certain phenomena reflected a need to introduce new knowledge to students with the FancyBook AR. A device-based or a teacher-led introduction regarding the explanation, pronunciation, and application of unfamiliar vocabularies and grammar is required to facilitate the learning procedure.

With both narrative MathMythosAR2, teachers spent little time instructing students with learning content directly. Since children of this age were already proficient in mathematical operations, and the narrative is not a necessary component of mathematical knowledge, children's incomplete understanding of the external narrative did not directly affect their ability to perform operations. Most cases requiring scaffolding are when students make repeated calculation errors. While instructing students, teachers often address arithmetic concepts outside of the narrative framework. Research suggests effective instruction for low achievers requires a context that facilitates teachers' explicit instructions and students' idea-sharing [73]. A feasible solution to create a context that facilitates certain teacher-student interactions is to design storyline branches and a separate scene that allow teachers to conduct explicit instruction for those who repeatedly make mistakes. Further strategies for addressing certain issues while situating learning in AR fantasy of game-based learning will be discussed in response to RQ3.

5.3. How to Situate Learning in AR Fantasy of Game-Based Learning in Consideration of Learners' Engagement, Experience, and the Teachers' Scaffolding?

Although endogenous fantasy is often regarded as an essential way of improving players' engagement, situating learning in AR fantasy of game-based learning does not necessarily require turning every learning goal into a fantasy representation and integrating each part endogenously. In combination with our result, we suggest designers firstly consider whether the narrative is intrinsically tied to the learning goals. When a narrative is not necessarily a fixed body of the learning content, designers can consider incorporating fantasy as portraying abstract concepts for narrative-independent knowledge to activate the players' imagination, further enhancing players' motivation and immersion, especially in repetitive practices. In situations such as second-language learning, when the narrative serves as the fixed body of the learning contents, designers first should examine whether the additional cognitive burden imposed by the fantasy narrative would impede or facilitate students' comprehension of the knowledge. Design should consider activating users' analogy for real-life with narrative-related knowledge, facilitating understanding where fantasy can be applied as a next step after students have established a good understanding of explicit knowledge. To leverage endogenous fantasy's positive effects, designers can partly integrate learning content with fantasy. Designers can create fantasy and real-life mixed settings, e.g., an alien visitor settles down to regular life on Earth, allowing students to understand the learning context with imagination and interest.

To further support teachers' scaffolding on gameplay interaction, narrative, and knowledge understanding, we synthesize several strategies for situating learning in the AR fantasy of game-based learning. Regarding interaction, we suggest designing games with simple and consistent ways of interaction. Complex AR interaction for the first time needs real-world instruction. While digital instruction is ineffective regardless of the game's storyline, we suggest designers consider reducing complicated interaction and giving teachers a role in the story. It is feasible to design a pedagogical agent that enables instructors to take on the role of a scaffolder. To assist teachers in scaffolding new knowledge introduction and providing more instructions to low-achieving learners, we advise designing introduction scenes using real-world settings to activate analogy and offer supplementary instruction scenes for explicit explanations.

5.4. Limitations

There is a possible weakness in our study groups from various locations since we could not invite participants from different regions to the same laboratory due to the local COVID-19 policy restrictions. Influence from different lab settings was not balanced out in our research. The two games differ in the specific play steps and details due to the different learning methods of the two subjects, which we did not eliminate as an effect on the experimental results. We could not eliminate the chance that quantitative research with children might involve over-scored situations and fail to represent their thinking effectively. We could only give more comprehensive perspectives by adding qualitative analysis. For experimental purposes, we only set magic story narratives as the fantasy condition and school-home life narratives as the real-life condition. In actual game design, fantasy covers a wider range of categories, and its application is often integrated with real-life conditions, which requires further exploration and research.

6. Conclusions

We answered three research questions regarding situating learning in AR fantasy of game-based learning through experiments and analysis with qualitative and quantitative data. We discovered that fantasy narratives increase students' engagement in mathematics learning while disengaging students in second-language learning. They perceived stronger immersion with AR fantasy for mathematics learning than AR fantasy for second-language learning. Fantasy narrative activated players' imagination while real-life narrative enables their minds to analogy. We found the basis of developing affective incentives to fantasy is related to participants' comprehension of the contexts. They were more open to imagination and fantasy when there was less challenge in learning. We found that teachers' scaffolding mainly focused on complicated interaction in AR with MathMythosAR2 for learning math. With FancyBookAR for learning English, teachers' scaffolding mainly focused on narrative interpretation and knowledge explanation. We further contextualized and explained the phenomena we found and synthesized several strategies for situating learning in the AR fantasy of game-based learning. (1) When the narrative is not intrinsically tied to the learning goal, design can employ fantasy narrative to represent abstract concepts, activating effective incentives to immerse and engage students more, especially in repetitive practice. (2) When the story is intrinsically tied to knowledge to be learned, we recommend starting with real-life narratives that engage the user's analogy mind to aid interpretation, then progressing to fantasy narratives as the next level of challenge. (3) We also recommend mixing fantasy and real-life settings to achieve positive effects of imagination and analogy mental states in the actual design. (4) We recommend a design with simple AR interaction such as card tracking, which is effort-saving for teachers and task-concentrating for students. (5) It is feasible to design a pedagogical agent that enables teachers to take on the role of scaffolding seamlessly at any time. (6) Design should also prioritize analogy activation when constructing scenes for device-based or teacher-led presentations of new information for beginners, as well as clear explanations for low-achieving learners.

Author Contributions: Conceptualization, T.Z., E.V.d.S., J.H., M.B.; software, T Z., E.V.d.S.; validation, J.J.; formal analysis, T.Z., E.V.d.S., M.B., J.H.; investigation, J.J.; resources, J.J.; data curation, T.Z., J.J.; writing—original draft preparation, T.Z. writing—review and editing, E.V.d.S., M.B.; visualization, T.Z.; supervision, E.V.d.S., M.B., J.H.; project administration, E.V.d.S., M.B., J.H.; funding acquisition, T.Z., J.H. All authors have read and agreed to the published version of the manuscript.

Funding: This work was supported by the Chinese Scholarship Council [grant numbers 201806790091].

Institutional Review Board Statement: The study was conducted in accordance with the Declaration of Helsinki, and approved by the Institutional Review Board of Eindhoven University of Technology (protocol code ERB2020ID165 and date of approval 2 November 2020).

Informed Consent Statement: Informed consent was obtained from all subjects involved in the study.

Acknowledgments: We would like to express our deep and sincere gratitude to Tao who helped conduct the experiment in Qingdao, and Yixuan Bian who designed the game FancyBook AR. We are grateful to Jingya Li who offered support on the experiment measures.

Conflicts of Interest: The authors declare no conflict of interest.

References

1. Taufiq, M.; Nuswowati, M.; Widiyatmoko, A. Study of the Applicability Level of Merge Cube Augmented Reality Media on Junior Hight School Science Teachers. *Unnes Sci. Educ. J.* **2021**, *10*, 132–136.
2. Radu, I. Augmented Reality in Education: A Meta-Review and Cross-Media Analysis. *Pers. Ubiquitous Comput.* **2014**, *18*, 1533–1543. [CrossRef]
3. Lave, J.; Wenger, E. *Situated Learning: Legitimate Peripheral Participation*; Cambridge University Press: Cambridge, UK, 1991; ISBN 0521423740.
4. Dunleavy, M.; Dede, C.; Mitchell, R. Affordances and Limitations of Immersive Participatory Augmented Reality Simulations for Teaching and Learning. *J. Sci. Educ. Technol.* **2009**, *18*, 7–22. [CrossRef]
5. Fotaris, P.; Pellas, N.; Kazanidis, I.; Smith, P. A systematic review of augmented reality in stem education. *Stud. Appl. Econ.* **2017**, *39*, 1.
6. Li, J.; Spek, E.D.; van der Feijs, L.; Feng, W.; Hu, J. *Augmented Reality Games for Learning: A Literature Review*; Springer: Cham, Switzerland, 2017; Volume 10291, ISBN 9783319586960.
7. Pellas, N.; Fotaris, P.; Kazanidis, I.; Wells, D. Augmenting the Learning Experience in Primary and Secondary School Education: A Systematic Review of Recent Trends in Augmented Reality Game-Based Learning. *Virtual Real.* **2019**, *23*, 329–346. [CrossRef]
8. Chen, C.H. Impacts of Augmented Reality and a Digital Game on Students' Science Learning with Reflection Prompts in Multimedia Learning. *Educ. Technol. Res. Dev.* **2020**, *68*, 3057–3076. [CrossRef]
9. Birk, M.V.; Atkins, C.; Bowey, J.T.; Mandryk, R.L. Fostering Intrinsic Motivation through Avatar Identification in Digital Games. In Proceedings of the 2016 CHI Conference on Human Factors in Computing Systems, San Jose, CA, USA, 7 May 2016; pp. 2982–2995. [CrossRef]
10. Charsky, D. From Edutainment to Serious Games: A Change in the Use of Game Characteristics. *Games Cult.* **2010**, *5*, 177–198. [CrossRef]
11. Alessi, S.M.; Trollip, S.R. *Multimedia for Learning: Methods and Development*; Allyn and Bacon: Boston, MA, USA, 2001; ISBN 0205276911.
12. Zuo, T.; Spek, E.D.; Hu, J.; Birk, M.V. Investigating the motivational effect of fantasy and similarity through avatar identification in AR game-based learning. In Proceedings of the 13th EAI International Conference, INTETAIN 2021, Virtual, 3–4 December 2021; Springer: Cham, Switzerland, 2022; pp. 279–294.
13. Job, V.; Brandstätter, V. Get a Taste of Your Goals: Promoting Motive–Goal Congruence through Affect-focus Goal Fantasy. *J. Pers.* **2009**, *77*, 1527–1560. [CrossRef]
14. Malone, T.W. What Makes Things fun to learn? Heuristics for designing instructional computer games. In Proceedings of the 3rd ACM SIGSMALL Symposium and The First SIGPC Symposium on SMALL Systems, Palo Alto, CA, USA, 18 September 1980; Volume 162, pp. 162–169. [CrossRef]
15. Zuo, T.; Feijs, L.; Van Der Spek, E.D.; Hu, J. A classification of fantasy in serious games. In *The Extended Abstracts of the Annual Symposium on Computer-Human Interaction in Play Companion Extended Abstracts*; ACM, Inc.: New York, NY, USA, 2019; pp. 821–828. [CrossRef]
16. Connolly, J.A.; Doyle, A.B. Relation of Social Fantasy Play to Social Competence in Preschoolers. *Dev. Psychol.* **1984**, *20*, 797–806. [CrossRef]
17. Zuo, T.; Birk, M.V.; Van Der Spek, E.D.; Hu, J. Exploring fantasy play in mathmythos AR. In *The Extended Abstracts of the 2020 Annual Symposium on Computer-Human Interaction in Play*; ACM, Inc.: New York, NY, USA, 2020; pp. 413–417. [CrossRef]
18. Parker, L.E.; Lepper, M.R. Effects of Fantasy Contexts on Children's Learning and Motivation: Making Learning More Fun. *J. Personal. Soc. Psychol.* **1992**, *62*, 625–633. [CrossRef]
19. Lee, J. Effects of Fantasy and Fantasy Proneness on Learning and Engagement in a 3D Educational Game. Ph.D. Thesis, University of Texas at Austin, Austin, TX, USA, 2015.
20. Van Der Spek, E.D.; Sidorenkova, T.; Porskamp, P.; Matthias, R. The Effect of Familiar and Fantasy Aesthetics on Learning and Experience of Serious Games. *Lect. Notes Comput. Sci.* **2014**, *8770*, 133–138. [CrossRef]
21. Brown, J.S.; Collins, A.; Duguid, P. Situated Cognition and the Culture of Learning. *Educ. Res.* **1989**, *18*, 32–42. [CrossRef]
22. Belland, B.R. Scaffolding: Definition, Current Debates, and Future Directions. In *Handbook of Research on Educational Communications and Technology*; Springer: Berlin/Heidelberg, Germany, 2014; pp. 505–518.

23. Van, R. *Gaming and Cognition: Theories and Practice from the Learning Sciences*; IGI Global: Hershey, PA, USA, 2010; ISBN 9781615207176.

24. Eraut, M. Non-formal Learning and Tacit Knowledge in Professional Work. *Br. J. Educ. Psychol.* **2000**, *70*, 113–136. [CrossRef]

25. Hughes, C.E.; Stapleton, C.B.; Hughes, D.E.; Smith, E.M. Mixed Reality in Education, Entertainment, and Training. *IEEE Comput. Graph. Appl.* **2005**, *25*, 24–30. [CrossRef]

26. Klopfer, E.; Yoon, S. Developing Games and Simulations for Today and Tomorrow's Tech Savvy Youth. *TechTrends* **2005**, *49*, 33–41. [CrossRef]

27. Fotaris, P.; Pellas, N.; Kazanidis, I.; Smith, P. A Systematic Review of Augmented Reality Game-Based Applications in Primary Education. In Proceedings of the 11th European Conference on Games Based Learning (ECGBL 2017), Graz, Austria, 5–6 October 2017; pp. 181–190.

28. Li, J.; Van Der Spek, E.D.; Yu, X.; Hu, J.; Feijs, L. Exploring an augmented reality social learning game for elementary school students. In Proceedings of the Interaction Design and Children Conference, London, UK, 21–24 June 2020; pp. 508–518. [CrossRef]

29. Dunleavy, M.; Dede, C. Augmented Reality Teaching and Learning. In *Handbook of Research on Educational Communications and Technology*; Springer: New York, NY, USA, 2014; pp. 735–745.

30. Rodrígues, P.; Bidarra, J. Transmedia Storytelling and the Creation of a Converging Space of Educational Practices. *Int. J. Emerg. Technol. Learn.* **2014**, *9*, 42–48. [CrossRef]

31. Malone, T.W.; Lepper, M.R. *Making Learning Fun, A Taxonomy of Intrinsic Motivations for Learning*; Routledge: London, UK, 1987.

32. Huizinga, J. *Homo Ludens Ils 86*; Routledge: London, UK, 2014; ISBN 1315824167.

33. Montola, M. Exploring the Edge of the Magic Circle: Defining Pervasive Games. *Proc. DAC* **2005**, *1966*, 16–19.

34. Stapleton, C.B.; Hughes, C.E.; Moshell, J.M. MIXED FANTASY: Exhibition of Entertainment Research for Mixed Reality. In Proceedings of the Second IEEE and ACM International Symposium on Mixed and Augmented Reality, ISMAR, Tokyo, Japan, 27 October 2003; pp. 354–355. [CrossRef]

35. Heater, C. Being There: The Subjective Experience of Presence. *Presence Teleoper. Virtual Environ.* **1992**, *1*, 262–271. [CrossRef]

36. Doherty, K.; Doherty, G. Engagement in HCI. *ACM Comput. Surv.* **2018**, *51*, 1–39. [CrossRef]

37. Lee, J.; Shute, V.J. Personal and Social-Contextual Factors in K-12 Academic Performance: An Integrative Perspective on Student Learning. *Educ. Psychol.* **2010**, *45*, 185–202. [CrossRef]

38. Bouta, H.; Retalis, S. Enhancing Primary School Children Collaborative Learning Experiences in Maths via a 3D Virtual Environment. *Educ. Inf. Technol.* **2013**, *18*, 571–596. [CrossRef]

39. Filsecker, M.; Kerres, M. Engagement as a Volitional Construct: A Framework for Evidence-Based Research on Educational Games. *Simul. Gaming* **2014**, *45*, 450–470. [CrossRef]

40. Zuo, T.; Birk, M.V.; van der Spek, E.D.; Hu, J. The Mediating Effect of Fantasy on Engagement in an AR Game for Learning. *Entertain. Comput.* **2022**, *42*, 100480. [CrossRef]

41. Sedano, C.I.; Leendertz, V.; Vinni, M.; Sutinen, E.; Ellis, S. Hypercontextualized Learning Games: Fantasy, Motivation, and Engagement in Reality. *Simul. Gaming* **2013**, *44*, 821–845. [CrossRef]

42. Mozelius, P. Game Based Learning-a Way to Stimulate Intrinsic Motivation. In Proceedings of the International Conference on e-Learning, Valparaiso, Chile, 26–27 June 2014; Academic Conferences International Limited: Oxfordshire, UK, 2014; pp. 272–278.

43. Council, N.R.; Suzanne, M.; John, D.; James, W. *How People Learn: Bridging Research and Practice*; National Academies Press: Washington, DC, USA, 1999. Available online: http://www.nap.edu/catalog/9457.html (accessed on 23 July 2022)ISBN 030918584X.

44. Lombardi, M.M.; Oblinger, D.G. Authentic Learning for the 21st Century: An Overview. *Educ. Learn. Initiat.* **2007**, *1*, 1–12.

45. Ladley, P. *Games Based Situated Learning: Games-ED Whole Class Games and Learning Outcomes*; The Pixel Foundation Ltd.: London, UK, 2010.

46. Newmann, F.M.; Marks, H.; Gamoran, A. *Authentic Pedagogy: Standards That Boost Student Performance*; Issues in Restructuring Schools; Center on Organization and Restructuring of Schools, University of Wisconsin-Madison: Madison, WI, USA, 1995; 17p.

47. van de Pol, J.; Volman, M.; Beishuizen, J. Scaffolding in Teacher–Student Interaction: A Decade of Research. *Educ. Psychol. Rev.* **2010**, *22*, 271–296. [CrossRef]

48. Wouters, P.; Van Oostendorp, H. Overview of Instructional Techniques to Facilitate Learning and Motivation of Serious Games. In *Instructional Techniques to Facilitate Learning and Motivation of Serious Games*; Springer: Berlin/Heidelberg, Germany, 2017; pp. 1–16.

49. Young, M.F. Instructional Design for Situated Learning. *Educ. Technol. Res. Dev.* **1993**, *41*, 43–58. [CrossRef]

50. Garrett, T.; The, S.; Interaction, C.; Case, A. Student-Centered and Teacher-Centered Classroom Management: A Case Study of Three Elementary Teachers Student-Centered and Teacher-Centered of Three Classroom Teachers Management: Study Elementary. *J. Classr. Interact.* **2015**, *43*, 34–47.

51. Wouters, P.; van Oostendorp, H. *Instructional Techniques to Facilitate Learning and Motivation of Serious Games*; Springer: Cham, Switzerland, 2016; pp. 1–16.

52. Ryan, R.M.; Rigby, C.S.; Przybylski, A. The Motivational Pull of Video Games: A Self-Determination Theory Approach. *Motiv. Emot.* **2006**, *30*, 347–363. [CrossRef]

53. Yuen, S.C.-Y.; Yaoyuneyong, G.; Johnson, E. Augmented Reality: An Overview and Five Directions for AR in Education. *J. Educ. Technol. Dev. Exch.* **2011**, *4*, 11. [CrossRef]

54. Pivec, M.; Dziabenko, O.; Schinnerl, I. Aspects of Game-Based Learning. In Proceedings of the 3rd International Conference on Knowledge Management, Graz, Austria, 2–4 July 2003; pp. 216–225.
55. Vygotsky, L.S.; Cole, M. *Mind in Society: Development of Higher Psychological Processes*; Harvard University Press: Cambridge, MA, USA, 1978; ISBN 0674576292.
56. Xin, Y.P.; Jitendra, A.K. The Effects of Instruction in Solving Mathematical Word Problems for Students with Learning Problems: A Meta-Analysis. *J. Spec. Educ.* **1999**, *32*, 207–225.
57. Taylor, W.L. "Cloze Procedure": A New Tool for Measuring Readability. *J. Q.* **1953**, *30*, 415–433. [CrossRef]
58. Rigby, S.; Ryan, R. *The Player Experience of Need Satisfaction (PENS) Model*; Immersyve Inc.: Celebration, FL, USA, 2007; pp. 1–22.
59. Johnson, D.; Gardner, M.J.; Perry, R. Validation of Two Game Experience Scales: The Player Experience of Need Satisfaction (PENS) and Game Experience Questionnaire (GEQ). *Int. J. Hum. Comput. Stud.* **2018**, *118*, 38–46. [CrossRef]
60. Scott, J. Children as Respondents: The Challenge for Quantitative Methods. *Res. Child. Perspect. Pract.* **2000**, *2*, 87–108.
61. Li, J.; Van Der Spek, E.D.; Hu, J.; Feijs, L. Turning Your Book into a Game: Improving Motivation through Tangible Interaction and Diegetic Feedback in an AR Mathematics Game for Children. In Proceedings of the Annual Symposium on Computer-Human Interaction in Play, Barcelona, Spain, 22–25 October 2019; pp. 73–85. [CrossRef]
62. Plain, C. Build an Affinity for KJ Method. *Qual. Prog.* **2007**, *40*, 88.
63. Piaget, J.; Cartalis, E.; Escher, S.; Hanhart, U.; Hahnloser, L. *Le Jugement et le Raisonnement Chez L'enfant*; Delachaux et Niestlé: Neuchâtel, Switzerland; Paris, France, 1947; ISBN 2242000659.
64. Gaffield-Vile, N. Content-Based Second Language Instruction at the Tertiary Level. *Elt J.* **1996**, *50*, 108–114. [CrossRef]
65. Vollmer, H.J. Language across the Curriculum. In Proceedings of the Conference of Languages in Education, Ljubljana, Slovenia, 25–26 September 2008; pp. 27–39.
66. O'malley, J.M.; O'Malley, M.J.; Chamot, A.U.; O'Malley, J.M. *Learning Strategies in Second Language Acquisition*; Cambridge University Press: London, UK, 1990; ISBN 052135837X.
67. Nieuwdorp, E. The Pervasive Interface; Tracing the Magic Circle. In Proceedings of the DiGRA 2005: Changing Views: Worlds in Play, 2005 International Conference, Vancouver, BC, Canada, 16–20 June 2005; 2005.
68. Atkinson, R.C.; Bower, G.H.; Crothers, E.J. *Introduction to Mathematical Learning Theory*; Wiley: New York, NY, USA, 1965.
69. Steffe, L.P.; Nesher, P.; Cobb, P.; Sriraman, B.; Greer, B. *Theories of Mathematical Learning*; Routledge: London, UK, 2013; ISBN 1136485546.
70. Egan, K. Young Children's Imagination and Learning: Engaging Children's Emotional Response. *Young Child.* **1994**, *49*, 27–32.
71. Egan, K. Memory, Imagination, and Learning: Connected by the Story. *Phi Delta Kappan* **1989**, *70*, 455–459.
72. Barzilai, S.; Blau, I. Scaffolding Game-Based Learning: Impact on Learning Achievements, Perceived Learning, and Game Experiences. *Comput. Educ.* **2014**, *70*, 65–79. [CrossRef]
73. Pfister, M.; Moser Opitz, E.; Pauli, C. Scaffolding for Mathematics Teaching in Inclusive Primary Classrooms: A Video Study. *ZDM* **2015**, *47*, 1079–1092. [CrossRef]

MDPI

Article

Development of a Virtual Object Weight Recognition Algorithm Based on Pseudo-Haptics and the Development of Immersion Evaluation Technology

Eunjin Son, Hayoung Song, Seonghyeon Nam and Youngwon Kim *

Korea Electronics Technology Institute, Gwangju 61011, Korea; ejson@keti.re.kr (E.S.); shy1230@keti.re.kr (H.S.); sadguest@keti.re.kr (S.N.)
* Correspondence: kimforever920@keti.re.kr

Abstract: In this work, we propose a qualitative immersion evaluation technique based on a pseudo-haptic-based user-specific virtual object weight recognition algorithm and an immersive experience questionnaire (IEQ). The proposed weight recognition algorithm is developed by considering the moving speed of a natural hand tracking-based, user-customized virtual object using a camera in a VR headset and the realistic offset of the object's weight when lifting it in real space. Customized speeds are defined to recognize customized weights. In addition, an experiment is conducted to measure the speed of lifting objects by weight in real space to obtain the natural object lifting speed weight according to the weight. In order to evaluate the weight and immersion of the developed simulation content, the participants' qualitative immersion evaluation is conducted through three IEQ-based immersion evaluation surveys. Based on the analysis results of the experimental participants and the interview, this immersion evaluation technique shows whether it is possible to evaluate a realistic tactile experience in VR content. It is predicted that the proposed weight recognition algorithm and evaluation technology can be applied to various fields, such as content production and service support, in line with market demand in the rapidly growing VR, AR, and MR fields.

Keywords: virtual reality; pseudo-haptic; human-centered computing; mixed and augmented reality

Citation: Son, E.; Song, H.; Nam, S.; Kim, Y. Development of a Virtual Object Weight Recognition Algorithm Based on Pseudo-Haptics and the Development of Immersion Evaluation Technology. *Electronics* **2022**, *11*, 2274. https://doi.org/10.3390/electronics11142274

Academic Editor: Dah-Jye Lee

Received: 14 June 2022
Accepted: 19 July 2022
Published: 21 July 2022

Publisher's Note: MDPI stays neutral with regard to jurisdictional claims in published maps and institutional affiliations.

1. Introduction

Research is continuing to increase the utilization of VR devices as VR accessibility and the usability of ordinary users increase due to the light weight and convenience of VR devices [1]. VR devices are applied in various industries such as education, medical care, and entertainment, but there are still areas to be solved in order to commercialize them [2]. Among them, the most important thing is to provide users with realistic stimuli and experiences [3]. To this end, a realistic haptic experience should be provided in a virtual environment to induce a higher sense of immersion, and appropriate interaction should be possible. In order to achieve this purpose, various studies [4–6] are continuing to provide a realistic tactile experience in the VR environment. Virtual object manipulation used by users to interact in a VR environment can be said to be a key factor in increasing the immersion of virtual reality content [7]. Representative interactions with virtual objects are mainly actions that involve lifting objects, such as lifting, catching, throwing, picking up, and rolling. The weight of a virtual object that a user feels in lifting an object, one of the basic interactions, is typically felt through visual and haptic feedback, a technique that simulates tactile sensations in a virtual environment using visual feedback and the properties of human visual-touch perception called pseudo-haptics [8]. Pseudo-haptics has mainly been conducted with studies that provide the right sense when catching virtual objects using haptic interfaces. Haptic interfaces such as haptic gloves are attached to the user's arms, hands, and fingers, and they are adjusted to provide weight by giving the user a force appropriate to the weight of the virtual object in the process of lifting the object [9–11]. In addition, research is being conducted to allow users to feel the weight of several objects by varying the degree of vibration and pressure felt on the skin according to the weight of each virtual object [12–14]. Weight recognition in VR environments is one of the most important tactile properties that contributes to realistic interactions with virtual objects and immersive experiences, and most methods of weighting virtual objects tend to rely on manipulating physical senses or visual information [15]. The physical sensation was made to feel the weight by utilizing the vibration of an electric muscle stimulator or controller, and visual information was for detecting the weight by applying an offset between the actual and virtual hands [3,16,17]. Although this method is used to increase immersion in a VR environment, other devices (controllers, electric muscle stimulation devices, etc.) require high costs and have difficulty interacting with users, thereby lowering immersion. To address this problem, we propose a pseudo-haptic-based, user-customized virtual object weight recognition algorithm using natural hand tracking and visual illusion effects with only cameras in VR headsets.

In this study, we propose a method to provide the weight of a virtual object and a method to qualitatively measure the user's content commitment, focusing on the actual weight recognition applied with a camera-based offset without a separate haptic device in the VR headset environment. The implementation of a pseudo-haptic-based virtual object weight recognition algorithm induces the user-customized virtual object to feel its weight without a controller and is designed to implement a more realistic offset considering the result value of the object lifting speed measurement experiment in real space to set its natural offset threshold. In addition, we intend to measure the weight recognition and content commitment of the proposed algorithm by requiring experimental participants to complete a questionnaire for evaluation.

Our proposed pseudo-haptic-based virtual object weight recognition method considers the weights of virtual objects when lifting virtual objects by tracking real hands in a VR environment. By creating a deliberate error in the distance between the mapped 3D hand model and the lifted virtual object, we intend to create a visual illusion effect of weight and qualitatively measure the immersion in a given environment. A visual illusion effect means that by using the algorithm developed in this paper, decoupling of the user's hand position in XR content and the user's hand position in real space creates a visual illusion that makes the actual hand position different from that shown to the real user. We expect that this study will provide algorithms and pilot contents for pseudo-haptics technology based on the experimental visual illusion effect, which can lead to the implementation of more detailed research in the area and contribute to the development of a cultural content industry based on increased immersion.

Our main suggestions for weight sensing are as follows:

- Development of decoupling algorithms for user hand positions in VR environments and user hand positions in real space;
- A weight recognition software approach that takes into account the volume, density, and speed of a virtual object based on hand tracking using a camera in a VR headset;
- Develop a customized virtual object weight algorithm based on the speed of lifting standard objects in real space;
- After conducting an experiment to measure the speed of lifting an object according to weight in real space, consider the resulting value as a weight;
- Because the user feels a different individual weight depending on the experience and degree of immersion in VR, the user lifts a golf ball in the real space and measures this speed to define it as a customized standard speed;
- A study showing the subject's perceived weight using the proposed approach;
- Development of qualitative evaluation technology based on an immersive experience questionnaire (IEQ) within a VR environment;
- Qualitative user experience evaluation based on developed pseudo-haptics technology and analysis of the results.

2. Related Work

Multi-sensory feedback, such as a sense of experience with a virtual object in a VR environment, provides a high degree of immersion to the user [3,18,19]. Among them, tactile sense has features that can explore texture, hardness, temperature, and weight, a major factor contributing to improving user immersion in VR content [20]. Accordingly, many studies related to tactile sensation have been developed, and similar tactile feedback is typically being studied in various applications [4,21]. Similar tactile feedback helps one navigate VR content, such as feeling the size, texture, and weight of an object [22]. There are two main similar tactile feedback study methods: one is to detect the actual weight by adding separate equipment, and the other is to add visual elements [2,23]. In this study, a pseudo-haptic-based virtual object weight recognition algorithm is implemented to induce a sense of weight for a virtual object, and to set a natural offset threshold according to weight, a more realistic offset is considered.

2.1. Pseudo-Haptics

Rietzler et al. [16] implemented virtual object weight through the occurrence of intentional offsets with virtual objects in a virtual environment. Through experiments quantifying the perceptual threshold of the offset, it was suggested that experimenters could recognize the object weight according to the visual volume. However, we used VR headset controllers to track the hands, which reduced immersion, and did not consider volume or speed in the implementation of virtual object weights. Kim et al. [24], in a related study, showed that the effect of visual pseudo-haptic feedback occurred differently depending on the size of the virtual object. Another study by Kim et al. [3] studied the multi-sensory-like haptic effect that combines the control if display ratio manipulation and electrical muscle stimulation. These studies have shown that multi-sensory-like haptic feedback can produce more powerful simulated senses but tends to be somewhat device-dependent. These leading studies were based on equipment, including controllers, as the start stage of university-centered technology development. In addition, although the content of the study is weight recognition using offset-based visual illusion effects, the purpose of the study is to quantify the offset threshold in weight recognition, so it does not provide a systematic algorithm according to the weight and moving speed of the virtual object. In addition, we show that the threshold of the offset, which feels a sense of weight, is different for each experimental participant, but it has a limitation in that it does not provide a customized weight algorithm.

2.2. Muscle Strength for One Hand

According to Mo et al. [25], the maximum weight that does not cause fatigue during one-handed lifting is appropriate, and men had a relatively higher recommended weight than women. It is recommended that the weight be reduced by 30% for repetitive motion. According to Kim [26], the static maximum muscle strength when raising with one hand differs according to the height of exertion, and the dynamic maximum muscle strength is not statistically significant between one hand and both hands or the difference between the left and right hands. Based on this research, in order to set a natural offset threshold according to weight, an experiment was conducted to measure the speed of lifting an object according to the weight of an object in real space, and the resulting value of the experiment was considered as a weight to implement a more realistic offset.

3. Designing a Virtual Object Weight Algorithm

This chapter describes the design method of the virtual object weight algorithm considering the density, volume, and speed weights of virtual objects in a VR environment.

The algorithm we propose is based on pseudo-haptics-based visual illusion effects. The user lifts the virtual object in a VR environment. At this time, the difference in weight of different virtual objects is shown as an offset, which is the distance between the virtual object and the user's hand. The proposed algorithm sets the speed at which standard objects (golf balls) present in real space are lifted to the user's standard speed. Based on this, when lifting objects of different weights, it deliberately controls the speed at which the virtual object is lifted, thereby inducing custom weight recognition. In addition, in order to consider the perception of real-life weights, we experimented with measuring the speed of lifting an object according to the weight of the object in real space, and the resulting values were used as weights to guide one to define a realistic offset. Considering that immersion can be weakened by excessive offsets during relatively long interactions, it is designed to slow down the moving speed as the distance between the virtual object and the hand increases. The proposed algorithm ultimately derives the moving speed $v_{final\_virtual}$ of the object, dependent on the weight of the virtual object.

Figure 1 shows the basic driving principle of the pseudo-haptic customized virtual object weight algorithm based on visual illusion effects. There are two main types of speeds for personalized weight recognition: v_{real}, which refers to the actual movement speed of the user's hand, and v_{vr_obj}, which refers to the speed at which the virtual object is raised according to the weight of the virtual object. Users feel different weights for each individual depending on the presence or absence of VR experience and the degree of immersion. It induces user-customized weight recognition based on the speed of lifting a standard weight object (golf ball) existing in real space, and this speed affects the rate at which a virtual object with a weight other than the standard weight is lifted (see Figure 1).

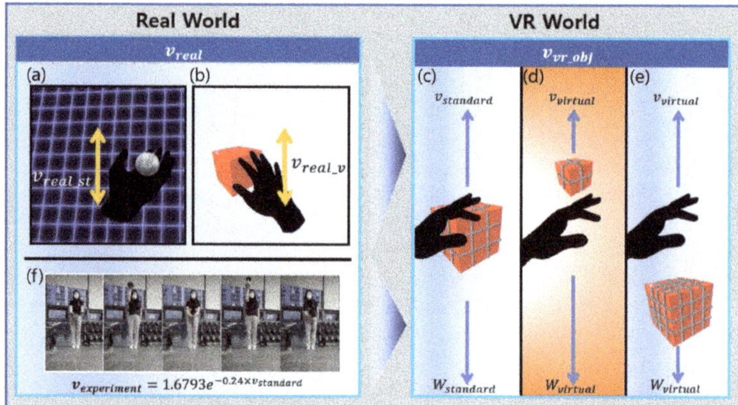

Figure 1. Based on the visual illusion effect, fundamental driving principles of pseudo-haptic customized virtual object weight algorithms. (**a**) the speed of lifting a golf ball existing in a real space, which refers to the y-axis movement speed of a 3D hand model (v_{real_st}), (**b**) the speed of a 3D hand model that lifts a virtual object other than the standard weight (v_{real_v}), and (**c**) the driving principle of an object with a standard weight. In order to recognize a user-customized weight, it is necessary to lift an object having the same weight existing in real space. The weight of an object with a standard weight is called $\mathcal{W}_{standard}$ and is based on the actual weight of the golf ball. The lifting speed of this object is called $v_{standard}$, which is defined as the standard speed of raising the standard weight, so it can be considered to be the same as the speed v_{real_st} of the 3D hand model. In addition, Figre 1 also shows (**d,e**) a situation in which the weight of a virtual object is lighter or heavier than a golf ball having a standard weight, where $\mathcal{W}_{virtual}$ refers to the weight of a virtual object that has a different weight than the standard weight and $v_{virtual}$ refers to the speed at which a virtual object with a weight of $\mathcal{W}_{virtual}$ is lifted, which is the moving speed of the virtual object and not the speed of the 3D hand model, (**d**) that if the weight of the virtual object is lighter than $\mathcal{W}_{standard}$, the virtual object is higher even if the object is lifted at a speed of $v_{standard}$, (**e**) that if the weight of the virtual object is heavier than $\mathcal{W}_{standard}$, the virtual object is lower even if the object is lifted at a speed of $v_{standard}$, and (**f**) the equation according to the experiment of lifting an object according to a weight in real space ($v_{experiment}$). An experiment was conducted in real space to set a natural offset threshold in a VR environment.

3.1. Ideas and Algorithms

When lifting an object existing in real space, there is a weight (W) according to the mass (m) of the object and the gravitational acceleration (g) and a physical strength ($F_{physical}$), which is a force originating from the muscle used to lift it. However, in a VR environment, there is no force for a camera-tracked hand to lift a virtual object. The implementation of the corresponding force to be similar in a VR environment is called a virtual force ($F_{virtual}$). In real space, the density (ρ) and volume (V) are considered for the mass (m) of an object, and the moving speed of the object is determined by the force of lifting the object but not considered in a VR environment. Therefore, we propose a virtual object weight algorithm that takes into account the weight of a virtual object, customized travel speed, and a realistic offset according to the weight of the object when lifting the object in real space.

For customized virtual object weight recognition, users were asked to lift objects of the same weight to define the y-axis movement speed as a customized speed, and the speed of the virtual object was controlled to be proportional to the volume and density of the virtual object. In order to set the natural offset threshold according to the weight, it was designed to implement a more realistic offset by considering the resulting value through the experiment of measuring the speed of lifting an object according to the weight of an object in real space.

The confinement conditions of our virtual object weight algorithm were as follows:

- The virtual object can be lifted only on the y-axis in a three-dimensional VR environment;
- The location of each virtual object is represented by local coordinates and moves in a positive space greater than or equal to 0 (the y-axis coordinates of a virtual object cannot be negative);
- The unit time for measuring the average speed of a virtual object is defined as 1 s;
- The densities of the virtual objects compared are the same.

The proposed virtual object weight recognition algorithm was implemented so that when two virtual objects with different weights were lifted in a VR environment, the movement distances of objects with larger weights were small even if the actual hand was raised to the same height. In addition, even if we moved our hands at the same height when lifting two virtual objects with the same weight, we implemented that the movement of the two fast-moving virtual objects was greater, considering the instantaneous acceleration of the 3D hand model lifting the virtual object:

$$\mathcal{W} = m \times g \tag{1}$$

In Equation (1), m is the mass considering the density and volume of the object, g is the acceleration of gravity, and $\mathcal{W}$ is the weight of the object.

When lifting a virtual object, considering the weight of an object in a VR environment, an object with a standard weight ($\mathcal{W}$) that can be lifted at the same speed as the tracked 3D hand model is required:

$$\mathcal{W}_{standard} = 0.045 \text{ kg} \times 9.80665 \text{ m/s}^2 \tag{2}$$

In Equation (2), $\mathcal{W}_{standard}$ is the standard weight of a virtual object required to measure the force of lifting by the user, and the gravitational acceleration was considered based on the weight of a general golf ball (0.045 kg). To consider the speed of lifting a virtual object, a standard speed of a virtual object that could move simultaneously in the same position as the 3D hand model was required. There was a deviation in the speed of lifting virtual objects for each user, so there needed to be a standard speed for each user. The initial speed at which a user lifts a virtual object with a weight of $\mathcal{W}_{standard}$ with a force of $F_{v\_standard}$ is called $v_{standard}$:

$$F_{v\_standard} = \mathcal{W}_{standard} \times v_{real\_st} \tag{3}$$

Equation (3) shows the relationship between the standard weight $\mathcal{W}_{standard}$ and v_{real_st} according to the customized speed $v_{standard}$ of lifting the object. The customized standard speed $v_{standard}$ for lifting a virtual object with a standard weight $\mathcal{W}_{standard}$, defined in Equation (2), determines the v_{real_st} for lifting the standard weight object:

$$v_{virtual} = v_{real\_v} \times (\mathcal{W}_{standard} \times \mathcal{W}_{virtual}) \tag{4}$$

Equation (4) is a formula for obtaining the speed $v_{virtual}$ of lifting a virtual object other than the standard weight. The corresponding equation was cross-verified through comparison to $\mathcal{W}_{standard}$ and v_{real_v}.

The cases of cross-verification are as follows:

1. The virtual object is lighter in weight and faster in speed;
2. If the weight is lighter and the speed is slower;
3. If the weight is heavier and the speed is the same;
4. If the weight is lighter and the speed is the same;
5. If the weight is heavier and the speed is also slow.

According to Equation (5), the moving speed $v_{virtual}$ of the virtual object in consideration of the weight of each virtual object may be derived. However, the speed at which this is derived is not realistic. Therefore, a more realistic offset was realized through an experiment of measuring the speed of lifting an object according to its weight:

$$v_{experiment} = 1.6793e^{-0.24 \times \mathcal{W}_{virtual}}, \tag{5}$$

Equation (5) is a formula derived from the experiment of measuring the one-handed lifting speed according to the weight in real space, and it shows the change in the speed of lifting a realistic object, while $v_{experiment}$ shows the velocity of the actual object derived according to the actual weight:

$$v_{final\_virtual} = \frac{v_{virtual} + v_{experiment}}{2}, \tag{6}$$

Equation (6) is a formula that adjusts the weight of the $v_{virtual}$ speed derived from Equation (5) and the speed of $v_{experiment}$ derived from the experiment. Through this equation, an implement offset was induced, which is the distance between a more realistic hand and a virtual object. Here, the speed of the user-customized virtual object considering the weight was determined by the derived speed of $v_{final\_virtual}$. However, considering the weakening of the sense of immersion due to an excessive offset during a relatively long interaction, it was designed to slow down the moving speed as the distance between the virtual object and the hand increased.

If the distance between the center point of the virtual object and the tracked 3D hand model was greater than 0.2 m during the virtual object lift operation, the speed $v_{final\_virtual}$ obtained from Equation (6) was corrected in a gradual deceleration manner. After checking the distance between the virtual object and the hand, it was corrected by moving $v_{final\_virtual} + 0.02$ (m) faster on the y-axis if the object was lighter than $\mathcal{W}_{standard}$ and moving -0.02 slower if the object was heavier than the standard weight.

3.2. Experiment to Measure the Speed of a One-Handed Lift According to the Weight in Real Space

In order to set a natural offset according to the weight of an object in a VR environment, an experiment was conducted to measure the speed of one hand according to the weight in real space.

3.2.1. Participants and Experimental Tools

A simple experiment was conducted with one woman in her 20 s. The experiment was conducted using five exercise dumbbells with different weights. According to the study, the maximum weight that did not cause fatigue during the one-handed lifting operation of Mo et al. [19] was set at 12.25 N, 19.60 N, 24.50 N, 39.20 N, and 49.00 N by setting the maximum weight to 5 kg (see Figure 2).

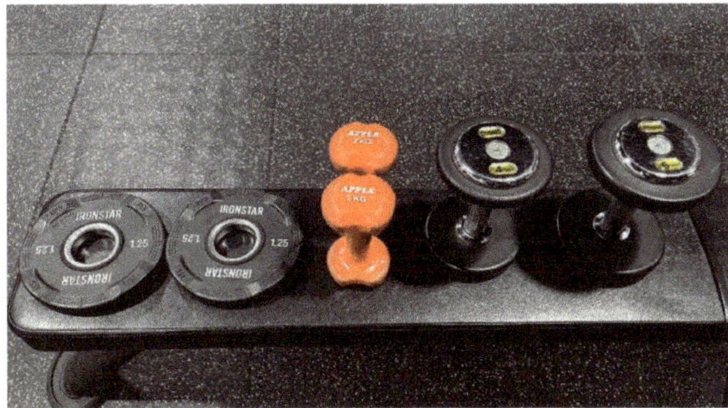

Figure 2. Five-step object image used in the experiment.

3.2.2. Experimental Procedures and Methods

The measurement of the moving speed of an object according to the static muscle ability of the palm center height to the top of the head was measured at 78.00 cm (approximately 30.71 in) in height for the general palm joint height and elbow height, taking into account the simplicity of the participant's experimental procedure. In addition, the maximum dynamic muscle strength during the lifting operation was measured according to the five weights of these palm-centered to elbow-height tasks: 12.25 N, 19.60 N, 24.50 N, 39.20 N, and 49.00 N. At this time, the speed measurement was performed by using the right hand, which was the main hand of the participant. The operation of lifting an object with a weight of five steps in a real space was repeated six times. The experiment for the purpose of this study was organized as follows (see Figure 3):

1. The experimenter set the camera position and angle in consideration of the height of the participant;
2. The position and angle of the camera filming the participant were fixed;
3. Participants lifted an object weighing 12.25 N using one hand;
4. The lifting of an object of the same weight was repeated six times;
5. The experimenter measured the movement time of the object from the center of the participant's palm to the top of the head;
6. Repeated lifting of objects weighing 19.60 N, 24.50 N, 39.20 N, and 49.00 N (steps 2–5);
7. The experimenter used the measured data to calculate the speed of lifting objects by weight;
8. The user derived a formula suitable for the speed of lifting objects by weight.

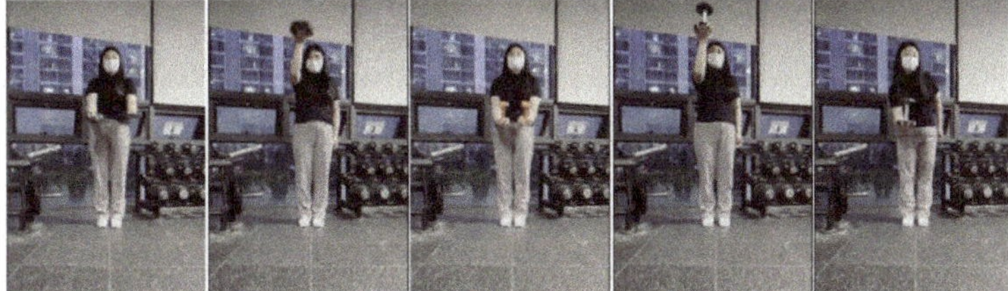

Figure 3. Experimental scene of measuring the speed of one-hand raising by weight in real space.

3.2.3. Speed Measurement Experiment Results

The resulting value of the measurement of the moving speed of the object for the object lifting task according to the dynamic muscle strength ability was averaged according to the weight to derive an appropriate equation. As the relatively light object was lifted, the deviation of the object's lift speed was large, so it was easier to eliminate outliers by using the average value of the speed according to each weight rather than using the value of each population as a whole (see Figure 4).

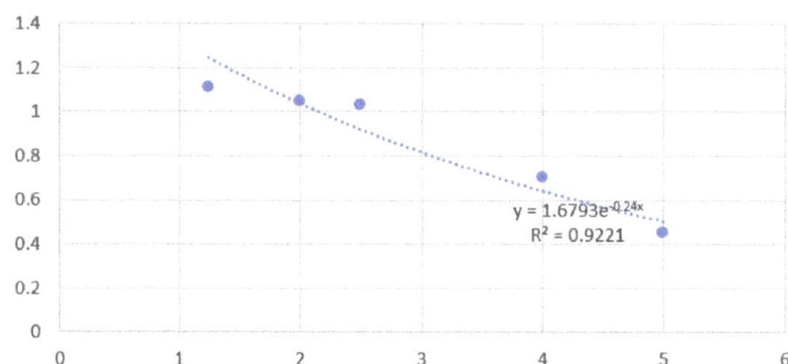

Figure 4. Experimental results for measuring the lifting speed of objects by weight in real space (from 12.25 N to 49 N).

The resulting value of the experiment is as shown in Equation (5).

3.3. Exclude Inertia

If only the weight due to gravity and the muscle force, which is the force to lift an object, are applied to the VR environment, maximum inertia occurs, and it takes a relatively long time for the object to stop, causing the virtual object to shake. Therefore, we removed the falling inertia that occurred at the moment of holding the object to reduce the maximum inertia of the object. When a tracked 3D hand model passes through a weighted virtual object in a VR environment, the tracked 3D hand model has a force F to lift the object at the same time as it passes so that the virtual object does not fall to the floor and moves the virtual object using the Y-axis position value of the tracked hand.

3.4. Hand Tracking

In this study, the controller was not used for the natural user interface (NUI) for natural interaction. Hand tracking using a camera in a VR headset was used for interaction. Hand tracking using the Oculus Interaction Software Development Kit (SDK) can improve user immersion in a VR environment.

4. Design and System Implementation of Simulated Content

In this chapter, we describe the construction of a virtual environment for virtual object weight recognition algorithm-based simulation content and the method of hand tracking through a camera within a VR headset. We also describe an experimental simulation for immersion evaluation based on the proposed algorithm.

4.1. Building a Virtual Environment

We used an Oculus Quest 2, one of the most popular VR headsets, as the equipment for running the proposed virtual object weight recognition algorithm simulation software. Our authoring environment was developed with Unity 2019.4.18f1. The virtual environment includes tasks implemented to show interactions with virtual objects.

Based on the pseudo-haptic-based virtual object weight recognition algorithm proposed in this study, we implemented a virtual experience space to implement simulation content. A 3D table model for placing a 3D virtual experience space and a virtual object were constructed with simple objects supported by the Unity Asset Store (see Figure 5).

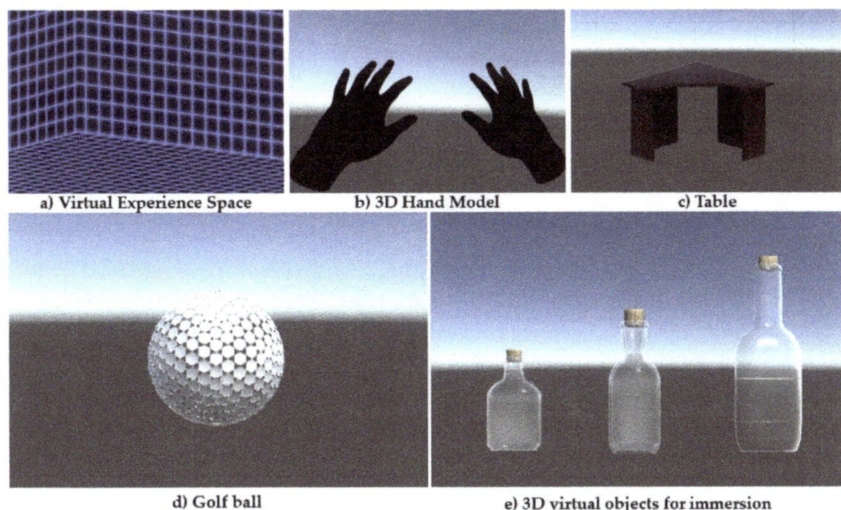

a) Virtual Experience Space   b) 3D Hand Model   c) Table

d) Golf ball   e) 3D virtual objects for immersion
measurement experiments

Figure 5. Virtual object weight recognition algorithm-based simulation content's virtual environment. (**a**) A virtual experience space where the proposed algorithm is simulated, (**b**) a 3D hand model that receives and maps posture information from the user's hand using the camera in Oculus Quest 2, (**c**) a table model that places virtual objects, (**d**) 3D Virtual Objects for Weight Recognition Experiments, and (**e**) 3D virtual objects for immersion measurement experiments.

4.2. User and 3D Virtual Object Interaction

The Oculus Interaction SDK version 38.0 provided by Meta, a manufacturer of the Oculus Quest 2, recognized the user's hand movements. Using a camera within a VR headset, the hand was recognized without additional equipment and mapped to a VR environment hand model. It was used to interact with 3D virtual objects to which the weighting algorithm was applied.

When the content started, three virtual objects floated in a virtual reality (VR) environment, and a 3D hand model appeared, in which the recognized hand was mapped by the camera of the VR headset. The three virtual objects floating in the VR environment consisted of virtual objects of different sizes, and the weights and renderings of the virtual objects may have been different depending on the density inside each virtual object. The user could lift two or more with both hands, and the virtual object moved in the Y-axis direction. To experience an algorithm that recognized weights, one had to lift each virtual object at least once. One of the three virtual objects was an object with a standard weight, so it moved with the hand model. However, when lifting a virtual object heavier than the standard weight, the object fell short of the position of the 3D hand model.

4.3. System Implementation and Enforcement

Figure 6 shows the proposed algorithm simulation content driving scene (see Figure 6).

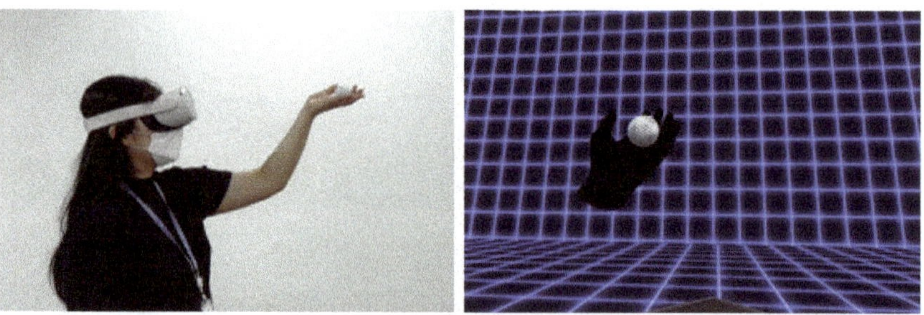

a) **Proposed Personalized Standard Speed Measurement Scene**

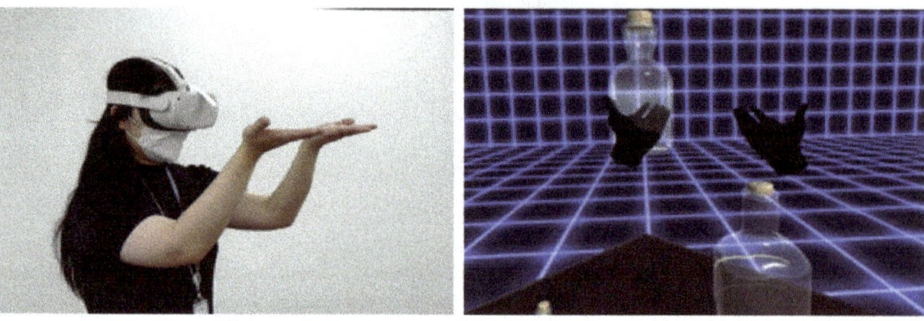

b) **Virtual Object Weight Algorithm Scene**

Figure 6. Simulation scene based on virtual object weight algorithm. (**a**) Because the user feels a different individual weight depending on the experience and degree of immersion in VR, the user lifts a golf ball in the real space and measures this speed to define it as a customized standard speed, (**b**) Virtual Object Weight Algorithm.

5. Questionnaire Development

As various media such as games and videos appear, questionnaires measuring immersion that fit the characteristics of each media are being developed. Jennett [27] has prepared a questionnaire for an experiment to measure game content immersion. Three experiments were conducted to measure game immersion, and the goal was to measure immersion so that the participants could better understand how to participate in the game. Overall, the research results showed that immersion can be measured not only by objective factors (e.g., task completion time and eye movement) but also by subjective factors (e.g., survey papers). Rigby [28] wrote a questionnaire to measure the degree of immersion of viewers when watching a video. Based on the verified game immersion questionnaire of Jennett, four factors were found through exploratory factor analysis, and the questionnaire was constructed by modifying the content of the questionnaire to be suitable for video viewing immersion measurement.

The questionnaire design method used in this study was as follows. It is important to clarify the objectives in developing the questionnaire. First of all, the field to be proven through the experimental results was selected. We judged that the weight perception of virtual objects in VR environments can affect immersion enhancement through existing leading studies [16,17] and designed a questionnaire to find out the user's weight perception and degree of it in VR environments. In the first experiment, we developed a questionnaire to see if a user could detect the weight of an object in a virtual environment, and in the second experiment, we developed a questionnaire to measure the user's immersion. In the second experiment, the immersion measurement section, a separate item was also added so that participants could evaluate their own immersion level. When developing a questionnaire, a resulting value should be considered, and in order to reduce a slight error in the process of rounding the result value, the questionnaire was selected in 10 units. Due to the characteristics of the two experiments, the sensitivity of the weight felt by the user and immersion in the content were measured. In addition, in order to select the right question, it was necessary to select the question item in consideration of the respondent's position. First of all, technical terms should be excluded, and terms that are easy for respondents to understand and intuitive words should be constructed. Each question should also focus on obtaining specific information, but only one question should be asked so that respondents are not confused. The contents of each question used in this study were selected to measure the weight and immersion by reflecting the characteristics of the experiment, and questions were selected by mixing positive and negative questions to secure consistency in the user answers. Using the characteristics of the pseudo-haptic content, the questionnaire items were divided into weight detection measurement (Part 1), immersion measurement (Part 2), and immersion self-evaluation (Part 3), (Appendices A and B). Whether or not the weight was detected could be measured in consideration of visual or touch, object movement speed, virtual or real hand position, and weight comparison. In the immersion measurement, the items for immersion measurement were selected in consideration of the experimenter's concentration, interest, time flow oblivion, and sense of reality. Finally, through self-evaluation, the items were selected so that the participants could measure how much they were immersed in the content themselves.

Specifically, in the measurement of weight detection and immersion, the participants were asked to what extent they agreed with each statement describing their experience participating in the experiment. The answers to each question could be measured by dividing them into a five-point scale, and in the self-evaluation of immersion, it was composed of questionnaire items asking how much immersion there was overall on a scale from 1 to 10. In addition, among some items, questions regarding mean negation were calculated by reversing the scale. In the first experiment, we demonstrated a virtual object weight recognition algorithm implemented using a weight detection measurement questionnaire, and in the second experiment, we used an immersion measurement and immersion self-assessment questionnaire to measure the user's immersion state when weight was given to a virtual reality through a validated virtual object weight recognition algorithm.

6. Experiment 1

6.1. Procedure

Using our developed virtual object weight recognition algorithm, we conducted an experiment to find out whether the weight of a virtual object could be detected in user VR content. The algorithm in the experiment induced user-customized virtual object weights based on the speed of lifting standard objects in real space, such as in the proposed algorithm. However, in the final stage of the algorithm, no correction was made according to the center point of the virtual object and the distance of the tracked 3D hand model. In addition, for the composition of the experiment, the part where the density of the virtual object, one of the algorithm's limitations, was the same was modified. An experiment was conducted to measure whether or not the weights of three rectangular parallels having different densities of the same size were detected (see Figure 7). It is necessary to implement wooden boxes of the same size but different densities in a general background and to recognize that each object has different weights by lifting it with both hands without a controller. Twelve participants were recruited, and the experiment was conducted. It consisted of various age groups from 23 to 39 years old with VR experience, with 7 male and 5 female participants. The participants were asked to wear VR headsets (Oculus Quest 2), and the controller was not used due to the nature of the content, so the experiment was conducted after informing them how to use hand tracking. The experiment was conducted one by one, and the questionnaire was filled out after experiencing the content for 5 min. As for the questionnaire, a questionnaire related to weight detection was prepared. At this time, the user was not informed of the density of the object.

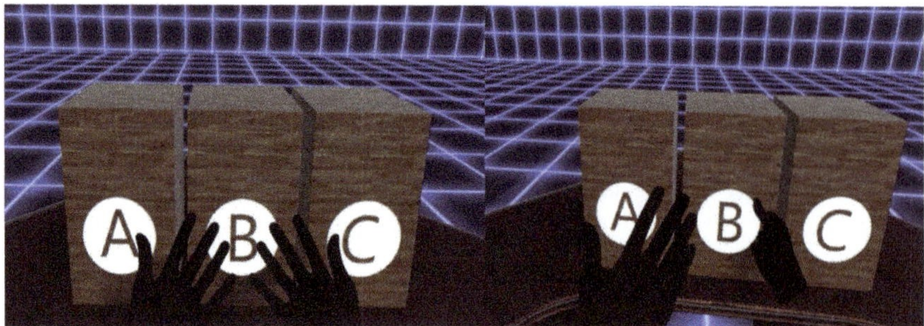

Figure 7. Screenshot of the scene in Experiment 1. The experiment was constructed so that the weight felt different by applying different offsets to objects A, B and C. Participants may think the weight was heavy or light in the order in which the objects were listed, so B was set as the heaviest weight.

6.2. Results

To verify the virtual object weight recognition algorithm, the participants conducted weight detection experiments and analyzed the questionnaire prepared. The answers to each question in the weight detection questionnaire were prepared by dividing the degree to which the participants agreed with the question on a five-point scale. The SA was calculated as 5 points and the SD as 1 point, and Q4 and Q7 for the mean negation were calculated by reversing the score. The results of calculating the average of the questionnaire prepared by the participants are shown in Table 1. This means that most of the experimental participants felt the weight when lifting a virtual object in a virtual environment. In addition, 9 out of 12 participants in the experiment mentioned one object (B) and gave an accurate answer when asked which object was the heaviest in the interview that followed the questionnaire. In addition, when asked about their preference for applying offsets to virtual objects, 10 out of 12 said that applying offsets to virtual objects helped sense the weight. On the other hand, one participant said that he felt the buffering of the screen during the operation with hand tracking, and another participant said that it was uncomfortable because they were unfamiliar with the operation of hand tracking. One of the participants in the experiment, who responded positively during the experiment, expressed surprise that the weight of the virtual object was felt even when the controller was not used.

Table 1. Results of the weight sensing experiment questionnaire.

P1	P2	P3	P4	P5	P6
4.6	5.0	3.8	3.2	4.3	4.8
P7	**P8**	**P9**	**P10**	**P11**	**P12**
5.0	4.7	2.9	3.7	4.5	4.9

7. Experiment 2

7.1. Procedure

The second experiment was conducted to measure the user's immersion state when weight was given to a virtual object in virtual reality through a verified virtual object weight recognition algorithm. Before conducting this experiment, a preliminary experiment was conducted to find out the difference in weight between the object in the virtual world and the object in the real world. This was an experiment to find out the relationship between the virtual object and the weight of the real space object. Preliminary experiments showed that the weight of the water bottle felt in the virtual environment was similar to the weight felt when the actual water bottle was lifted. We conducted the second experiment in reference to this. In the second experiment, the degree of immersion was measured by placing water bottles commonly encountered in daily life by size so that the participants could easily distinguish and judge experiences in the virtual environment and in the real world (see Figure 8). The algorithm of the ongoing experiment induced user-customized virtual object weights based on the speed of lifting standard objects in real space, as suggested by the method in the first experiment. However, in the final stage of the algorithm, no correction was made according to the center point of the virtual object and the distance of the tracked 3D hand model. The participants should have recognized that each object had a different weight and was similar to the difference in weight in the real world by lifting the water bottle with both hands without a controller. The experiment was conducted with 12 participants consisting of various age groups from 23 to 39. Eight were male and four were female participants. The participants were asked to wear VR headsets (Oculus Quest 2), and the controller was not used due to the nature of the content, so the experiment was conducted after informing them how to operate hand tracking. The experiment was conducted one by one, and the questionnaire was filled out after experiencing the content for 5 min. As for the questionnaires, an immersion measurement questionnaire and an immersion evaluation questionnaire were prepared. At this time, the users were not informed that the weight was reflected in each object.

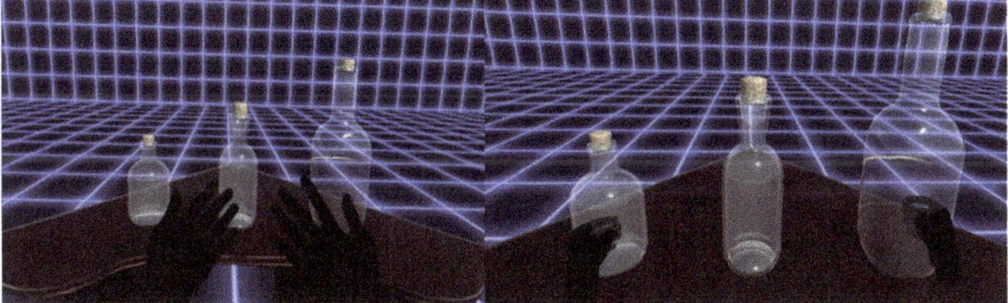

Figure 8. Screenshot of scene from Experiment 2.

7.2. Results

We conducted an experiment to measure user immersion when weight was given to a virtual object through content reflecting the virtual object weight recognition algorithm that we previously verified. At this time, the immersion measurement and immersion evaluation questionnaire form were utilized to measure the immersion state of the user. The immersion measurement questionnaire consisted of 10 questions, and the degree to which each item was agreed upon was divided into a five-point scale. The participants wrote down the extent to which each item corresponded to them on the questionnaire. We calculated five points for closer to "a lot" and one point for the closer to "not at all". Q3, Q5, Q8, and Q9 were calculated by inverse calculation as questions using negative words to secure user consistency. The calculated the survey averages created by the participants are as shown in Table 2, indicating that the majority of the participants in the experiment had higher immersion in the virtual environments with virtual object weight recognition algorithms. In addition, the degree of immersion self-assessment questionnaire for the participants themselves evaluating the degree of immersion was composed of a single question, and the degree of immersion was measured by using the degree of immersion as a 10-point scale. According to Table 3, the immersion level evaluated directly by the participants was high, and the participants' responses were positive in the subsequent interviews. Most people said that they felt the weight of lifting a water bottle in the real world, even though it had no effect on their hands when holding a water bottle with the contents embodied in virtual reality. In addition, the dominant evaluation was that applying the offset to the weights of virtual objects had a significant impact on high immersion in the content. On the other hand, one person said he was unable to immerse himself due to VR sickness. Nevertheless, 11 people with VR experience answered that the content used in the experiment was more immersive than the content they had previously experienced.

Table 2. Results of the immersion measurement questionnaire.

P1	P2	P3	P4	P5	P6
3.8	4.2	3.1	4.6	4.7	2.1
P7	**P8**	**P9**	**P10**	**P11**	**P12**
3.5	4.8	4.7	4.2	4.3	4.6

Table 3. The results of the immersion self-assessment.

P1	P2	P3	P4	P5	P6
7	10	8	10	10	5
P7	**P8**	**P9**	**P10**	**P11**	**P12**
8	9	8	8	9	9

8. Discussion

Experiment 1 attempted to verify the algorithm developed through an experiment to find out whether the weight of a virtual object was detected in a virtual environment. The questionnaire completed by participants after conducting the weight detection experiment showed that most of the participants felt the weight when they lifted the virtual object in their virtual environment. However, some participants answered that they were uncomfortable because they were unfamiliar with the process of manipulation with hand tracking. In Experiment 2, which was conducted for the second time, we tried to measure the user's immersion when weight was given to a virtual object through content that reflected the virtual object weight recognition algorithm verified in advance. It was confirmed that most of the participants had a high degree of immersion in the experience in the virtual environment to which the virtual object weight recognition algorithm was applied. While most people said they felt the weight when they picked up the object in the real world, even though it was not doing anything to their hands when they held it in a virtual environment, one person said that they could not immerse themselves due to motion sickness in virtual reality. Nevertheless, 11 people with VR experience answered that the content used in the experiment was more immersive than the previously experienced content. We believe that inducing a tactile experience through Experiments 1 and 2 contributed to increasing the immersion of the VR content, and in future studies, we intend to implement more stable content by upgrading the virtual object weight recognition algorithm.

9. Conclusions

In this paper, to improve VR headset environment immersion, we produced pseudo-haptic and simulation contents based on visual illusion effects and proposed qualitative evaluation techniques using an IEQ. Experiment 1 demonstrated its own virtual object weight recognition algorithm, and Experiment 2 demonstrated that the virtual object weight recognition algorithm was effective at increasing the user's content commitment. In other words, it was confirmed that the visual illusion effect of the user using the weight and speed of the virtual object helped to recognize the weight of the object in the VR environment. This is a standalone software-based approach that facilitates the tracking of VR interactions. It is not dependent on other equipment because it uses only VR headsets without using controllers among VR devices. It can also be seen as an advantage that the user's burden can be reduced by using a single device. Through the virtual object weight recognition algorithm and IEQ-based qualitative evaluation technology proposed in this paper, algorithms and pilot contents for pseudo-haptic technologies can be developed to provide more detailed research in the field, and they can be applied to various fields.

On the other hand, the proposed algorithm only applies to the method of lifting an object on one axis—the y-axis—and has the limitation that the contents of the deceleration in travel speed according to height for high-weight objects are not considered in the algorithm.

As a future study, it is intended to find a way to add an algorithm for speed deceleration according to the movement in the xyz direction and the height applied in the real space.

Author Contributions: Conceptualization, E.S.; methodology, Y.K.; software, E.S.; validation, E.S.; formal analysis, H.S.; investigation, H.S. and S.N.; data curation, H.S.; writing—original draft preparation, H.S. and E.S.; writing—review and editing, E.S., H.S. and S.N.; visualization, E.S.; supervision, Y.K.; project administration, H.S. and Y.K. All authors have read and agreed to the published version of the manuscript.

Funding: This research was supported by the Culture, Sports and Tourism R&D Program through the Korea Creative Content Agency grant funded by the Ministry of Culture, Sports and Tourism in 2022 (Project Name: The expandable Park Platform technology development and demonstration based on Metaverse·AI convergence, Project Number: R2021040269, Contribution Rate: 100%).

Informed Consent Statement: Informed consent was obtained from all subjects involved in the study.

Acknowledgments: Thank you to everyone who supported this study.

Conflicts of Interest: The authors declare no conflict of interest.

Appendix A. Immersion Questionnaire Used in Experiment 1

(Part1) Weight Detected

How much do you agree with the following items?

SD = strongly disagree; D = disagree; N = neutral; A = agree; SA = strongly agree.

1. When I lifted an object in a virtual environment, I felt like lifting an object in reality.

    SD D N A SA

2. When I lifted the object, I felt like one or more objects were moving slower than I thought.

    SD D N A SA

3. At the same time, the weight of the object to be lifted felt differently.

    SD D N A SA

4. The weight of each object did not feel different.

    SD D N A SA

5. I felt like the lightest object was moving fast.

    SD D N A SA

6. When the actual hand felt higher than expected, the weight of the object felt heavier.

    SD D N A SA

7. I didn't feel the difference in weight for each object.

    SD D N A SA

8. The actual hand was not manipulated much, but I could feel the strength in my arm.

    SD D N A SA

9. The slower the object, the heavier it felt.

    SD D N A SA

10. I felt like my actual hand was holding an object directly.

    SD D N A SA

Appendix B. Immersion Questionnaire Used in Experiment 2

(Part2) Measurement of immersion (5 = A lot; 1 = Not at all)

1. I easily understood the content.

Not at all 1 2 3 4 5 A lot

2. Due to the weight of objects felt in the virtual environment, the experience in the virtual world felt similar to the real world.

Not at all 1 2 3 4 5 A lot

3. Interacting in the virtual world didn't feel as real as the real world.

Not at all 1 2 3 4 5 A lot

4. I didn't know what was happening around me.

Not at all 1 2 3 4 5 A lot

5. I found it boring to execute the content.

Not at all 1 2 3 4 5 A lot

6. While proceeding with the content, I felt cut off from the outside.

Not at all 1 2 3 4 5 A lot

7. While proceeding with the content, I was able to fully immerse myself in the given situation.

Not at all 1 2 3 4 5 A lot

8. My daily thoughts and worries still remained in my head.

Not at all 1 2 3 4 5 A lot

9. Even though I was participating in the content, I felt like I was still in the real world.

Not at all 1 2 3 4 5 A lot

10. I felt that a very short time had passed during the content.

Not at all 1 2 3 4 5 A lot

(Part3) Self-assessment of immersion (10 = Very much so; 0 = Not at all)

Q. How immersed do you think you felt?

1 2 3 4 5 6 7 8 9 10

References

1. MacIsaac, D. Google Cardboard: A virtual reality headset for $10? *AAPT* **2015**, *53*, 125. [CrossRef]
2. Praveena, P.; Rakita, D.; Mutlu, B.; Gleicher, M. Supporting Perception of Weight through Motion-induced Sensory Conflicts in Robot Teleoperation. In Proceedings of the ACM/IEEE International Conference on Human-Robot Interaction, Online, 9 March 2020; pp. 509–517.
3. Kim, J.; Kim, S.; Lee, J. The Effect of Multisensory Pseudo-Haptic Feedback on Perception of Virtual Weight. *IEEE Access* **2022**, *10*, 5129–5140. [CrossRef]
4. Collins, K.; Kapralos, B. Pseudo-haptics: Leveraging cross-modal perception in virtual environments. *Senses Soc.* **2019**, *14*, 313–329. [CrossRef]
5. Lécuyer, A.; Coquillart, S.; Kheddar, A.; Richard, P.; Coiffet, P. Pseudo-haptic feedback: Can isometric input devices simulate force feedback? In Proceedings of the IEEE Virtual Reality 2000, New Brunswick, NJ, USA, 18–22 March 2000; pp. 83–90.
6. Punpongsanon, P.; Iwai, D.; Sato, K. Softar: Visually manipulating haptic softness perception in spatial augmented reality. *IEEE Trans. Vis. Comput. Graph.* **2015**, *21*, 1279–1288. [CrossRef] [PubMed]
7. Lopes, P.; You, S.; Cheng, L.P.; Marwecki, S.; Baudisch, P. Providing haptics to walls & heavy objects in virtual reality by means of electrical muscle stimulation. In Proceedings of the 2017 CHI Conference on Human Factors in Computing Systems, Online, 2 May 2017; pp. 1471–1482.
8. Culbertson, H.; Schorr, S.B.; Okamura, A.M. Haptics: The present and future of artificial touch sensation. *Annu. Rev. Control. Robot. Auton. Syst.* **2018**, *1*, 385–409. [CrossRef]
9. Borst, C.W.; Indugula, A. Realistic virtual grasping. In Proceedings of the IEEE Virtual Reality, Bonn, Germany, 12–16 March 2005; pp. 91–98.
10. Ott, R.; De Perrot, V.; Thalmann, D.; Vexo, F. MHaptic: A haptic manipulation library for generic virtual environments. In Proceedings of the 2007 International Conference on Cyberworlds (CW'07), Hannover, Germany, 24–26 October 2007; pp. 338–345.
11. Ott, R.; Vexo, F.; Thalmann, D. Two-handed haptic manipulation for CAD and VR applications. *Comput. Des. Appl.* **2010**, *7*, 125–138. [CrossRef]
12. Fisch, A.; Mavroidis, C.; Melli-Huber, J.; Bar-Cohen, Y. Haptic devices for virtual reality, telepresence, and human-assistive robotics. Biologically inspired intelligent robots. *Biol. Inspired Intell. Robot.* **2003**, *73*, 1–24.
13. Turner, M.L.; Gomez, D.H.; Tremblay, M.R.; Cutkosky, M.R. Preliminary tests of an arm-grounded haptic feedback device in telemanipulation. *ASME* **1998**, *15861*, 145–149.

14. Giachritsis, C.D.; Garcia-Robledo, P.; Barrio, J.; Wing, A.M.; Ferre, M. Unimanual, bimanual and bilateral weight perception of virtual objects in the master finger 2 environment. In Proceedings of the 19th International Symposium in Robot and Human Interactive Communication, Viareggio, Italy, 13–15 September 2010; pp. 513–519.
15. Hummel, J.; Dodiya, J.; Wolff, R.; Gerndt, A.; Kuhlen, T. An evaluation of two simple methods for representing heaviness in immersive virtual environments. In Proceedings of the 2013 IEEE Symposium on 3D User Interfaces (3DUI), Orlando, FL, USA, 16–17 March 2013; pp. 87–94.
16. Rietzler, M.; Geiselhart, F.; Gugenheimer, J.; Rukzio, E. Breaking the tracking: Enabling weight perception using perceivable tracking offsets. In Proceedings of the 2018 CHI Conference on Human Factors in Computing Systems, Online, 19 April 2018; pp. 1–12.
17. Rietzler, M.; Geiselhart, F.; Frommel, J.; Rukzio, E. Conveying the perception of kinesthetic feedback in virtual reality using state-of-the-art hardware. In Proceedings of the 2018 CHI Conference on Human Factors in Computing Systems, Online, 21 April 2018; pp. 1–13.
18. Lécuyer, A.; Burkhardt, J.M.; Coquillart, S.; Coiffet, P. "Boundary of illusion": An experiment of sensory integration with a pseudo-haptic system. In Proceedings of the IEEE Virtual Reality 2001, Yokohama, Japan, 13–17 March 2001; pp. 115–122.
19. Lopes, P.; Baudisch, P. Interactive systems based on electrical muscle stimulation. *Computer* **2017**, *50*, 28–35. [CrossRef]
20. Azmandian, M.; Hancock, M.; Benko, H.; Ofek, E.; Wilson, A.D. Haptic retargeting: Dynamic repurposing of passive haptics for enhanced virtual reality experiences. In Proceedings of the 2016 Chi Conference on Human Factors in Computing Systems, Online, 7 May 2016; pp. 1968–1979.
21. Lopes, P.; You, S.; Ion, A.; Baudisch, P. Adding force feedback to mixed reality experiences and games using electrical muscle stimulation. In Proceedings of the 2018 CHI Conference on Human Factors in Computing Systems, Online, 21 April 2018; p. 446.
22. Choi, I.; Culbertson, H.; Miller, M.R.; Olwal, A.; Follmer, S. Grabity: A wearable haptic interface for simulating weight and grasping in virtual reality. In Proceedings of the 30th Annual ACM Symposium on User Interface Software and Technology, Online, 20 October 2017; pp. 119–130.
23. Park, C.H.; Kim, H.T. Induced pseudo-haptic sensation using multisensory congruency in virtual reality. In Proceedings of the HCI Society of Korea Conference, Jeju-si, Korea, 14 February 2017; pp. 1055–1059.
24. Kim, J.; Lee, J. The Effect of the Virtual Object Size on Weight Perception Augmented with Pseudo-Haptic Feedback. In Proceedings of the 2021 IEEE Conference on Virtual Reality and 3D User Interfaces Abstracts and Workshops (VRW), Lisbon, Portugal, 27 March–1 April 2021; pp. 575–576.
25. Mo, S.; Kwag, J.; Jung, M.C. Literature Review on One-Handed Manual Material Handling. *J. Ergon. Soc. Korea* **2010**, *29*, 819–829. [CrossRef]
26. Kim, H.K. Comparison of Muscle Strength for One-hand and Two-hands Lifting Activity. *J. Ergon. Soc. Korea* **2007**, *26*, 35–44.
27. Jennett, C.; Cox, A.L.; Cairns, P.; Dhoparee, S.; Epps, A.; Tijs, T.; Walton, A. Measuring and defining the experience of immersion in games. *Int. J. Hum. Comput. Stud.* **2008**, *66*, 641–661. [CrossRef]
28. Rigby, J.M.; Brumby, D.P.; Gould, S.J.; Cox, A.L. Development of a questionnaire to measure immersion in video media: The Film IEQ. In Proceedings of the 2019 ACM International Conference on Interactive Experiences for TV and Online Video, Online, 4 June 2019; pp. 35–46.

electronics

Systematic Review

Virtual/Augmented Reality for Rehabilitation Applications Using Electromyography as Control/Biofeedback: Systematic Literature Review

Cinthya Lourdes Toledo-Peral [1,2], Gabriel Vega-Martínez [1,2], Jorge Airy Mercado-Gutiérrez [1,2], Gerardo Rodríguez-Reyes [1], Arturo Vera-Hernández [2], Lorenzo Leija-Salas [2] and Josefina Gutiérrez-Martínez [1,*]

[1] Instituto Nacional de Rehabilitacion Luis Guillermo Ibarra Ibarra, Mexico City 14389, Mexico; phd.toledo@outlook.com (C.L.T.-P.); gvegam@outlook.com (G.V.-M.); mercado.jorgea@gmail.com (J.A.M.-G.); grodriguezreyes@gmail.com (G.R.-R.)

[2] Centro de Investigación y de Estudios Avanzados del Instituto Politécnico Nacional, Mexico City 07360, Mexico; arvera@cinvestav.mx (A.V.-H.); lleija@cinvestav.mx (L.L.-S.)

* Correspondence: josefina_gutierrez@hotmail.com

Citation: Toledo-Peral, C.L.; Vega-Martínez, G.; Mercado-Gutiérrez, J.A.; Rodríguez-Reyes, G.; Vera-Hernández, A.; Leija-Salas, L.; Gutiérrez-Martínez, J. Virtual/Augmented Reality for Rehabilitation Applications Using Electromyography as Control/Biofeedback: Systematic Literature Review. *Electronics* **2022**, *11*, 2271. https://doi.org/10.3390/ electronics11142271

Academic Editors: Jorge C. S. Cardoso, André Perrotta, Paula Alexandra Silva and Pedro Martins

Received: 19 June 2022
Accepted: 18 July 2022
Published: 20 July 2022

Publisher's Note: MDPI stays neutral with regard to jurisdictional claims in published maps and institutional affiliations.

Abstract: Virtual reality (VR) and augmented reality (AR) are engaging interfaces that can be of benefit for rehabilitation therapy. However, they are still not widely used, and the use of surface electromyography (sEMG) signals is not established for them. Our goal is to explore whether there is a standardized protocol towards therapeutic applications since there are not many methodological reviews that focus on sEMG control/feedback. A systematic literature review using the PRISMA (preferred reporting items for systematic reviews and meta-analyses) methodology is conducted. A Boolean search in databases was performed applying inclusion/exclusion criteria; articles older than 5 years and repeated were excluded. A total of 393 articles were selected for screening, of which 66.15% were excluded, 131 records were eligible, 69.46% use neither VR/AR interfaces nor sEMG control; 40 articles remained. Categories are, application: neurological motor rehabilitation (70%), prosthesis training (30%); processing algorithm: artificial intelligence (40%), direct control (20%); hardware: Myo Armband (22.5%), Delsys (10%), proprietary (17.5%); VR/AR interface: training scene model (25%), videogame (47.5%), first-person (20%). Finally, applications are focused on motor neurorehabilitation after stroke/amputation; however, there is no consensus regarding signal processing or classification criteria. Future work should deal with proposing guidelines to standardize these technologies for their adoption in clinical practice.

Keywords: artificial intelligence; classification; control; motor rehabilitation; prosthesis; stroke; surface electromyography signals; user interface

1. Introduction

Rehabilitation therapies currently include a variety of techniques and approaches that have allowed specialized care of impaired patients up to personalized therapy, which is becoming a leading strategy in public health [1]. Among them, a new category of rehabilitation systems and virtual environments for therapy has arisen, from telemedicine [2], alterations of user interfaces and videogame controllers [3], to serious games [4], virtual reality (VR), and augmented reality (AR) [5]. Conventional physical therapy (CPT) and VR/AR therapies are believed to have a symbiotic relationship, where the latter increase patients' engagement and help them immerse in therapy, while CPT stimulates tactile and proprioceptive paths by means of mobilization, strengthening, and stretching. Therefore, the combination of both approaches could be beneficial to patients, bringing a more comprehensive and integrated treatment that can be clinically useful [5]. VR/AR environments for rehabilitation are mainly used for stroke aftermath rehabilitation therapy and prosthesis control training.

Furthermore, the control of the interface and the feedback received by the user are crucial to stimulate neurological pathways that aid in cases like neuromotor rehabilitation [6,7], or in learning how to use a new prosthetic device [8].

The use of biological signals such as surface electromyography (sEMG), which is the electrical representation of muscle activity, additionally improves the biofeedback benefits of therapy with advantages such as avoiding fatigue [9,10].

A virtual environment is used by means of an interface; VR interfaces should comply with certain conditions, i.e., no significant lag time to perceive it as a real time interaction, have seamless digitalization, use a behavioral interface (sensorial and motor skills), and have an effective immersion as close as possible to reality [11]. The use of virtual reality technologies for rehabilitation purposes has recently increased [12–15], especially for motor rehabilitation applications [16,17]. In the literature, it has been found that the main uses of these technologies fall into two main areas: motor neurological rehabilitation [15,18,19] and training for prosthesis control [20,21].

Feedback is very important during rehabilitation of new neuromotor pathways since it helps the user to correct the direction of the movement or intention towards the right track [15,20,22–25]. Instant feedback tells the brain and the body how to recalibrate in the same way that it learned it the first time [22]. Moreover, as the user interaction in rehabilitation systems grows towards a closed loop approach there is a need for a wider variety of feedback strategies, whether in the form of visual and audio-visual [26], tactile [27], or haptic [28]. Some studies centered on feedback, report closed-loop [29,30] and open-loop algorithms, among other combinations.

sEMG signals have been widely used as a control signal for rehabilitation systems and applications for a long time now [31]. This approach has several advantages regarding signal acquisition which allows the user to move freely depending on the type of hardware used, including wearable arrays, wireless systems, and even implantable electrodes [32–34]. Additionally, there is a variety of electrodes and electrode types, shapes, and arrays that can suit different applications and needs [35,36].

The use of sEMG signals as a control strategy has been widely explored in the myoelectric prosthesis research area, but it is not until recent years that these control techniques have migrated towards other therapy applications, for example, in the control of computer interfaces and environments such as VR and AR [37]. sEMG signals bring about the possibility of a complex multichannel/multiclass type of control algorithm that enables, in turn, the implementation of more intuitive user interfaces for the patient to perform [38]. This is important as it closes the loop of control/feedback interaction, and this special quality promotes neuroplasticity pathways to emerge [39]. However, the use of biosignals in VR/AR applications require more effort than other control strategies that involve other sensor measurements or motion analysis.

On one hand, sEMG signals have been widely explored for control purposes [31] and, on the other hand, VR/AR interfaces have been explored to improve the outcomes of physical rehabilitation therapies [40,41]. Furthermore, as mentioned above, several feedback techniques have been considered recently, but mainly as sensors that record a variable and return a quantitative measure to the experimenter or provide the user feedback that might feel unnatural [26–28].

Literature suggests that using sEMG signals to control VR/AR interfaces can provide better outcomes when paired to CPT. In 2018, Meng et al. [9] performed a 20-day follow-up experiment to prove the effectiveness of a rehabilitation training system based on sEMG feedback and virtual reality that showed that this system has a positive effect on recovery and evaluation of upper limb motor function of stroke patients. Then, in 2019, Dash et al. [42] carried out an experiment with healthy subjects and post-stroke patients to increase their grip strength. Both groups showed an improvement in task-related performance score, physiological measures (using sEMG features), and readings from a dynamometer; from the latter, both groups gained at least twice their grip ability. Later, in 2021, Hashim et al. [43] found a significant correlation of training time and the Box and

Block Test score when testing healthy subjects and amputee patients in 10 sessions during a 4-week period using a videogame-based rehabilitation protocol. They demonstrated improved muscle strength, coordination, and control in both groups. Moreover, these features added to induced neuroplasticity and enabled a better score in this test, which is related to readiness to use a myoelectric prosthesis. More recently, in 2022, Seo et al. [44] proposed to determine feasibility of training sEMG signals to control games with the goal of improving muscle control. They found improvement in completion times of the daily life activities proposed; however, interestingly, they report no significant changes in the Box and Block Test. The contrasting results found should be further investigated for specific clinical instruments or experimental settings.

Literature supports the hypothesis that sEMG signals can be a robust biofeedback method for VR/AR interfaces that can potentially boost therapy effects. On top of an increased motivation and adherence from patient to complete rehabilitation therapies [43], the method also yields a different type of awareness to the patient of their own rehabilitation progress. Furthermore, this type of therapy offers quantitative data to the therapist, potentially allowing a better understanding of patient progress, which brings certainty to the process.

Nonetheless, sEMG signals as control or feedback of a VR/AR interface merge has not been investigated thoroughly, and even less so in the form of a systematic literature review (SLR). sEMG signals can be better interpretated by patients as control and the visual feedback completes the natural pathway they lost and are trying to get back through rehabilitation therapy.

There are some SLRs that have analyzed VR and AR interfaces used for hand rehabilitation, but some lack an adequate inclusion of feedback techniques [45], whereas others include feedback and focus on the similarity of techniques among VR interfaces for rehabilitation therapy and CPT [5]. Some studies use a computer screen interface to address virtual rehabilitation therapies [46]. Although there are several articles reporting individually the use of sEMG signals and VR/AR interfaces [9,43,44,47], there are no SLRs focused on the use of for these interfaces.

Although some studies show sEMG signals paired with VR/AR environments, just a few discuss if there are advantages in clinical results compared to conventional therapy groups, and there is a shortage of standardized protocols when sEMG signals are used for rehabilitation therapy purposes [37,48,49]. Hence, there is not enough information in the literature to determine if these biosignals used as control or feedback of VR/AR systems improve the positive outcomes of neuromotor rehabilitation therapies and whether they promote, e.g., neuroplasticity or support training of myoelectric control for prosthesis fitting. It is also important to know if the hardware used is commercially available or proprietary/developed, and if the signal processing techniques used are similar enough to be compared. Likewise, it is important to learn the rehabilitation target to which this technology has been applied to and if they have been tested with healthy subjects or patients, and if this technology is aimed for a clinical environment or only for research protocols.

To address these matters, our goal and contribution to the field is to explore and analyze if sEMG signals can be used as control and/or biofeedback for VR/AR interfaces, and to find out if the proposed techniques converge on a standard of care protocol, since there are not many methodological reviews to date that focus on sEMG control/feedback. Therefore, we considered it essential and necessary to carry out an exhaustive review of the published scientific literature regarding this topic. In this paper we applied the PRISMA (preferred reporting items for systematic reviews and meta-analyses) methodology [50] for a systematic literature review (SLR) to find out how AR and VR interfaces are used in rehabilitation applications that are controlled through sEMG signals. To complete this task, we collected relevant articles dealing with state-of-the-art AR and VR environments used for rehabilitation purposes based on sEMG signals used as control or biofeedback.

2. Materials and Methods

The PRISMA methodology was followed to conduct the SLR search [50]. A set of six academic and scientific databases were searched: PubMed/Medline, IEEE Xplore, Science Direct, Scopus, EBSCO, and Google Scholar. The search included titles, abstracts, and keywords of articles written in the English language. The search was conducted from January 2017 to March 2022.

Search query and selection criteria—The aim of the SLR was to find and analyze the state-of-the-art of motor neurological rehabilitation based on VR/AR interfaces, focusing on those using sEMG control covering feature extraction and classification algorithms. The search query was performed in three steps (Figure 1). Step 1—Identification: from the articles resulting from the Boolean search of keywords in databases, titles, abstracts, and keywords; they are looked over to eliminate duplicates and unrelated articles. Step 2—Screening eligibility: articles were selected if dealing with any form of VR/AR interfaces for rehabilitation controlled by sEMG, while excluding those that cannot be retrieved, the ones that aim at other research focus, those that do not use sEMG as control or feedback, the ones that use sEMG for assessment purposes, and those that do not include a virtual interface. Step 3—Including: the filtered articles are selected for analysis after full text reading.

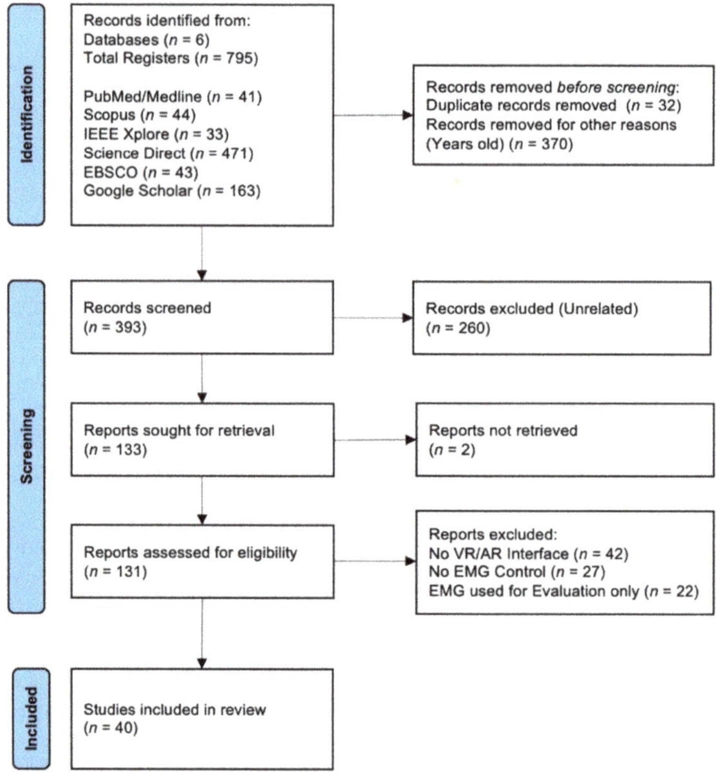

Figure 1. PRISMA flow diagram for the systematic literature review.

Research questions—The main information to be extracted from the SLR, to find out the use of sEMG control for rehabilitation using VR/AR application, is summarized in the following series of research questions (RQ):

RQ1: What is the share in the use of VR and AR interfaces in rehabilitation?
RQ2: Which is the target anatomical region aimed to be rehabilitated?

RQ3: What type of rehabilitation therapy is the interface used for?
RQ4: What are the characteristics of VR/AR interfaces when used for rehabilitation?
RQ5: How are sEMG signals used to interact with VR/AR interfaces for rehabilitation?
RQ6: What hardware is used for signal acquisition?

Inclusion and exclusion criteria—The keywords used for a Boolean search through the databases were: ((Virtual Reality) OR (Augmented Reality)) AND (Rehabilitation) AND (Surface Electromyography) AND (Control OR Feedback). Articles from peer-reviewed conference proceedings, indexed scientific journals, books, and chapters are included. After this examination, articles that are duplicated or unrelated to the scope of this paper were removed. The remaining articles were explored for other related keywords such as: interface, videogame, stroke, and prosthesis. Those which were considered relevant and belong to recent advances in the techniques of interest were selected for analysis in the SLR.

Data extraction and analysis—This section describes the proposed classification for the selected articles, including original and review articles (Figure 2). The articles were filtered into three classes: first class consisted of sEMG control algorithms and was subdivided into pattern recognition and direct control; the second class was the mode of rehabilitation application that can be either for neurological rehabilitation (e.g., stroke) or for amputation rehabilitation in the form of training for prosthesis control. Finally, the third class took up the categories of VR/AR interface interaction, including the training scene model, first-person mode, and videogame interfacing.

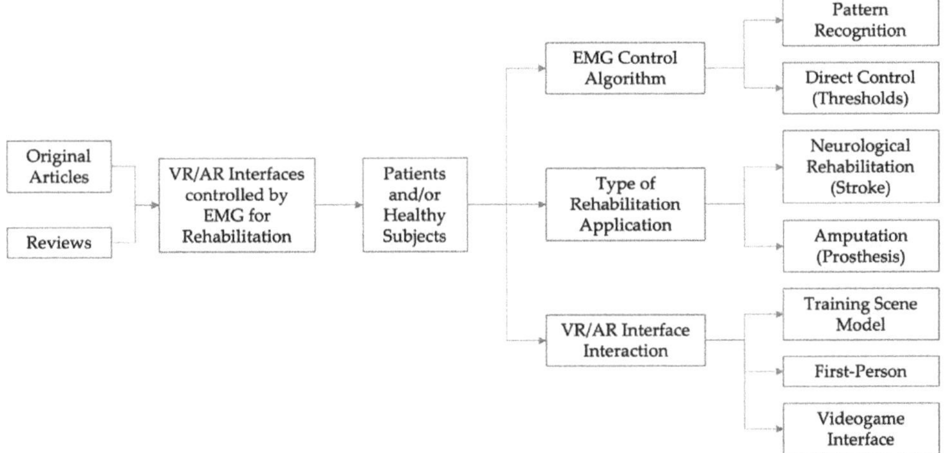

Figure 2. Taxonomy of the SLR: sEMG control algorithms for VR/AR interfaces for motor rehabilitation.

3. Results

The Boolean search of keywords in electronic databases accounted for 795 studies (Step 1—Identification). Articles dealing with any form of VR/AR interfaces for rehabilitation controlled by sEMG signals were selected (Figure 3). First, irrelevant articles were removed, including those eliminated for being duplicated and those older than January 2017, summing up to 573 articles.

Then, articles unrelated to the focus topic and those that could not be retrieved were eliminated (Step 2—Screening eligibility), subtracting 104 additional articles.

Finally, those that do not use sEMG as control or feedback (27), the ones that use sEMG signals just for evaluation (22), and those that do not include a virtual or augmented reality interface (42) were excluded (Step 3—Including), leaving us with 40 articles for full text reading and analysis.

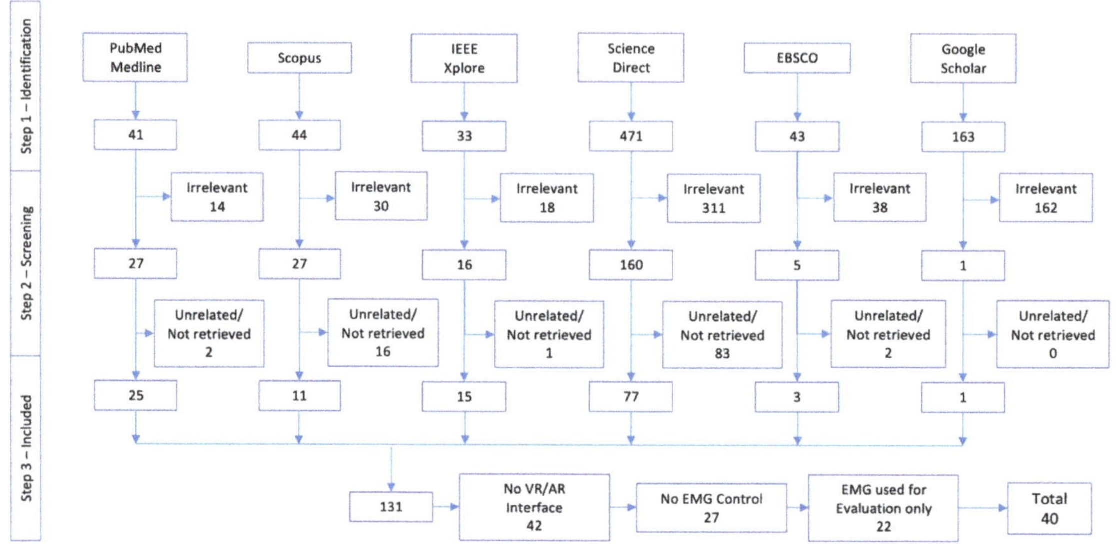

Figure 3. Number of records identified from each database for the systematic literature review.

From the 40 works analyzed, 2 were review articles, 2 were book chapters, and the remaining (36) were original articles. Some 82.5% (33) of literature articles were oriented to upper limb, and 17.5% (7) to lower limb. Patients were included in 15 articles (37.5%), with 7 concerning amputee patients and 8 concerning post-stroke patients. A total of 67.5% (27) of the articles included abled-bodied healthy subjects in their trials. All the articles (40) used visual feedback through the VR or AR interfaces, but a few used a second type of feedback, such as 2 articles that included fatigue and closed-loop feedback to regulate intensity [51,52], while another 2 papers used audio feedback [47,53], 2 relied on tactile feedback [53–55], 1 had haptic feedback, and 1 asked the subject to think of the movement (to be detected through electroencephalography (EEG)) as well as to perform it [56]. Just 3 articles mentioned exoskeletons for movement assistance triggered by sEMG signals [56–58], and 1 article used functional electrical stimulation (FES) for movement assistance [59]; all 4 of them belong to neurorehabilitation applications.

3.1. RQ1: What Is the Share in the Use of VR and AR Interfaces in Rehabilitation?

From the analyzed articles, we found that 57.5% (23) of them use a VR interface environment for rehabilitation purposes (Table 1). Some 27.5% (11) of the articles propose a virtual interface that operates as a computer interface (CI) (Table 2). Meanwhile, 4 (10%) of them use AR interfaces as biofeedback (Table 3).

Mostly, VR and CI interfaces show an environment to be controlled by the user to complete an action or different movements repetitions. There are three main variations: videogame interface (11 for VR, 7 for CI), imitation tasks named training scene model (2 articles for both cases VR and CI), and first person, e.g., outreach tasks (10 for VR and 11 for CI). In the case of AR interfaces, 3 correspond to a videogame or serious games interfaces [60–62], and 1 to a training scene model [63].

In total, 18 (52.94%) of the articles concerning VR or CI interfaces use a videogame or a serious game as interface with 7 of them, for CI, showing tests performed by patients with positive performance results [42,43,47,51,64–66], and 3 for VR interfaces [53,67,68].

There were 2 VR and 2 CI articles presenting interfaces based on a training scene model, with only 2 of them showing results for patient use [69,70]. Meanwhile, 5 (13.8%) articles use a first-person approach for their interface, however, only 3 show results with patients and do not report performance metrics [29,56,70–72].

Only [61] reports the use of the AR interface with patients (five amputees). Melero et al. [60] use a Microsoft Kinect to locate upper limbs of three abled-bodied subjects; when sEMG signals show a perfect performance of the activity, the subject scores a point. They report a 77% accuracy in hand gesture classification.

There is a widespread conception of what a VR interface is implied to have and look like. However, most articles use a conventional videogame computer interface, and only 8 (22.2%) report using a headset [61,69,72–74], an environment [65,68], or immersive VR [75].

3.2. RQ2: Which Is the Target Anatomical Region Aimed to Be Rehabilitated?

There are two types of pathologies to which the VR/AR interfaces are targeted: post-stroke paresis rehabilitation and training for myoelectric prosthesis use. From the articles that include patients in their studies (40%), these are mainly upper limb amputees (50%) [6,43,56,61,65,67,68,70], post-stroke with hemiparesis patients (43.75%) (which may need both upper and/or lower limb rehabilitation) [42,47,51,53,64,66,69], and there is 1 study where authors tested their environment with a patient that presented a bilateral upper-limb congenital transverse deficiency [72].

Most of the developments are focused on upper limb rehabilitation (82.5%), which include all the AR interfaces described above. Even though there are more cases of lower limb amputations and paresis than upper limb amputations [76], upper limb disability has been reported as a larger burden than lower limb impairment or loss [77].

3.3. RQ3: What Type of Rehabilitation Therapy Is the Interface Used for?

Neurological motor rehabilitation is the goal of 28 (70%) of the analyzed articles, while 12 (30%) present interfaces used for training the amputee patient for future myoelectric prosthesis use.

3.4. RQ4: What Are the Characteristics of VR/AR Interfaces When Used for Rehabilitation?

We found the interfaces can be divided into three types: videogame, first-person, and training scene model.

The less common interface is the training scene model (15% of analyzed articles), only used in 2 (8.69%) VR interfaces, in 2 (18.18%) CI interfaces, and 2 (50%) AR interfaces. Here, the user is shown an arm and/or hand that performs the movements the user is sending for control or biofeedback. The next type of interface is first-person with 16 (40%) interfaces found. This type of interface is trying to embed the user in the environment, as if they were going through it; most of the times the user can see either their arms and hands or some tool used to attain the goal of the game. Finally, the videogame interface is based on the movement of a character to perform a given task within a designed environment, and each virtual movement is related to a real movement from the impaired hand. The articles show 11 (27.5%) videogame type for VR interfaces, and 7 (17.5%) for CI interfaces; no videogame interfaces were reported for AR interfaces in the analyzed articles.

3.5. RQ5: How Are sEMG Signals Used to Interact with VR/AR Interfaces for Rehabilitation?

Regarding the user interaction with the interface, the main aspect to be described is sEMG control, which is based on acquiring and classifying muscle activity to detect volitional activity, and in some cases, which type of hand gesture is being performed. To accomplish this, several types of classifiers are presented within the analyzed articles.

Regarding signal processing algorithms, Li et al. [55] reported the use of the wavelet transform to process sEMG signals and to extract features for further classification using support vector machines (SVM). However, 13 articles (32.5%) reported the use of pattern/gesture recognition to differentiate among hand grasps.

Support vector machine is used in 4 (10%) articles, only 3 of them report performance (with 96.3%, 95%, and 99.5% (healthy subjects)/94.75% (stroke patients), respectively) [9,51,52,55]. Another 4 (10%) articles use neural networks, and 1 reports a 97.5% performance using a convolutional neural network [63], while another 1 uses a deep learning model [73], and

1 more a probabilistic neural network [64]. Only 1 article uses linear discriminant analysis for classification of grasps [67]. The Myo Software® is used for classification of hand grasps in 2 articles [59,61], and [6] reports the use of a Kalman filter-base decode algorithm. The above-mentioned techniques are state-of-the-art classification methods for sEMG control, which is based on pattern/gesture recognition and accounts for 32.5% (13) of the upper limb prosthesis training papers analyzed.

Furthermore, classic control approaches such as proportional control and threshold-based classification are found within the analyzed articles, with 2 articles for the former [47,60] which also considers the strength of the muscle contraction, and 4 (10%) concerning the latter [54,56,62,78]. Additionally, 4 (10%) articles [53,68,78,79] consider the intention of motion to generate a control or activation signal. The remaining ones either do not specify or are unclear or ambiguous regarding their classification method.

3.6. RQ6: What Hardware Is Used for Signal Acquisition?

For sEMG signal acquisition, 9 (22.5%) of the analyzed articles report the use of a Myo Armband (Thalmic Labs, Kitchener, Canada) with all applications related to the use and training for upper limb prosthesis; this accounts for 75% of the upper limb prosthesis applications reported. Another 4 (10%) articles used a model of Delsys© sEMG acquisition system (Delsys Inc., Natick, MA, USA); both hardware systems are considered among the three best acquisition systems regarding the quality of their signals and the high classification accuracy achieved with them [80]. A further 7 (17.5%) research articles present applications using proprietary hardware. Refs. [73,74] use the Leap Motion (Ultraleap, San Francisco, CA, USA) hardware to acquire arm/hand movements as an extra input for system control. Melero, et al. used the Microsoft® Kinect as a second acquisition input for control [60].

Finally, exoskeletons are used by [56–58], triggered by the events detected from sEMG signals, to promote correct trajectories during rehabilitation therapy.

For biofeedback, visual interfaces are used in all cases, but some also incorporate other types of feedback. For instance, Wang et al. [51] use fatigue to adapt the level of difficulty of the videogame interface; Dash et al. [47] present an audiovisual stimulus to the user, as do as Llorens et al. [53], where they incorporate tactile user feedback to the audiovisual modality; Li et al. [29] use electrotactile feedback for a closed-loop control application with a VR environment; Ruiz-Olaya et al. [58] use visual and haptic feedback, whereas Covaciu et al. [81] use visualization of the functional movement through the VR interface as feedback.

The following tables shows the most relevant characteristics of the articles analyzed. When the article did not include information regarding a certain topic the slot is left blank. The first table shows results for VR interfaces (Table 1), the next one presents the summary of articles dealing with AR interfaces (Table 2), and finally, the last one regards computer interfaces found among the analyzed articles (Table 3).

Table 1. VR interfaces used for motor rehabilitation based on sEMG Control.

| First Author, Year | Type of Rehabilitation | Type of Interface | Interaction | Subjects | | Anatomical Region | Acquisition Hardware | Feature Extraction | Classification Algorithm and Performance | Feedback |
				Healthy Subject	Patient					
Mazzola, S., 2020 [73]	Neurological Motor Rehabilitation	VR 3D Upper Limbs Precision-based block staking task	Comparison of sEMG signals by means of RMS Voltage	24	-	Upper limb/ flexor carpi radialis, extensor digitorum, biceps brachii, triceps brachii bilateral	Vive Pro HMDLeap Motion Delsys Trigno wireless Electrodes 4-channels	RMS from sEMG [+]	Compare RMS [+] level from sEMG Signal (with and without the VR Interface)	Visual feedback
Lydakis, A., 2017 [69]	Neurological Motor Rehabilitation	VR Videogame Interface Movement Imitation	3D Avatar (1) Assessment experiment (2) Action observation (3) Combined motor imagery and action observation	-	4 post-stroke	Upper limb/ Musculi flexor pollicis longus, flexor digitorum superficialis and flexor carpi radialis	Myo Armband+ R7 AR Glasses + IMU g.Hlamp	RMS from sEMG [+]	Thresholds	Visual feedback
Woodward, R.B., 2019 [67]	Prosthesis Training	VR Virtual forest virtual crossbow in real-time	sEMG control of real-time hand grasps	16	4 amputees (3 transradial, 1 wrist dislocation)	Upper limb/ Forearm Hand gestures (no motion, hand open, hand close, wrist pronation, wrist supination, wrist flexion, and wrist extension)	Custom-fabricated sEMG acquisition armbands included six pairs of stainless-steel dome electrodes TI ADS1299 bioin-strumentation chip	Movement velocity (advanced proportional control algorithm) Speed (smoothed) MRV, WVL, ZC SSC, and ARF from sEMG [+]	Pattern recognition 3D Target Achievement Control Test LDA ***	Visual feedback
Summa, S., 2019 [74]	Neurological Motor Rehabilitation	VR Robotic platform (Dynamic Oriented Rehabilitative Integrated System–DORIS) + motion analysis + sEMG	Training of equilibrium and gait Game experiences for VR	-	-	Lower limb Core Equilibrium and gait	Unreal VR Headset + Leap Motion + Vicon/sEMG Server – DORIS	-	-	Visual feedback
Kluger, D.T., 2019 [6]	Prosthesis Training	VR Virtual Modular Prosthetic Limb	Closed-loop virtual task	-	2 amputees (transradial)		19 contact sensors at the hand	MAV [+] from sEMG	Modified Kalman-filter-based decode	Visual feedback

Table 1. *Cont.*

First Author, Year	Type of Rehabilitation	Type of Interface	Interaction	Subjects		Anatomical Region	Acquisition Hardware	Feature Extraction	Classification Algorithm and Performance	Feedback
				Healthy Subject	Patient					
Meng, Q., 2019 [9]	Neurological Motor Rehabilitation	VR Rehabilitation Game	Virtual rehabilitation game	8	-	Upper limb/ Wrist flexion–extension	Property Design	Moving average window Autoregressive model (AR)parameter model in time domain	SVM *** Recognition of action 96.3%	Visual feedback
Nissler, C., 2019 [68]	Prosthesis Training	VR Environment Serious Games (Unity)	Box and Block Test At virtual living room and kitchen	15	1 amputee (uses prosthesis)	Upper limb	Myo Armband	-	Intent detection	Visual feedback
Covaciu, F., 2021 [81]	Neurological Motor Rehabilitation	VR Collect Apples	Foot movements	10	-	Lower limb/ Ankle	Gyroscope Accelerometer Myoware	-	KNN *** 5-fold cross-validation 81.35%	Visual/functional feedback
Llorens, R., 2021 [53]	Neurological Motor Rehabilitation	VR Pick up Apples that grow before they disappeared	Intention of action while administering transcranial direct current stimulation	-	29	Upper limb/ brachioradialis, palmaris longus, and flexors and extensors of the fingers	Myo Armband	-	Intention of action while administering transcranial direct current stimulation	Audiovisual and tactile feedback
Li, K., 2019 [29]	Neurological Motor Rehabilitation	VR Environment Virtual Hand	Control with sEMG Electrotactile stimulation module Force proportional to intensity	10	-	Upper limb	Multichannel sEMG Acquisition System Elonxi Ltd.	sEMG intensity	-	Visual Feedback: Numerical indicators of force and deformation ElectrotactileStimulation Closed-loop
Cardoso, V.F., 2019 [75]	Neurological Motor Rehabilitation	VR Immersive Serious Game	EEG sEMG Robotic Monocycle	8 (5 males)	-	Lower limb	Property sEMG acquisition 4-channels	-	-	Visual feedback
Li, X., 2019 [55]	Neurological Motor Rehabilitation	VR Kitchen Scene (open door, clean table, ventilator, cut food)	Control with sEMG	4	-	Upper limb	Wireless acquisition module	MAV, RMS, SD from sEMG + MAV, singular values of wavelet coefficients	SVM, PNN *** 95% for wavelet coefficients	Visual feedback

Table 1. *Cont.*

First Author, Year	Type of Rehabilitation	Type of Interface	Interaction	Subjects		Anatomical Region	Acquisition Hardware	Feature Extraction	Classification Algorithm and Performance	Feedback
				Healthy Subject	Patient					
Bank, P., 2017 [78]	Neurological Motor Rehabilitation	VR Imitation Game	Control with sEMG	18	-	Upper limb/ wrist flexor carpi radialis and extensor carpi radialis	Porti7 22 bits A/D fs = 2000 Hz	MVC from sEMG [+]	Task Performance 96.6% / Effort 100% / Co-contraction 99.8%	Visuomotor tracking
Ruiz-Olaya, A.F., 2019 [58]	Neurological Motor Rehabilitation	VR Environments and/or Headsets	Control left/right position of virtual car / High-density surface sEMG EEG	-	-	Upper limb, lower limb, full body Exoskeletons	Several	-	.	Visual feedback Haptic Exoskeleton
Castellini, C., 2020 [82]	Neurological Motor Rehabilitation	VR/AR Avatar Upper Limb Interaction	Control with sEMG	-	-	Upper limb	-	MVC from sEMG [+]	ML *** Pattern Recognition	Visual feedback (Positive psychological effects)
Raz, G., 2020 [71]	Neurological Motor Rehabilitation	VR Environment	Headset Sit in real table Arms represented in virtual world	-	-	Upper limb	-	-	-	Visual feedback
Bhagat, N.A., 2020 [56]	Neurological Motor Rehabilitation	VR Outreach task	BMI detects motion intention from sEMG and EEG motor intent to trigger exoskeleton for assistance	-	10 chronic post-stroke	Upper limb/ biceps brachii, triceps brachii	Proprietary sEMG	RMS from sEMG [+]	sEMG threshold + EEG motor intent	Think of movement Visual feedback
Heerschop, A., 2020 [83]	Prosthesis Training	VR Serious Games:Control a grabber, free catching task, following task	Control from sEMG	43	-	Upper limb/ flexor-extensor of wrist	Otto Bock 13E200 Electrodes 2-channels	-	-	Visual feedback
Liew, S.L., 2022 [59]	Neurological Motor Rehabilitation	VR Serious Games	Control from sEMG to trigger FES *	-	-	Upper limb/ lower limb	Several	-	-	Visual feedback FES * activation

Table 1. *Cont.*

First Author, Year	Type of Rehabilitation	Type of Interface	Interaction	Subjects		Anatomical Region	Acquisition Hardware	Feature Extraction	Classification Algorithm and Performance	Feedback
				Healthy Subject	Patient					
Ida, H., 2022 [84]	Neurological Motor Rehabilitation	VR Videogame	Postural adjustment after perturbation (rReal vs VR) Single-leg obstacle avoidance task	10	-	Lower limb	Myopac RUN	Mean + SD from sEMG [+]	-	Visual feedback
Montoya-Vega, M.F., 2019 [52]	Prosthesis Training	VR Serious Games Force Defense	Change difficulty of videogame depending on fatigue	12	-	Upper limb/ biceps brachii	Myo Armband	Fatigue	Motor Learning	Muscle fatigue as feedback
Mazzola, S., 2020 [73]	Neurological Motor Rehabilitation	VR Gesture-level hand tracking	Stack blocks using dominant hand	24	-	Upper limb/ flexor carpi radialis, extensor digitorum, biceps brachii and triceps brachii Bilateral	Delsys Trigno wireless electrodes	Amplitude RMS from sEMG [+] Completion Task Time	-	Visual feedback
Kisiel-Sajewicz, K., 2020 [72]	Neurological Motor Rehabilitation	VR Headset Virtual Upper Extremity	Reaching task precision fine grasping	1	1	Upper limb	OTbioLab ELSCHO064LS	MVC, Sub-MVC (20% MVC) from sEMG [+]	-	Visual feedback

* IMU—inertial movement unit, FES—functional electrical stimulation. *** SVM—support vector machine, ML—machine learning, ANN—artificial neural network, LDA—linear discriminant analysis, DL—deep learning, KNN–K-nearest neighbor, PNN—probabilistic neural network, CNN—convolutional neural network. [+] Features acronyms: SSC–slope sign changes, ZC–zero crossings, RMS–root mean square, WL–waveform length, MDF–median frequency, MNF–mean frequency, MAV—mean average value, MPF—mean power frequency, SE—self-ordering entropy, MRV—mean relative value, WVL—waveform vertical length, ARF—auto-regressive features, MVC—maximum voluntary contraction, ASS—absolute value of the summation of square root, MSR—mean value of square root.

Table 2. CI interfaces used for motor rehabilitation based on sEMG control.

First Author, Year	Type of Rehabilitation	Type of Interface	Interaction	Subjects		Anatomical Region	Acquisition Hardware	Feature Extraction	Classification Algorithm and Performance	Feedback
				Healthy Subject	Patient					
Wang, L., 2017 [51]	Neurological Motor Rehabilitation	VR Videogame Interface	Training scene model	4	4 post-stroke	Upper limb/ hand gestures	eego™sports, Delsys	SE, MPF from sEMG +	SVM *** to identify action patterns – 99.5% healthy subjects (4) 94.75% stroke patients (4)	EEG and sEMGfatigue status to adapt level of difficulty
Lai, J., 2017 [79]	Prosthesis Training	VR Real-time interaction	Visual feedback real time response of sEMG control	1	-	Upper Limb/ Forearm	Danyang Prosthetic Electrodes 4-channel and UBS6351 NI	×5 20 repetitions trials	SVM *** Training and pattern recognition	Visual feedback
Dash, A., 2019 [47]	Neurological Motor Rehabilitation	VR Videogame Interface Basketball tower, 3 goal posts	sEMG biofeedback for strength inference and EDA for tonic mean	6	6 post-stroke hemiplegic	Upper limb/ flexor carpi radialis and extensor carpi radialis longus	Biopac MP150 fs = 1000 Hz	MAV from sEMG +	Levels of strength	Audio-visual feedback
Trifonov, A.A., 2020 [57]	Prosthesis Training	VR Movement Imitation	Replicates movements in VR, sEMG used as input of an exoskeleton that places the limb at given coordinates	1	-	Upper limb	Proprietary AD8232 2-channels – Exoskeleton	RMS, MAV from sEMG +	ANN *** (Two layers: Kohonen and Grossberg)	Visual feedback
Nasri, N. 2020 [64]	Neurological Motor Rehabilitation	VR (Unity) Serious Games	sEMG control	-	4	Upper limb/ hand gestures	Myo Armband	-	DL Model Conv-GRU architecture	Visual feedback
Dash, A., 2020 [42]	Neurological Motor Rehabilitation	VR (Unity) Fountains, Basketball court	sEMG control triggered grip exercise (move VR objects according to hand gesture)	8	12 post-stroke	Upper limb	Biopac MP150	MAV + from sEMG sEMG-controlled dynamic positioning of VR object	-	Visual feedback
Lukyanenko, P., 2021 [70]	Prosthesis Training	VR Representation of a prosthetic hand	Activate virtual hand using EMG	-	2	Upper limb	Chronically implanted EMG (ciEMG) electrodes Ripple Grapevine system collected ciEMG fs = 2000 Hz 15–350 Hz filter	MAV + from EMG	KNN mapping technique	Visual feedback

Table 2. *Cont.*

First Author, Year	Type of Rehabilitation	Type of Interface	Interaction	Subjects		Anatomical Region	Acquisition Hardware	Feature Extraction	Classification Algorithm and Performance	Feedback
				Healthy Subject	Patient					
Hashim, N., 2021 [43]	Neurological Motor Rehabilitation	VR Videogames Crate Whacker, Race the Sun, Fruit Ninja, andKaiju Carnage	1-h sessions 4-week rehabilitation program Box and Block Test sEMG Assessment	5	5 amputees Transradial	Upper limb/ forearm	Myo Armband	MVC $^+$ from sEMG to randomly select a game	-	Visual feedback Timer and score visible
Quinayás, C., 2019 [65]	Prosthesis Training	VR Environment to locate & grasp object (Unity)	Hand grasps: rest, open hand, power, and precision grip 20 trials	-	1	Upper limb/ forearm	Property sEMG Bracelet fs = 1000 Hz	ASS, MSR from sEMG $^+$	Online recognition of motion intention 86.6%	Visual feedback
Yassin, M.M., 2021 [85]	Neurological Motor Rehabilitation	VR Cellphone Apps (Patient + therapist) Car game (gauge & bar)	Control from sEMG	5	-	Upper limb	Property sEMG microcontroller—based on ARM Cortex 32- bit M3 architecture	RMS$^+$ from sEMG	-	Visual feedback (gauge and bar)
Ma, L., 2018 [66]	Neurological Motor Rehabilitation	VR Videogame Interface (Hamster, Flappy Bird)	Picture guidance Gesture recognition of hand movement generates game character movement	6 validation only	9 post-stroke (pre-, mid-, post-rehab) (5–right hemiplegia)	Upper limb/ hand grasps (relax, open hand, close hand)	Delsys 4-channels, dry electrode, fs = 2000 Hz	SSC, ZC, RMS, WL, MDF, and MNF from sEMG $^+$	ML *** 2-fold model fusion of Stacking – 95% in healthy subjects 90% 2 post-rehab patients' hemiplegic side	Visual feedback

*** SVM—support vector machine, ML—machine learning, ANN—artificial neural network, LDA—linear discriminant analysis, DL—deep learning, KNN—K-nearest neighbor, PNN—probabilistic neural network, CNN—convolutional neural network. $^+$ Features acronyms: SSC—slope sign changes, ZC—zero crossings, RMS—root mean square, WL—waveform length, MDF—median frequency, MNF—mean frequency, MAV—mean average value, MPF—mean power frequency, SE—self-ordering entropy, MRV—mean relative value, WVL—waveform vertical length, ARF—auto-regressive features, MVC—maximum voluntary contraction, ASS—absolute value of the summation of square root, MSR—mean value of square root.

Table 3. AR Interfaces used for motor rehabilitation based on sEMG control.

First Author, Year	Type of Rehabilitation	Type of Interface	Interaction	Subjects	Anatomical Region	Acquisition Hardware	Feature Extraction	Classification Algorithm and Performance	Feedback
Melero, M., 2019 [60]	Prosthesis Training	AR Visualization of muscle activity Dance game Imitation	Score when movements are performed correctly Perform choreographed dance containing hand gestures involved in upper limb rehabilitation therapy	3	-	Upper limb/ hand gestures	Wired intramuscular sEMG recording implant 4-channels Myo Armband Microsoft Kinect	10 trials	Myo Armband Software 77% Accuracy Hand Gesture Classification	Visual feedback
Gazzoni, M., 2021 [62]	Neurological Motor Rehabilitation	AR Superimposed muscles Smartglasses	Control from sEMG	1	-	Upper limb/ lower limb	Due 14-channels	RMS from sEMG [+]	Threshold sEMG	
Liu, L., 2020 [63]	Prosthesis Training	AR Imitation game	Exercise finger movements	100	-	Upper limb/ aand	Myo Armband	Spectogram	CNN * Pattern Recognition 10 gestures 97.8%	Visual feedback
Palermo, F., 2019 [61]	Prosthesis Training	AR Portable Environment	AR Environment renders a table, a hand and bottle, a screwdriver, tennis ball, pen, can Control with sEMG	5	5 amputees Transradial	Upper limb	Microsoft HoloLens Myo Armband	-	Pattern recognition with Myo Software	Visual feedback

* CNN—Convolutional Neural Network. [+] Features acronyms: SSC—slope sign changes, ZC—zero crossings, RMS—root mean square, WL—waveform length, MDF—median frequency, MNF—mean frequency, MAV—mean average value, MPF—mean power frequency, SE—self-ordering entropy, MRV—mean relative value, WVL—waveform vertical length, ARF—auto-regressive features, MVC—maximum voluntary contraction, ASS—absolute value of the summation of square root, MSR—mean value of square root.

4. Discussion

VR/AR technologies can be used as a visual guide to perform an activity, or to immerse in a different environment, but they can also be controlled using a variety of sensors or biosignals where a natural movement of the body generates a response in the environment displayed as movement or control of an avatar [11]. A simple example would be to adapt the environment so that when the subjects walk, it moves too, and they can explore it. Several rehabilitation strategies can emerge from these interactions [2,7,9,17,86,87]. In this paper we presented the analysis of 40 articles that deal with the use of sEMG signals to control or feedback a VR/AR interface for rehabilitation purposes, which provides a global framework of the most common applications found.

VR is characterized as being an immersive interface or environment [11]. Despite this, it was found that 11 out of the 34 articles (32.35%) proposing VR interfaces interact with the user through a computer interface. The effects of these type of interfaces should be further investigated since they present advantages (less expensive, ready to use, needs less space to be used) and disadvantages (lack of immersion, allows distractions).

The application of AR technology has more challenging requirements to emulate virtual items over real life environments, which could be a room, furniture, or an open field. This visualization is commonly made through a screen that shows a virtual object projected over a real time image or video of the experimentation room [88]. A way to overcome this issue requires technology that can be as advanced as a holographic projector. On the other hand, this type of interface is much more immersive; as its name suggests, it is closer to reality, trying to erase the limits between the virtual and the real world. So, an AR interface could have a potentially higher impact on the user's brain and consequently on rehabilitation therapy [5,41,88].

Videogame and first-person interfaces engage the patient by allowing them to train actively, compared to traditional rehabilitation therapy where monotonous repetitions are typical. This type of therapy approach is copied by training scene model interfaces.

On the other hand, VR/AR technologies have great potential, since they can completely change the perception of the user's own motor functions, potentially restructuring body proprioception, vital for neurorehabilitation applications, neuroplasticity, and motor rehabilitation in general. For example, Osumi et al. found that VR therapy helped alleviate phantom limb pain effectively, compared to CPT [87]. The interaction achieved with these technologies highly improve patient engagement with therapy, adherence to treatment, and excitement to come back [9,42,43,82]. Specifically, Castellini et al. [82] mentioned a positive psychological effect from VR/AR interfaces used in rehabilitation.

Acquisition hardware is a sensitive subject because is the first link to the user, a mistake here can cause chaos in the system. Melero et al. [60] and Palermo et al. [61] use the Myo Armband for signal acquisition which allows them to have a more compact and portable system. In total, 9 articles report the use of this band. Different models of the Delsys acquisition system are reported in other articles [51,66,73]. Both systems are not only in the top 3 devices for sEMG signals acquisition [80], but are also very small, portable, and convenient to use, which translates into an easier way to use this technology in a clinical environment, consequently involving more patients in the tryouts. Several authors chose to develop their own hardware, which come with advantages (specific design to fulfill specific needs) and disadvantages (manufacture can be problematic, especially to miniaturize the electronics).

Surprisingly, 32.5% of the articles did not mention the processing algorithms, type of classifier used, classification performance, etc., and a few more mentioned it but were very ambiguous—they neither reported the protocol followed for therapy, the evaluations performed to the technique selected, or the effectiveness of the technique for therapy purposes. Reporting this data is highly important disregarding the clinical section of the results. The use of sEMG signals and their processing aiming rehabilitation applications is still scattered and heterogenous, and there is no consensus to select the methodology for processing algorithms, signal features, classification approach, and performance evaluation.

We consider it would be very important for authors to state the signal processing and classification algorithm designed or used, and their performance metrics, which could be among the technical guidelines that could be proposed to homogenize the protocols.

All these environments are controlled using sEMG signals and provide feedback through a visual, tactile, or functional stimulus. It becomes obvious that virtual therapy is also based on repetitions and practice, like CPT, but the manner the patient can interact with it is what makes it as engaging and addictive as ludic videogames.

Motor rehabilitation and neurorehabilitation are both intended to help the brain adjust to a new way to function, to re-learn how to control an impaired limb, and even to generate new paths to communicate with the limb, a process best known as neuroplasticity [89]. Neuroplasticity is based on principles such as goal-oriented practice, multisensorial stimulation, explicit feedback, implicit knowledge of performance, and action observation, among others [89]. These qualities are implicit to the use of VR/AR interfaces [5], and when they are aggregated to the improvement of muscle control, coordination, and control of movements or contractions [43], therapy can take an upturn in the best interest of the patient. An outstanding aspect is that patient betterment will have quantitative recordings that could ultimately yield specific changes in their therapy to target the aspects that need the most attention.

Some authors [16] propose sEMG signals as a popular form of biofeedback, nevertheless, there have been developments where they combine two biosignals [51,90,91]. Electroencephalography (EEG) and sEMG are non-invasive biopotentials that offer plenty of information regarding brain and muscle activity in clinical and daily life contexts. Interestingly, sEMG signals are commonly considered an undesired noise source in EEG recordings. Cortico-muscular (EEG–sEMG) coherence is a new analysis tool that studies the functional connection between the brain (EEG) and muscle (sEMG) electrical activity. EEG–sEMG coherence has been used for assessment of neuronal recovery [91] in rehabilitation applications, including those based on virtual reality [92]. Moreover, it has been shown that EEG–sEMG coherence, measured from a single EEG and a single sEMG channel, can be used as a control signal for distinction of hand movements [93] with potential for rehabilitation applications. Furthermore, simultaneous recording and analysis of multiple sEMG and EEG signals in key body and scalp zones can help to evaluate potential effects and interrelations between types and parameters of AR/VR during rehabilitation interventions, on the activity of central and peripheral central nervous system structures related to movement control, planning, and execution. Moreover, since VR/AR technologies are oriented to visual simulation during body movements, EEG–sEMG coherence or other combined parameters could be an alternative to evaluate relations between visual attention to objects in the VR/AR interface and visual information processing in the brain and motor responses in the body.

Considering the above evidence, it is not rare that only a few developments have reached a commercial environment and therefore been applied in the clinic [94]. All the analyzed articles are still in a research and development stage and do not mention their use for therapy; on the contrary, they propose larger studies as future work, meaning that even though this is a promising technology, more and larger studies are necessary to prove its efficiency.

To have access to a wide variety of human movements and dynamic interaction, through VR/AR therapy, is an additional benefit that has the potential to generate new solutions in rehabilitation. This information could be useful, especially if there were specific guidelines or protocols to standardize the acquisition of other sensors or signals, as there are for sEMG signals [95], as well as other data used for control and feedback. This would allow to propose the design of a database to house standardized reports for documentation and filing regarding signal recordings, acquisition hardware, environmental or user conditions, and experimentation.

Finally, there is a major gap regarding the standard of care protocols or guidelines to perform VR/AR therapies for rehabilitation. To further evaluate the advantages of these developments, a structured methodology should be proposed and followed. It could include session time and frequency, maximum number of activations per limb, proper guides to use external aids such as exoskeletons, FES, orthosis, or prosthetics, type of movements to be commanded according to the pathology or target therapeutic application, along with a larger list of requirements.

Future Directions

Future directions in sEMG-based control of VR/AR interfaces for rehabilitation applications include several features. Hardware implementations of acquisition systems that come closer to a wearable device where most of the system could be integrated will be significant. Hybrid multisignal inputs accompanied by signal processing algorithms that incorporate the contributions of several systems of the body being analyzed simultaneously could be the first proposal of a novel approach for a complex and robust control that adapts to the patient dexterity level and moves up and down with them through difficulty levels.

It will be very important to go beyond the current widespread interfaces with visual feedback for these systems–one option could be to incorporate tactile and haptic feedback, based on information from gyroscope and accelerometer sensors. These hardware systems combined with VR/AR interfaces will promote a richer environment to develop rehabilitation therapies, where several metrics related to the patient's movement could be monitored and used as feedback to promote motor rehabilitation and neuroplasticity. Some applications have shown that using exoskeletons or FES can be beneficial to help the patient in training muscles and neural pathways to practice the correct movement trajectories during therapy. Moreover, there is a need for more cohesive technologies (hardware and software) that allows the user, patient, and care provider to perform this type of therapy in a real-life environment. For this technology to become a regular therapy it must be integrated and ready to use, without the complications of too many wires or lengthy donning and doffing procedures.

Personalized therapy is also within reach by means of VR/AR technologies, since these interfaces can adapt the complexity level to patient performance and be updated as the patient improves their control over the impaired limb. In this paper, we have examined examples where researchers use biofeedback to adjust the complexity of the task, e.g., fatigue, correct position (proprioception), or performance of repetitions, e.g., the TAC test proposed by Simon et al. [96]. Personalization includes videogame difficulty levels for the VR/AR interfaces which can be controlled as in a regular ludic videogame, except in therapy the user can downgrade levels. This characteristic could be very useful in case of muscular fatigue, which is very common during therapies. This little detail might allow patients to complete more repetitions or to endure larger therapy sessions; also, changes in sEMG signals during therapy can be considered too, i.e., retraining the control algorithm mid-session to lower the patient's muscular strength demand.

5. Conclusions

This SLR provides a global framework of the most common application of sEMG signals for control/feedback of VR/AR interfaces. Nowadays, the use of these signals for rehabilitation is still scattered and heterogenous. There is no consensus regarding the selection methodology of sEMG signal processing algorithms, signal features, the classification approach, the performance evaluation, and even less about its use in applications for rehabilitation. There are no reports of these interfaces being adopted in clinical practice. Future work should be targeted to propose a set of guidelines to standardize these technologies for clinical therapies.

Author Contributions: Conceptualization, C.L.T.-P. and J.G.-M.; methodology, C.L.T.-P., G.V.-M., J.A.M.-G. and J.G.-M.; software, C.L.T.-P.; validation, C.L.T.-P., G.V.-M., J.A.M.-G. and J.G.-M.; formal analysis, C.L.T.-P., G.V.-M. and J.A.M.-G.; investigation, C.L.T.-P., G.V.-M. and J.A.M.-G.; resources, C.L.T.-P., G.V.-M., J.A.M.-G., G.R.-R., A.V.-H., L.L.-S. and J.G.-M.; data curation, C.L.T.-P., G.V.-M., J.A.M.-G., G.R.-R., A.V.-H., L.L.-S. and J.G.-M.; writing—original draft preparation, C.L.T.-P., G.V.-

M., J.A.M.-G., G.R.-R., A.V.-H., L.L.-S. and J.G.-M.; writing—review and editing, C.L.T.-P., G.V.-M., J.A.M.-G., G.R.-R., A.V.-H., L.L.-S. and J.G.-M.; visualization, C.L.T.-P., G.V.-M., J.A.M.-G., G.R.-R., A.V.-H., L.L.-S. and J.G.-M.; supervision, G.R.-R., A.V.-H., L.L.-S. and J.G.-M.; project administration, L.L.-S. and J.G.-M.; funding acquisition, L.L.-S. and J.G.-M. All authors have read and agreed to the published version of the manuscript.

Funding: This research was funded by CYTED-DITECROD-218RT0545, Proyecto IV-8 call AMEXcid-AUCI 2018-2022.

Conflicts of Interest: The authors declare no conflict of interest.

References

1. Stucki, G. Advancing the Rehabilitation Sciences. *Front. Rehabil. Sci.* **2021**, *1*, 617749. [CrossRef]
2. Reddy, N.P.; Unnikrishnan, R. EMG Interfaces for VR and Telematic Control Applications. *IFAC Proc. Vol.* **2001**, *34*, 443–446. [CrossRef]
3. Mejia, J.A.; Hernandez, G.; Toledo, C.; Mercado, J.; Vera, A.; Leija, L.; Gutierrez, J. Upper Limb Rehabilitation Therapies Based in Videogames Technology Review. In Proceedings of the 2019 Global Medical Engineering Physics Exchanges/Pan American Health Care Exchanges (GMEPE/PAHCE) 2019, Buenos Aires, Argentina, 26–31 March 2019; pp. 1–5.
4. González-González, C.S.; Toledo-Delgado, P.A.; Muñoz-Cruz, V.; Torres-Carrion, P.V. Serious games for rehabilitation: Gestural interaction in personalized gamified exercises through a recommender system. *J. Biomed. Inform.* **2019**, *97*, 103266. [CrossRef] [PubMed]
5. Cerritelli, F.; Chiera, M.; Abbro, M.; Megale, V.; Esteves, J.; Gallace, A.; Manzotti, A. The Challenges and Perspectives of the Integration Between Virtual and Augmented Reality and Manual Therapies. *Front. Neurol.* **2021**, *12*, 700211. [CrossRef] [PubMed]
6. Kluger, D.T.; Joyner, J.S.; Wendelken, S.M.; Davis, T.S.; George, J.A.; Page, D.M.; Hutchinson, D.T.; Benz, H.L.; Clark, G.A. Virtual Reality Provides an Effective Platform for Functional Evaluations of Closed-Loop Neuromyoelectric Control. *IEEE Trans. Neural Syst. Rehabil. Eng.* **2019**, *27*, 876–886. [CrossRef]
7. Huang, J.; Lin, M.; Fu, J.; Sun, Y.; Fang, Q. An Immersive Motor Imagery Training System for Post-Stroke Rehabilitation Combining VR and EMG-based Real-Time Feedback. In Proceedings of the 2021 43rd Annual International Conference of the IEEE Engineering in Medicine & Biology Society (EMBC), Guadalajara, Mexico, 1–5 November 2021; pp. 7590–7593.
8. Muri, F.; Carbajal, C.; Echenique, A.M.; Fernández, H.; López, N.M. Virtual reality upper limb model controlled by EMG signals. *J. Phys. Conf. Ser.* **2013**, *477*, 012041. [CrossRef]
9. Meng, Q.; Zhang, J.; Yang, X. Virtual Rehabilitation Training System Based on Surface EMG Feature Extraction and Analysis. *J. Med. Syst.* **2019**, *43*, 48. [CrossRef] [PubMed]
10. Montoya, M.F.; Munoz, J.E.; Henao, O.A. Enhancing Virtual Rehabilitation in Upper Limbs With Biocybernetic Adaptation: The Effects of Virtual Reality on Perceived Muscle Fatigue, Game Performance and User Experience. *IEEE Trans. Neural Syst. Rehabil. Eng.* **2020**, *28*, 740–747. [CrossRef]
11. Thériault, L.; Robert, J.-M.; Baron, L. Virtual Reality Interfaces for Virtual Environments. Virtual Reality International Conference. Available online: https://www.researchgate.net/publication/259576863 (accessed on 20 April 2022).
12. Liang, Y.; Wu, D.; Ledesma, D.; Davis, C.; Slaughter, R.; Guo, Z. Virtual Tai-Chi System: A smart-connected modality for rehabilitation. *Smart Health* **2018**, *9–10*, 232–249. [CrossRef]
13. Chen, P.-J.; Penn, I.-W.; Wei, S.-H.; Chuang, L.-R.; Sung, W.-H. Augmented reality-assisted training with selected Tai-Chi movements improves balance control and increases lower limb muscle strength in older adults: A prospective randomized trial. *J. Exerc. Sci. Fit.* **2020**, *18*, 142–147. [CrossRef]
14. Muñoz, J.E.; Montoya, M.F.; Boger, J. *From Exergames to Immersive Virtual Reality Systems: Serious Games for Supporting Older Adults*, 1st ed.; Elsevier Inc.: Amsterdam, The Netherlands, 2021; pp. 141–204.
15. Barrett, A.M.; Oh-Park, M.; Chen, P.; Ifejika, N.L. Neurorehabilitation: Five new things. *Neurol. Clin. Pract.* **2013**, *3*, 484–492. [CrossRef]
16. Giggins, O.M.; Persson, U.M.; Caulfield, B. Biofeedback in rehabilitation. *J. Neuroeng. Rehabil.* **2013**, *10*, 60. [CrossRef] [PubMed]
17. Dosen, S.; Markovic, M.; Somer, K.; Graimann, B.; Farina, D. EMG Biofeedback for online predictive control of grasping force in a myoelectric prosthesis. *J. Neuroeng. Rehabil.* **2015**, *12*, 55. [CrossRef] [PubMed]
18. Ogourtsova, T.; Archambault, P.S.; Lamontagne, A. Let's do groceries-a novel virtual assessment for post-stroke unilateral spatial neglect Effects of virtual scene complexity and knowledge translation initiatives. In Proceedings of the 2017 International Conference on Virtual Rehabilitation (Icvr), Montreal, QC, Canada, 19–22 June 2017.
19. Tao, G.; Archambault, P.S.; Levin, M.F. Evaluation of Kinect skeletal tracking in a virtual reality rehabilitation system for upper limb hemiparesis. In Proceedings of the 2013 International Conference on Virtual Rehabilitation, ICVR 2013, Philadelphia, PA, USA, 26–29 August 2013; pp. 164–165.
20. Wada, T.; Takeuchi, T. A Training System for EMG Prosthetic Hand in Virtual Environment. In Proceedings of the Human Factors and Ergonomics Society Annual Meeting, New York, NY, USA, 22–26 September 2008; Volume 52, pp. 2112–2116.
21. Sime, D.W. Potential Application of Virtual Reality for Interface Customisation (and Pre-training) of Amputee Patients as Preparation for Prosthetic Use. *Adv. Exp. Med. Biol.* **2019**, *1120*, 15–24. [CrossRef] [PubMed]

22. Clemente, F.; D'Alonzo, M.; Controzzi, M.; Edin, B.B.; Cipriani, C. Non-Invasive, Temporally Discrete Feedback of Object Contact and Release Improves Grasp Control of Closed-Loop Myoelectric Transradial Prostheses. *IEEE Trans. Neural Syst. Rehabil. Eng.* **2016**, *24*, 1314–1322. [CrossRef] [PubMed]
23. Markovic, M.; Schweisfurth, M.A.; Engels, L.F.; Bentz, T.; Wüstefeld, D.; Farina, D.; Dosen, S. The clinical relevance of advanced artificial feedback in the control of a multi-functional myoelectric prosthesis. *J. Neuroeng. Rehabil.* **2018**, *15*, 28. [CrossRef]
24. Thomas, G.P.; Jobst, B.C. *Feedback-Sensitive and Closed-Loop Solutions*; Elsevier Inc.: Amsterdam, The Netherlands, 2017.
25. Svensson, P.; Wijk, U.; Björkman, A.; Antfolk, C. A review of invasive and non-invasive sensory feedback in upper limb prostheses. *Expert Rev. Med. Devices* **2017**, *14*, 439–447. [CrossRef]
26. Earnshaw, R.; Liggett, S.; Excell, P.; Thalmann, D. *Technology, Design and the Arts-Opportunities and Challenges*; Springer International Publishing: Cham, Switzerland, 2020.
27. Casellato, C.; Ambrosini, E.; Galbiati, A.; Biffi, E.; Cesareo, A.; Beretta, E.; Lunardini, F.; Zorzi, G.; Sanger, T.D.; Pedrocchi, A. EMG-based vibro-tactile biofeedback training: Effective learning accelerator for children and adolescents with dystonia? A pilot crossover trial. *J. Neuroeng. Rehabil.* **2019**, *16*, 150. [CrossRef] [PubMed]
28. Gutiérrez, Á.; Sepúlveda-Muñoz, D.; Gil-Agudo, Á.; de los Reyes Guzman, A. Serious Game Platform with Haptic Feedback and EMG Monitoring for Upper Limb Rehabilitation and Smoothness Quantification on Spinal Cord Injury Patients. *Appl. Sci.* **2020**, *10*, 963. [CrossRef]
29. Li, K.; Boyd, P.; Zhou, Y.; Ju, Z.; Liu, H. Electrotactile Feedback in a Virtual Hand Rehabilitation Platform: Evaluation and Implementation. *IEEE Trans. Autom. Sci. Eng.* **2019**, *16*, 1556–1565. [CrossRef]
30. Markovic, M.; Varel, M.; Schweisfurth, M.A.; Schilling, A.F.; Dosen, S. Closed-Loop Multi-Amplitude Control for Robust and Dexterous Performance of Myoelectric Prosthesis. *IEEE Trans. Neural Syst. Rehabil. Eng.* **2020**, *28*, 498–507. [CrossRef] [PubMed]
31. Parker, P.; Englehart, K.; Hudgins, B. Myoelectric signal processing for control of powered limb prostheses. *J. Electromyogr. Kinesiol.* **2006**, *16*, 541–548. [CrossRef] [PubMed]
32. Youn, W.; Kim, J. Development of a compact-size and wireless surface EMG measurement system. In Proceedings of the ICCAS-SICE 2009-ICROS-SICE International Joint Conference 2009, Fukuoka, Japan, 18–21 August 2009; pp. 1625–1628.
33. Lowery, M.; Weir, R.; Kuiken, T. Simulation of Intramuscular EMG Signals Detected Using Implantable Myoelectric Sensors (IMES). *IEEE Trans. Biomed. Eng.* **2006**, *53*, 1926–1933. [CrossRef] [PubMed]
34. Reategui, J.; Callupe, R. Surface EMG multichannel array using active dry sensors for forearm signal extraction. In Proceedings of the 2017 IEEE 24th International Congress on Electronics, Electrical Engineering and Computing, INTERCON 2017, Cusco, Peru, 15–18 August 2017; pp. 1–4.
35. Drost, G.; Stegeman, D.F.; van Engelen, B.G.; Zwarts, M.J. Clinical applications of high-density surface EMG: A systematic review. *J. Electromyogr. Kinesiol.* **2006**, *16*, 586–602. [CrossRef]
36. Xie, L.; Yang, G.; Xu, L.; Seoane, F.; Chen, Q.; Zheng, L. Characterization of dry biopotential electrodes. In Proceedings of the 2013 35th Annual International Conference of the IEEE Engineering in Medicine and Biology Society (EMBC), Osaka, Japan, 3–7 July 2013; pp. 1478–1481.
37. Roche, A.D.; Rehbaum, H.; Farina, D.; Aszmann, O.C. Prosthetic Myoelectric Control Strategies: A Clinical Perspective. *Curr. Surg. Rep.* **2014**, *2*, 44. [CrossRef]
38. Cordella, F.; Ciancio, A.L.; Sacchetti, R.; Davalli, A.; Cutti, A.G.; Guglielmelli, E.; Zollo, L. Literature Review on Needs of Upper Limb Prosthesis Users. *Front. Neurosci.* **2016**, *10*, 209. [CrossRef]
39. Pallavicini, F.; Ferrari, A.; Mantovani, F. Video Games for Well-Being: A Systematic Review on the Application of Computer Games for Cognitive and Emotional Training in the Adult Population. *Front. Psychol.* **2018**, *9*, 2127. [CrossRef]
40. Reilly, C.A.; Greeley, A.B.; Jevsevar, D.S.; Gitajn, I.L. Virtual reality-based physical therapy for patients with lower extremity injuries: Feasibility and acceptability. *OTA Int. Open Access J. Orthop. Trauma* **2021**, *4*, e132. [CrossRef]
41. Gil, M.J.V.; Gonzalez-Medina, G.; Lucena-Anton, D.; Perez-Cabezas, V.; Ruiz-Molinero, M.D.C.; Martín-Valero, R. Augmented Reality in Physical Therapy: Systematic Review and Meta-analysis. *JMIR Serious Games* **2021**, *9*, e30985. [CrossRef]
42. Dash, A.; Lahiri, U. Design of Virtual Reality-Enabled Surface Electromyogram-Triggered Grip Exercise Platform. *IEEE Trans. Neural Syst. Rehabil. Eng.* **2020**, *28*, 444–452. [CrossRef]
43. Hashim, N.A.; Razak, N.A.A.; Gholizadeh, H.; Osman, N.A.A. Video Game–Based Rehabilitation Approach for Individuals Who Have Undergone Upper Limb Amputation: Case-Control Study. *JMIR Serious Games* **2021**, *9*, e17017. [CrossRef] [PubMed]
44. Seo, N.J.; Barry, A.; Ghassemi, M.; Triandafilou, K.M.; Stoykov, M.E.; Vidakovic, L.; Roth, E.; Kamper, D.G. Use of an EMG-Controlled Game as a Therapeutic Tool to Retrain Hand Muscle Activation Patterns Following Stroke: A Pilot Study. *J. Neurol. Phys. Ther.* **2022**, *46*, 198–205. [CrossRef] [PubMed]
45. Pereira, M.F.; Prahm, C.; Kolbenschlag, J.; Oliveira, E.; Rodrigues, N.F. Application of AR and VR in hand rehabilitation: A systematic review. *J. Biomed. Inform.* **2020**, *111*, 103584. [CrossRef] [PubMed]
46. Merians, A.S.; Jack, D.; Boian, R.; Tremaine, M.; Burdea, G.C.; Adamovich, S.V.; Recce, M.; Poizner, H. Virtual Reality–Augmented Rehabilitation for Patients Following Stroke. *Phys. Ther.* **2002**, *82*, 898–915. [CrossRef] [PubMed]
47. Dash, A.; Yadav, A.; Lahiri, U. Physiology-sensitive Virtual Reality based Strength Training Platform for Post-stroke Grip Task. In Proceedings of the 2019 IEEE EMBS International Conference on Biomedical & Health Informatics (BHI), Chicago, IL, USA, 19–22 May 2019; pp. 1–4.

48. Peng, L.; Hou, Z.G.; Peng, L.; Luo, L.; Wang, W. Robot assisted upper limb rehabilitation training and clinical evaluation: Results of a pilot study. In Proceedings of the 2017 IEEE International Conference on Robotics and Biomimetics, ROBIO 2017, Macau, Macao, 5–8 December 2017; pp. 1–6.

49. Wei, X.; Chen, Y.; Jia, X.; Chen, Y.; Xie, L. Muscle Activation Visualization System Using Adaptive Assessment and Forces-EMG Mapping. *IEEE Access* **2021**, *9*, 46374–46385. [CrossRef]

50. Page, M.J.; McKenzie, J.E.; Bossuyt, P.M.; Boutron, I.; Hoffmann, T.C.; Mulrow, C.D.; Shamseer, L.; Tetzlaff, J.M.; Akl, E.A.; Brennan, S.E.; et al. The PRISMA 2020 statement: An updated guideline for reporting systematic reviews. *BMJ* **2021**, *372*, n71. [CrossRef]

51. Wang, L.; Du, S.; Liu, H.; Yu, J.; Cheng, S.; Xie, P. A virtual rehabilitation system based on EEG-EMG feedback control. In Proceedings of the 2017 Chinese Automation Congress (CAC), Jinan, China, 20–22 October 2017; pp. 4337–4340.

52. Vega, M.F.M.; Henao, O.A. Cross-validation of a classification method applied in a database of sEMG contractions collected in a body interaction videogame. *J. Phys. Conf. Ser.* **2019**, *1247*, 012049. [CrossRef]

53. Llorens, R.; Fuentes, M.A.; Borrego, A.; Latorre, J.; Alcañiz, M.; Colomer, C.; Noé, E. Effectiveness of a combined transcranial direct current stimulation and virtual reality-based intervention on upper limb function in chronic individuals post-stroke with persistent severe hemiparesis: A randomized controlled trial. *J. Neuroeng. Rehabil.* **2021**, *18*, 108. [CrossRef]

54. Li, Y.; Chen, J.; Yang, Y. A Method for Suppressing Electrical Stimulation Artifacts from Electromyography. *Int. J. Neural Syst.* **2019**, *29*, 1850054. [CrossRef]

55. Li, X.; Zhou, Z.; Liu, W.; Ji, M. Wireless sEMG-based identification in a virtual reality environment. *Microelectron. Reliab.* **2019**, *98*, 78–85. [CrossRef]

56. Bhagat, N.A.; Yozbatiran, N.; Sullivan, J.L.; Paranjape, R.; Losey, C.; Hernandez, Z.; Keser, Z.; Grossman, R.; Francisco, G.E.; O'Malley, M.K.; et al. Neural activity modulations and motor recovery following brain-exoskeleton interface mediated stroke rehabilitation. *Neuroimage Clin.* **2020**, *28*, 102502. [CrossRef] [PubMed]

57. Trifonov, A.A.; Kuzmin, A.A.; Filist, S.A.; Degtyarev, S.v.; Petrunina, E.v. Biotechnical System for Control to the Exoskeleton Limb Based on Surface Myosignals for Rehabilitation Complexes. In Proceedings of the 2020 IEEE 14th International Conference on Application of Information and Communication Technologies (AICT), Tashkent, Uzbekistan, 7–9 October 2020.

58. Ruiz-Olaya, A.F.; Lopez-Delis, A.; da Rocha, A.F. Upper and lower extremity exoskeletons. In *Handbook of Biomechatronics*; Elsevier: Amsterdam, The Netherlands; pp. 283–317.

59. Liew, S.-L.; Lin, D.J.; Cramer, S.C. *Interventions to Improve Recovery After Stroke, 7th ed*; Elsevier Inc.: Amsterdam, The Netherlands, 2022.

60. Melero, M.; Hou, A.; Cheng, E.; Tayade, A.; Lee, S.C.; Unberath, M.; Navab, N. Upbeat: Augmented Reality-Guided Dancing for Prosthetic Rehabilitation of Upper Limb Amputees. *J. Healthc. Eng.* **2019**, *2019*, 2163705. [CrossRef]

61. Palermo, F.; Cognolato, M.; Eggel, I.; Atzori, M.; Müller, H. An augmented reality environment to provide visual feedback to amputees during sEMG data acquisitions. In *Lecture Notes in Computer Science (Including Subseries Lecture Notes in Artificial Intelligence and Lecture Notes in Bioinformatics)*; Springer: Cham, Switzerland, 2019; Volume 11650, pp. 3–14.

62. Gazzoni, M.; Cerone, G.L. Augmented Reality Biofeedback for Muscle Activation Monitoring: Proof of Concept. In *IFMBE Proceedings*; Springer: Cham, Switzerland, 2021; Volume 80, pp. 143–150.

63. Liu, L.; Cui, J.; Niu, J.; Duan, N.; Yu, X.; Li, Q.; Yeh, S.-C.; Zheng, L.-R. Design of Mirror Therapy System Base on Multi-Channel Surface-Electromyography Signal Pattern Recognition and Mobile Augmented Reality. *Electronics* **2020**, *9*, 2142. [CrossRef]

64. Nasri, N.; Orts-Escolano, S.; Cazorla, M. An sEMG-Controlled 3D Game for Rehabilitation Therapies: Real-Time Time Hand Gesture Recognition Using Deep Learning Techniques. *Sensors* **2020**, *20*, 6451. [CrossRef] [PubMed]

65. Quinayás, C.; Barrera, F.; Ruiz, A.; Delis, A. Virtual Hand Training Platform Controlled Through Online Recognition of Motion Intention. In *Lecture Notes in Computer Science (including subseries Lecture Notes in Artificial Intelligence and Lecture Notes in Bioinformatics)*; Springer: Cham, Switzerland, 2019; Volume 11896, pp. 761–768.

66. Ma, L.; Zhao, X.; Li, Z.; Zhao, M.; Xu, Z. A sEMG-based Hand Function Rehabilitation System for Stroke Patients. In Proceedings of the 2018 3rd International Conference on Advanced Robotics and Mechatronics (ICARM), Singapore, 18–20 July 2018; pp. 497–502.

67. Woodward, R.B.; Hargrove, L.J. Adapting myoelectric control in real-time using a virtual environment. *J. Neuroeng. Rehabil.* **2019**, *16*, 11. [CrossRef]

68. Nissler, C.; Nowak, M.; Connan, M.; Büttner, S.; Vogel, J.; Kossyk, I.; Márton, Z.-C.; Castellini, C. VITA—an everyday virtual reality setup for prosthetics and upper-limb rehabilitation. *J. Neural Eng.* **2019**, *16*, 026039. [CrossRef]

69. Lydakis, A.; Meng, Y.; Munroe, C.; Wu, Y.N.; Begum, M. A learning-based agent for home neurorehabilitation. In Proceedings of the IEEE International Conference on Rehabilitation Robotics, London, UK, 17–20 July 2017; pp. 1233–1238.

70. Lukyanenko, P.; Dewald, H.A.; Lambrecht, J.; Kirsch, R.F.; Tyler, D.J.; Williams, M.R. Stable, simultaneous and proportional 4-DoF prosthetic hand control via synergy-inspired linear interpolation: A case series. *J. Neuroeng. Rehabil.* **2021**, *18*, 50. [CrossRef]

71. Raz, G.; Gurevitch, G.; Vaknin, T.; Aazamy, A.; Gefen, I.; Grunstein, S.; Azouri, G.; Goldway, N. Electroencephalographic evidence for the involvement of mirror-neuron and error-monitoring related processes in virtual body ownership. *NeuroImage* **2020**, *207*, 116351. [CrossRef]

72. Kisiel-Sajewicz, K.; Marusiak, J.; Rojas-Martínez, M.; Janecki, D.; Chomiak, S.; Mencel, J.; Mañanas, M.; Jaskólski, A.; Jaskólska, A. High-density surface electromyography maps after computer-aided training in individual with congenital transverse deficiency: A case study. *BMC Musculoskelet. Disord.* **2020**, *21*, 682. [CrossRef]

73. Mazzola, S.; Prado, A.; Agrawal, S.K. An upper limb mirror therapy environment with hand tracking in virtual reality. In Proceedings of the 2020 8th IEEE RAS/EMBS International Conference for Biomedical Robotics and Biomechatronics (BioRob), New York, NY, USA, 29 November–1 December 2020; pp. 752–758.

74. Summa, S.; Gori, R.; Castelli, E.; Petrarca, M. Development of a dynamic oriented rehabilitative integrated system. In Proceedings of the Annual International Conference of the IEEE Engineering in Medicine and Biology Society, EMBS, Berlin, Germany, 23–27 July 2019; pp. 5245–5250.

75. Cardoso, V.F.; Pomer-Escher, A.; Longo, B.B.; Loterio, F.A.; Nascimento, S.S.G.; Laiseca, M.A.R.; Delisle-Rodriguez, D.; Frizera-Neto, A.; Bastos-Filho, T. Neurorehabilitation platform based on EEG, sEMG and virtual reality using robotic monocycle. In *IFMBE Proceedings*; Springer: Singapore; Volume 70, pp. 315–321.

76. Braza, D.W.; Martin, J.N.Y. Upper Limb Amputations. In *Essentials of Physical Medicine and Rehabilitation: Musculoskeletal Disorders, Pain, and Rehabilitation*; Elsevier: Amsterdam, The Netherlands; pp. 651–657.

77. Alshehri, F.M.; Ahmed, S.A.; Ullah, S.; Ghazal, H.; Nawaz, S.; Alzahrani, A.S. The Patterns of Acquired Upper and Lower Extremity Amputation at a Tertiary Centre in Saudi Arabia. *Cureus* **2022**, *14*, 4. [CrossRef]

78. Bank, P.J.; Dobbe, L.R.; Meskers, C.G.; De Groot, J.H.; De Vlugt, E. Manipulation of visual information affects control strategy during a visuomotor tracking task. *Behav. Brain Res.* **2017**, *329*, 205–214. [CrossRef]

79. Lai, J.; Zhao, Y.; Liao, Y.; Hou, W.; Chen, Y.; Zhang, Y.; Li, G.; Wu, X. Design of a multi-degree-of-freedom virtual hand bench for myoelectrical prosthesis. In Proceedings of the 2017 2nd International Conference on Advanced Robotics and Mechatronics (ICARM), Hefei/Tai'an, China, 27–31 August 2017; pp. 345–350.

80. Pizzolato, S.; Tagliapietra, L.; Cognolato, M.; Reggiani, M.; Müller, H.; Atzori, M. Comparison of six electromyography acquisition setups on hand movement classification tasks. *PLoS ONE* **2017**, *12*, e0186132. [CrossRef]

81. Covaciu, F.; Pisla, A.; Iordan, A.-E. Development of a Virtual Reality Simulator for an Intelligent Robotic System Used in Ankle Rehabilitation. *Sensors* **2021**, *21*, 1537. [CrossRef] [PubMed]

82. Castellini, C. Design Principles of a Light, Wearable Upper Limb Interface for Prosthetics and Teleoperation. In *Wearable Robotics*; Elsevier: Amsterdam, The Netherlands, 2020; pp. 377–391.

83. Heerschop, A.; van der Sluis, C.K.; Otten, E.; Bongers, R.M. Performance among different types of myocontrolled tasks is not related. *Hum. Mov. Sci.* **2020**, *70*, 102592. [CrossRef] [PubMed]

84. Ida, H.; Mohapatra, S.; Aruin, A.S. Perceptual distortion in virtual reality and its impact on dynamic postural control. *Gait Posture* **2021**, *92*, 123–128. [CrossRef] [PubMed]

85. Yassin, M.M.; Saber, A.M.; Saad, M.N.; Said, A.M.; Khalifa, A.M. Developing a Low-cost, smart, handheld electromyography biofeedback system for telerehabilitation with Clinical Evaluation. *Med. Nov. Technol. Devices* **2021**, *10*, 100056. [CrossRef]

86. Galido, E.; Esplanada, M.C.; Estacion, C.J.; Migriño, J.P.; Rapisora, J.K.; Salita, J.; Amado, T.; Jorda, R.; Tolentino, L.K. EMG Speed-Controlled Rehabilitation Treadmill With Physiological Data Acquisition System Using BITalino Kit. In Proceedings of the 2018 IEEE 10th International Conference on Humanoid, Nanotechnology, Information Technology, Communication and Control, Environment and Management (HNICEM), Baguio City, Philippines, 29 November–2 December 2018; pp. 1–5.

87. Osumi, M.; Inomata, K.; Inoue, Y.; Otake, Y.; Morioka, S.; Sumitani, M. Characteristics of Phantom Limb Pain Alleviated with Virtual Reality Rehabilitation. In *Pain Medicine*; Oxford University Press: Oxford, UK, 2019; Volume 20, pp. 1038–1046.

88. Sousa, M.; Vieira, J.; Medeiros, D.; Arsénio, A.; Jorge, J. SleeveAR: Augmented reality for rehabilitation using realtime feedback. In Proceedings of the International Conference on Intelligent User Interfaces, Proceedings IUI, Sonoma, CA, USA, 7–10 March 2016; pp. 175–185.

89. Maier, M.; Ballester, B.R.; Verschure, P.F.M.J. Principles of Neurorehabilitation After Stroke Based on Motor Learning and Brain Plasticity Mechanisms. *Front. Syst. Neurosci.* **2019**, *13*, 74. [CrossRef]

90. Quitadamo, L.R.; Cavrini, F.; Sbernini, L.; Riillo, F.; Bianchi, L.; Seri, S.; Saggio, G. Support vector machines to detect physiological patterns for EEG and EMG-based human–computer interaction: A review. *J. Neural Eng.* **2017**, *14*, 011001. [CrossRef]

91. Brambilla, C.; Pirovano, I.; Mira, R.M.; Rizzo, G.; Scano, A.; Mastropietro, A. Combined Use of EMG and EEG Techniques for Neuromotor Assessment in Rehabilitative Applications: A Systematic Review. *Sensors* **2021**, *21*, 7014. [CrossRef]

92. Marin-Pardo, O.; Laine, C.M.; Rennie, M.; Ito, K.L.; Finley, J.; Liew, S.-L. A Virtual Reality Muscle–Computer Interface for Neurorehabilitation in Chronic Stroke: A Pilot Study. *Sensors* **2020**, *20*, 3754. [CrossRef]

93. Lou, X.; Xiao, S.; Qi, Y.; Hu, X.; Wang, Y.; Zheng, X. Corticomuscular Coherence Analysis on Hand Movement Distinction for Active Rehabilitation. *Comput. Math. Methods Med.* **2013**, *2013*, 908591. [CrossRef] [PubMed]

94. Kim, W.-S.; Cho, S.; Ku, J.; Kim, Y.; Lee, K.; Hwang, H.-J.; Paik, N.-J. Clinical Application of Virtual Reality for Upper Limb Motor Rehabilitation in Stroke: Review of Technologies and Clinical Evidence. *J. Clin. Med.* **2020**, *9*, 3369. [CrossRef] [PubMed]

95. Hermens, H.J.; Freriks, B. Guidelines for reporting SEMG data. In *The state of the Art on Sensors and Sensor Placement Procedures for Surface Electromyography: A Proposal for Sensor Placement Procedures*; Roessingh Research and Development: Enschede, The Netherlands, 1997.

96. Simon, A.M.; Hargrove, L.J.; Lock, B.A.; Kuiken, T.A.; Simon, A. The Target Achievement Control Test: Evaluating real-time myoelectric pattern recognition control of a multifunctional upper-limb prosthesis. *J. Rehabil. Res. Dev.* **2011**, *48*, 619. [CrossRef] [PubMed]

 *electronics*

Article

A 3D Image Registration Method for Laparoscopic Liver Surgery Navigation

Donghui Li and Monan Wang *

Mechanical & Power Engineering College, Harbin University of Science and Technology, Harbin 150080, China; 1810100014@stu.hrbust.edu.cn
* Correspondence: mnwang@hrbust.edu.cn

Abstract: At present, laparoscopic augmented reality (AR) navigation has been applied to minimally invasive abdominal surgery, which can help doctors to see the location of blood vessels and tumors in organs, so as to perform precise surgery operations. Image registration is the process of optimally mapping one or more images to the target image, and it is also the core of laparoscopic AR navigation. The key is how to shorten the registration time and optimize the registration accuracy. We have studied the three-dimensional (3D) image registration technology in laparoscopic liver surgery navigation and proposed a new registration method combining rough registration and fine registration. First, the adaptive fireworks algorithm (AFWA) is applied to rough registration, and then the optimized iterative closest point (ICP) algorithm is applied to fine registration. We proposed a method that is validated by the computed tomography (CT) dataset 3D-IRCADb-01. Experimental results show that our method is superior to other registration methods based on stochastic optimization algorithms in terms of registration time and accuracy.

Keywords: laparoscopic AR navigation; liver surgery; 3D image registration method; point cloud registration

Citation: Li, D.; Wang, M. A 3D Image Registration Method for Laparoscopic Liver Surgery Navigation. *Electronics* **2022**, *11*, 1670. https://doi.org/10.3390/electronics11111670

Academic Editors: Jorge C. S. Cardoso, André Perrotta, Paula Alexandra Silva and Pedro Martins

Received: 11 April 2022
Accepted: 20 May 2022
Published: 24 May 2022

Publisher's Note: MDPI stays neutral with regard to jurisdictional claims in published maps and institutional affiliations.

1. Introduction

During the operation, the information that doctors can obtain through laparoscopy is very limited. They can only obtain the image information of a part of the surface area, and cannot obtain the information inside the organs, which rely heavily on preoperative imaging [1]. In this case, doctors can only rely on their own experience to judge the location of the internal lesions, which has high requirements for doctors and may cause the wrong location of the lesions [2,3]. In 1986, Roberts et al. [4] and Kelly et al. [5] performed AR-assisted surgery in neurosurgery. Since then, with the development of AR applications in auxiliary surgery, AR surgery navigation can accurately match the preoperative anatomical structure information with the intraoperative information, and then present it to the doctor, which has been applied in neurosurgery and orthopedic surgery [6]. The image guidance function of laparoscopic AR navigation has also made much progress in hepatectomy and nephrectomy [7–9]. The realization methods of laparoscopic AR navigation mainly include medical image processing, graphic image rendering, image registration, and display technology [10]. The main challenge is the speed and accuracy of 3D image registration [11]. Laparoscopic images and preoperative CT images were obtained from different imaging devices. Due to the different imaging modes, they belong to multi-modal registration. In laparoscopic AR navigation, the speed and accuracy of registration are critical to the impact of surgery [12].

In this paper, the 3D image registration method of laparoscopic AR liver surgery navigation is studied. The registration process involves preoperative point cloud reconstruction, intraoperative point cloud reconstruction, and related registration methods. As the imaging principles of preoperative images and intraoperative laparoscopic images are different, and there is no same standard to match it [13], after studying the multi-modal image registration method, the 3D–3D point cloud registration method is selected. Here we only list the most relevant work. A binocular vision camera can provide doctors with images similar to laparoscopy, which can be used for surface reconstruction by matching features between images [14–16]. In this study, we used a binocular vision camera to obtain intraoperative information. The novelty of the 3D image registration method proposed in this paper is that a combination of rough registration and fine registration is used for multi-modal liver image registration. The rough registration uses the AFWA with adaptive amplitude, which replaces the amplitude operator in the enhanced fireworks algorithm, and the fine alignment uses the ICP algorithm improved by the k-dimensional tree (KD-tree). Our goal is to achieve fast and more accurate 3D image registration for laparoscopic AR liver surgery navigation. In particular, the main work of this study includes the following:

(1) A 3D reconstruction of the segmented preoperative CT images using the Marching Cubes algorithm on the VTK platform, and the 3D point cloud was generated after obtaining the 3D model of the liver;

(2) The laparoscopic (binocular vision camera) image was processed, and the 3D point cloud of the intraoperative liver image was generated;

(3) A two-step combined registration method through rough registration and fine registration is introduced. First, AFWA is applied to rough registration, and then the optimized ICP is applied to fine registration, which solves the problem that the ICP algorithm will fall into local extreme values during the iterative process;

(4) The registration method we proposed and other registration methods based on stochastic optimization algorithms are jointly tested in experiments. From the point cloud registration results, our method is better in terms of computation time and registration accuracy.

2. Background

Surgery navigation is to accurately overlay the patient's preoperative or intraoperative images and the patient's anatomical structure to assist the doctor in accurately locating the lesion, thereby making the operation more precise and safer. Image registration in surgery navigation is the process of optimally mapping one or more images to the target image, and it is also the core of laparoscopic AR navigation. As shown in Figure 1, using the surgery navigation system, doctors can see the AR overlay image in the virtual reality glasses or display of laparoscopy, which seems to build a map for surgery, so that doctors can accurately find the location of lesions.

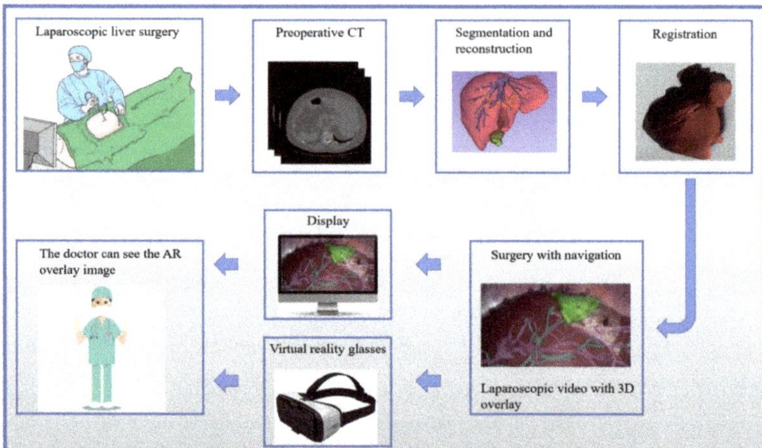

Figure 1. Overview of laparoscopic liver surgery navigation.

3. Related Work

Literature [17,18] reported the use of electromagnetic tracking to achieve the registration technology of preoperative CT and ultrasound imaging. Literature [19–22] has reported different registration techniques for image guidance in liver surgery. Fusaglia et al. [23] proposed a new registration method for liver surgery, which can register the intraoperative real-time reconstruction image with the preoperative image. Tam et al. [24] conducted a comprehensive investigation on rigid and non-rigid registration methods. In the process of 3D image registration, the ICP algorithm provides high precision and robustness and is widely used. Segal et al. [25] optimized the ICP algorithm and obtained higher robustness. Bentley et al. [26] proposed a KD-tree data structure, which provides a new space search idea. Liu et al. [27] used a KD-tree optimization algorithm to improve the original ICP, and the test results showed that the stability and registration speed were improved. It is worth noting that the application of ICP in 3D image registration also has certain drawbacks, such as a certain probability of falling into a local optimum. Li et al. [28] introduced an AFWA with high performance, and the experimental results proved that AFWA has high performance and does not take much time. Shi et al. [29] proposed a 3D point cloud registration method based on AFWA and ICP, which was verified by 3D point cloud registration of the physical model of the statue. The experimental results show that this method shows good calculation speed and accuracy, and can be applied in the field of cultural relics restoration. Chen et al. [30] proposed a new medical image registration method, which uses the fireworks algorithm to improve the coral reefs optimization algorithm for medical image registration. Through experimental tests, the method has a fast convergence speed and a significant improvement in computational performance. Zhang et al. [31] evaluated the LARN system they developed for application in liver surgery navigation. Through comparative analysis, the LARN system can help doctors to identify important anatomical structures during liver surgery, thus reducing surgery injuries. Pelanis et al. [32] tested and evaluated a liver surgery navigation system that provides an AR overlay on the laparoscopic camera view during laparoscopic liver surgery. The system can help doctors solve the difficulties associated with liver surgery, and thus perform safer liver surgery.

4. Materials and Methods

4.1. CT Data Preprocess

We used data from the publicly available 3D-IRCADb-01 dataset, which is provided by https://www.ircad.fr/research/3d-ircadb-01 (accessed on 3 January 2022). The CT dataset of three patients was selected, one of whom was a female patient, born in 1987, with a liver tumor located in the fifth zone. The CT voxel size is 0.78 mm $\times$ 0.78 mm $\times$ 1.6 mm, the pixels are 512 $\times$ 512 $\times$ 172, the average intensity of the liver in CT is 84, and the liver size is 20.1 cm $\times$ 16.9 cm $\times$ 15.7 cm. We use 3D Slicer as a tool for image segmentation. We import the patient's CT data into the 3D Slicer, use the segmentation module to segment the CT images, and extract the target area.

4.2. Preoperative Liver Point Cloud Generation

The 3D reconstruction of medical images has been extensively researched and is becoming increasingly mature [33], and it has contributed to the diagnosis of the patient's condition and 3D model printing. We choose to use the Marching Cubes algorithm in the VTK platform to perform a 3D reconstruction of the segmented CT images. The reconstructed models of liver, gallbladder, hepatic vena cava and portal vein, and liver tumor are shown in Figure 2a–d. After setting the transparency, these models are placed according to the original 3D space position, as shown in Figure 2e. At the same time, import the reconstructed model into the MeshLab software to generate a surface point cloud, which is shown in Figure 2f. The point cloud includes 7760 points.

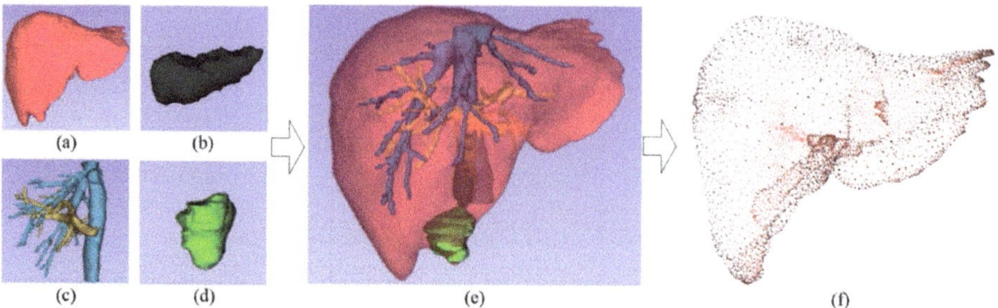

Figure 2. Three-dimensional reconstruction and surface point cloud generation, (**a**) liver model, (**b**) gallbladder model, (**c**) hepatic vena cava and portal vein model, (**d**) liver tumor model, (**e**) liver and internal tissue model, (**f**) preoperative liver point cloud.

4.3. Intraoperative Liver Point Cloud Generation

4.3.1. Calibration of Binocular Vision Camera

The MER-130-30UM binocular vision camera was used to simulate a stereo laparoscope. The installation and fixing of the binocular vision camera are shown in Figure 3a. Both cameras of the binocular vision camera are placed horizontally and fixed on the same reference plane, while the Y coordinates of the cameras must be horizontally aligned. The chessboard on the cardboard is placed in different positions such as far and near, up and down, left and right to take 20 pairs of images. We imported 20 pairs of chessboard images into Matlab (R2019a, America), and used the stereo vision calibration toolbox to obtain various parameters of the binocular vision camera through calculation. At the same time, 20 pairs of image feature points were matched, respectively, and the matching result of one pair of images is shown in Figure 3b.

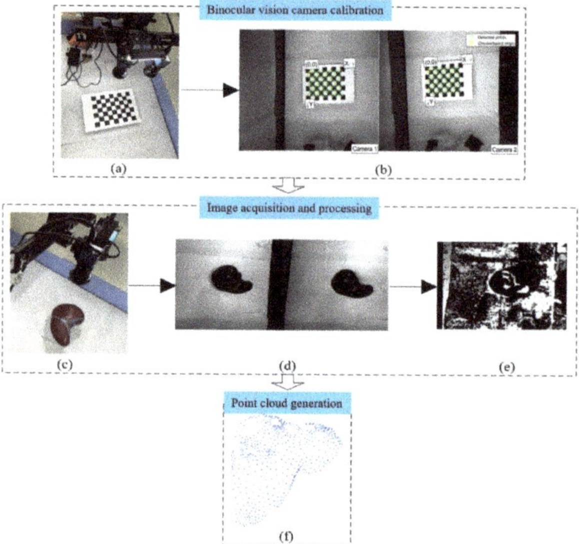

Figure 3. The liver model was photographed by binocular vision camera, and the point cloud was generated after image processing, (**a**) the installation and fixing of the binocular vision camera, (**b**) the matching result of one pair of images, (**c**) the calibrated binocular vision camera is used to photograph the liver model, (**d**) the result of using the Laplacian to sharpen the image, (**e**) the disparity image obtained by using the disparitySGM function, (**f**) the final result of the point cloud.

the surgical instruments with the liver. If this deformation is to be compensated, one can consider building a deformable biomechanical model of the liver, deformation guidance of the model by the surface motion of the liver, and a non-uniform deformation field. However, our proposed registration method compensates for liver deformation by taking into account improved calculation time and registration accuracy for initial registration and multiple intraoperative updates of the registration. This approach relies on a hybrid operating room with a real-time instrument to provide real-time images intraoperatively. During the surgery, the first registration is performed first, followed by multiple intraoperative registration updates. The CT image provided intraoperatively is used as the reference image and the image provided by the laparoscopic camera is used as the target image, thus performing multiple registration updates. After the first registration, the liver is deformed to varying degrees compared to its initial state due to a number of effects. Intraoperative registration updates can compensate for this deformation, and it still works even if the liver has undergone a large deformation.

Our proposed 3D image registration method will have a beneficial effect on surgery navigation systems, especially it will improve the registration accuracy and speed of surgery navigation systems. It is predictable that the improved surgery navigation system will help doctors quickly locate the lesion while bringing a good user experience to the doctor, so as to perform more accurate and safer surgery. Before our proposed method is applied in a real surgery setting, consideration should also be given to how to eliminate the effects caused by electrocautery during the procedure, such as smoke and liver bleeding. In future research, the better real-time registration of 3D images should be achieved through the improvement of computer hardware technology and the search for higher performance and faster 3D image registration methods, so as to provide a strong technical foundation for precision medicine and clinical application.

Author Contributions: D.L.: Conceptualization, methodology, software, writing—original draft; M.W.: Investigation, supervision, validation, and writing—review. All authors have read and agreed to the published version of the manuscript.

Funding: This study was supported by the National Natural Science Foundation of China (No. 61972117) and the Natural Science Foundation of Heilongjiang Province of China (ZD2019E007).

Data Availability Statement: Publicly available datasets were analyzed in this study. These data can be found here: https://www.ircad.fr/research/3d-ircadb-01 (accessed on 3 January 2022).

Conflicts of Interest: The authors declare no conflict of interest.

References

1. Winne, C.; Khan, M.; Stopp, F.; Jank, E.; Keeve, E. Overlay visualization in endoscopic ENT surgery. *Int. J. Comput. Assist. Radiol. Surg.* **2011**, *6*, 401–406. [CrossRef] [PubMed]
2. Collins, T.; Pizarro, D.; Bartoli, A.; Canis, M.; Bourdel, N. Computer-assisted laparoscopic myomectomy by augmenting the uterus with pre-operative MRI data. In Proceedings of the 2014 IEEE International Symposium on Mixed and Augmented Reality (ISMAR), Munich, Germany, 10–12 September 2014; IEEE: Munich, Germany, 2014; pp. 243–248.
3. Pascau, J. Image-guided intraoperative radiation therapy: Current developments and future perspectives. *Expert Rev. Med. Devices* **2014**, *11*, 431–434. [CrossRef]
4. Roberts, D.W.; Strohbehn, J.W.; Hatch, J.F.; Murray, W.; Kettenberger, H. A frameless stereotaxic integration of computerized tomographic imaging and the operating microscope. *J. Neurosurg.* **1986**, *65*, 545–549. [CrossRef] [PubMed]
5. Kelly, P.J.; Kall, B.A.; Goerss, S.; Earnest, F. Computer-assisted stereotaxic laser resection of intra-axial brain neoplasms. *J. Neurosurg.* **1986**, *64*, 427–439. [CrossRef]
6. Wengert, C.; Cattin, P.C.; Duff, J.M.; Baur, C.; Székely, G. Markerless endoscopic registration and referencing. In Proceedings of the International Conference on Medical Image Computing and Computer-Assisted Intervention, Copenhagen, Denmark, 1–6 October 2006; Springer: Berlin/Heidelberg, Germany, 2006; pp. 816–823.
7. Okamoto, T.; Onda, S.; Matsumoto, M.; Gocho, T.; Futagawa, Y.; Fujioka, S.; Yanaga, K.; Suzuki, N.; Hattori, A. Utility of augmented reality system in hepatobiliary surgery. *J. Hepato-Biliary Pancreat. Sci.* **2013**, *20*, 249–253. [CrossRef] [PubMed]
8. Teber, D.; Guven, S.; Simpfendörfer, T.; Baumhauer, M.; Güven, E.O.; Yencilek, F.; Gözen, A.S.; Rassweiler, J. Augmented reality: A new tool to improve surgical accuracy during laparoscopic partial nephrectomy? Preliminary in vitro and in vivo results. *Eur. Urol.* **2009**, *56*, 332–338. [CrossRef]

9. Kenngott, H.G.; Neuhaus, J.; Müller-Stich, B.P.; Wolf, I.; Vetter, M.; Meinzer, H.P.; Köninger, J.; Büchler, M.W.; Gutt, C.N. Development of a navigation system for minimally invasive esophagectomy. *Surg. Endosc.* **2008**, *22*, 1858–1865. [CrossRef]

10. Kersten-Oertel, M.; Jannin, P.; Collins, D.L. The state of the art of visualization in mixed reality image guided surgery. *Comput. Med. Imaging Graph.* **2013**, *37*, 98–112. [CrossRef]

11. Schneider, C.; Nguan, C.; Longpre, M.; Rohling, R. Motion of the Kidney between Preoperative and Intraoperative Positioning. *IEEE Trans. Biomed. Eng.* **2013**, *60*, 1619–1627. [CrossRef]

12. Haouchine, N.; Cotin, S.; Peterlik, I.; Dequidt, J.; Sanz-Lopez, M.; Kerrien, E.; Berger, M.O. Impact of soft tissue heterogeneity on augmented reality for liver surgery. *IEEE Trans. Vis. Comput. Graph.* **2014**, *21*, 584–597. [CrossRef]

13. Fischer, J.; Eichler, M.; Bartz, D.; Strasser, W. A hybrid tracking method for surgical augmented reality. *Comput. Graph.* **2007**, *31*, 39–52. [CrossRef]

14. Totz, J.; Thompson, S.; Stoyanov, D.; Gurusamy, K.; Davidson, B.R.; Hawkes, D.J.; Clarkson, M.J. Fast semi-dense surface reconstruction from stereoscopic video in laparoscopic surgery. In Proceedings of the International Conference on Information Processing in Computer-Assisted Interventions, Fukuoka, Japan, 28 June 2014; Springer: Cham, Switzerland, 2014; pp. 206–215.

15. Haouchine, N.; Dequidt, J.; Peterlik, I.; Kerrien, E.; Berger, M.O.; Cotin, S. Image-guided simulation of heterogeneous tissue deformation for augmented reality during hepatic surgery. In Proceedings of the 2013 IEEE International Symposium on Mixed and Augmented Reality (ISMAR), Adelaide, Australia, 1–4 October 2013; IEEE: Adelaide, Australia, 2013; pp. 199–208.

16. Chang, P.L.; Handa, A.; Davison, A.J.; Stoyanov, D.; Edwards, P.E. Robust real-time visual odometry for stereo endoscopy using dense quadrifocal tracking. In Proceedings of the International Conference on Information Processing in Computer-Assisted Interventions, Fukuoka, Japan, 28 June 2014; Springer: Cham, Switzerland, 2014; pp. 11–20.

17. Krücker, J.; Viswanathan, A.; Borgert, J.; Glossop, N.; Yang, Y.; Wood, B.J. An electro-magnetically tracked laparoscopic ultrasound for multi-modality minimally invasive surgery. In Proceedings of the International Congress Series, Madrid, Spain, 18–20 April 2005; Elsevier: Amsterdam, The Netherlands, 2005; Volume 1281, pp. 746–751.

18. Martens, V.; Besirevic, A.; Shahin, O.; Schlaefer, A.; Kleemann, M. LapAssistent-computer assisted laparoscopic liver surgery. In Proceedings of the Biomedizinischen Technik (BMT) Conference, Rostock, Germany, 5–8 October 2010.

19. Hammill, C.W.; Clements, L.W.; Stefansic, J.D.; Wolf, R.F.; Hansen, P.D.; Gerber, D.A. Evaluation of a minimally invasive image-guided surgery system for hepatic ablation procedures. *Surg. Innov.* **2014**, *21*, 419–426. [CrossRef] [PubMed]

20. Feuerstein, M.; Mussack, T.; Heining, S.M.; Navab, N. Intraoperative laparoscope augmentation for port placement and resection planning in minimally invasive liver resection. *IEEE Trans. Med. Imaging* **2008**, *27*, 355–369. [CrossRef] [PubMed]

21. Rauth, T.P.; Bao, P.Q.; Galloway, R.L.; Bieszczad, J.; Friets, E.M.; Knaus, D.A.; Kynor, D.B.; Herline, A.J. Laparoscopic surface scanning and subsurface targeting: Implications for image-guided laparoscopic liver surgery. *Surgery* **2007**, *137*, 229. [CrossRef] [PubMed]

22. Shekhar, R.; Dandekar, O.; Bhat, V.; Philip, M.; Lei, P.; Godinez, C.; Sutton, E.; George, I.; Kavic, S.; Mezrich, R. Live augmented reality: A new visualization method for laparoscopic surgery using continuous volumetric computed tomography. *Surg. Endosc.* **2010**, *24*, 1976–1985. [CrossRef] [PubMed]

23. Fusaglia, M.; Tinguely, P.; Banz, V.; Weber, S.; Lu, H. A Novel Ultrasound-Based Registration for Image-Guided Laparoscopic Liver Ablation. *Surg. Innov.* **2016**, *23*, 397–406. [CrossRef]

24. Tam, G.K.L.; Cheng, Z.Q.; Lai, Y.K.; Langbein, F.C.; Liu, Y.; Marshall, D.; Martin, R.R.; Sun, X.F.; Rosin, P.L. Registration of 3D point clouds and meshes: A survey from rigid to nonrigid. *IEEE Trans. Vis. Comput. Graph.* **2012**, *19*, 1199–1217. [CrossRef]

25. Segal, A.; Hhnel, D.; Thrun, S. Generalized-ICP. In Proceedings of the Robotics: Science and Systems V, Seattle, DC, USA, 28 June–1 July 2009.

26. Bentley, J.L. Multidimensional binary search trees used for associative searching. *Commun. ACM* **1975**, *18*, 509–517. [CrossRef]

27. Liu, J.; Zhu, J.; Yang, J.; Meng, X.; Zhang, H. Three-dimensional point cloud registration based on ICP algorithm employing KD tree optimization. In Proceedings of the Eighth International Conference on Digital Image Processing (ICDIP 2016), Chengdu, China, 20–22 May 2016; International Society for Optics and Photonics: Bellingham, WA, USA, 2016; Volume 10033, p. 100334D.

28. Li, J.; Zheng, S.; Tan, Y. Adaptive fireworks algorithm. In Proceedings of the 2014 IEEE Congress on Evolutionary Computation (CEC), Beijing, China, 6–11 July 2014.

29. Shi, X.J.; Liu, T.; Han, X. Improved Iterative Closest Point (ICP) 3D point cloud registration algorithm based on point cloud filtering and adaptive fireworks for coarse registration. *Int. J. Remote Sens.* **2020**, *41*, 3197–3220. [CrossRef]

30. Chen, Y.L.; He, F.Z.; Zeng, X.T.; Li, H.R.; Liang, Y.Q. The explosion operation of fireworks algorithm boosts the coral reef optimization for multimodal medical image registration. *Eng. Appl. Artif. Intell.* **2021**, *102*, 104252. [CrossRef]

31. Zhang, W.Q.; Zhu, W.; Yang, J.; Xiang, N.; Zeng, N.; Hu, H.Y.; Jia, F.C.; Fang, C.H. Augmented reality navigation for stereoscopic laparoscopic anatomical hepatectomy of primary liver cancer: Preliminary experience. *Front. Oncol.* **2021**, *11*, 996. [CrossRef] [PubMed]

32. Pelanis, E.; Teatini, A.; Eigl, B.; Regensburger, A.; Alzaga, A.; Kumar, R.P.; Rudolph, T.; Aghayan, D.L.; Riediger, C.; Kvarnström, N.; et al. Evaluation of a novel navigation platform for laparoscopic liver surgery with organ deformation compensation using injected fiducials. *Med. Image Anal.* **2021**, *69*, 101946. [CrossRef] [PubMed]

33. Verhey, J.T.; Haglin, J.M.; Verhey, E.M.; Hartigan, D.E. Virtual, augmented, and mixed reality applications in orthopedic surgery. *Int. J. Med. Robot. Comput. Assist. Surg.* **2020**, *16*, e2067. [CrossRef] [PubMed]

4.3.2. Image Acquisition and Image Processing

We used the processed liver CT data to obtain a 3D printed model to simulate the real liver. The calibrated binocular vision camera was used to photograph the liver model as shown in Figure 3c. The obtained images were corrected to remove distortions using the rectifyStereoImages function in Matlab. Figure 3d shows the result of using the Laplacian to sharpen the image.

4.3.3. Point Cloud Generation

The disparity image was generated using the SGM algorithm in Matlab, as shown in Figure 3e. The filtering operation is performed after reconstructing the liver model point cloud, and the final obtained point cloud is shown in Figure 3f. Since the computation time in the registration process is directly related to the number of points in the point cloud, it should be considered how to reasonably reduce the number of points in the point cloud. We select representative points in the point cloud through the filtering method to filter out unnecessary points and noise points [34]. The KD-tree algorithm is used to find the spatially neighboring point set of the point cloud, and to solve the average distance between the point cloud and the spatially neighboring point set, and the average and standard deviation of the global distance are calculated. After this, the points outside the range of the mean distance ± standard deviation are removed to obtain the filtered point cloud containing 1830 points.

4.4. Two-Step Combined Registration Method through Rough Registration and Fine Registration

As both binocular vision imaging and CT images can be regarded as 3D data, the 3D–3D registration method is used here. The registration of the 3D point cloud is used as the basis for the registration of CT images and binocular vision imaging, so as to register the obtained preoperative point cloud and intraoperative point cloud.

4.4.1. Rough Registration Process Based on AFWA

Before performing AFWA, it is necessary to determine the dataset, establish the fitness function, and determine the optimization goal. In the initial setting, the KD-tree can be used cleverly to determine the closest point. The preoperative model point cloud is stored in a KD-tree structure, and the K-nearest neighbor algorithm is used to search for the nearest neighbors of all points in the intraoperative point cloud in the KD-tree, and establish corresponding points. The point cloud generated from the intraoperative image was set as the target and set as P, while the point cloud generated from the preoperative model was set as the reference and set as Q. As the nearest neighbor point set of point cloud P, q can be obtained by searching in point cloud Q. For the sake of unity and convenience, we use p point set as a shorthand for point cloud P, and the fitness function is established

$$f(R, T) = \frac{1}{n} \sum_{i=1}^{n} \|q_i - (R \times p_i + T)\|^2 = \min \tag{1}$$

Among them, R is the rotation variable and T is the translation variable, including 3 rotation variables and 3 translation variables. Where n represents the number of points in the target p point set. After that, AWFA is used to realize rough registration. It is worth noting that adaptive explosion radius is the core mechanism of AWFA. In addition, in the AWFA, the fitness function is established to calculate the fitness value of each spark, so as to produce different numbers of sparks at different explosion radii. Figure 4a shows the rough registration process based on AFWA.

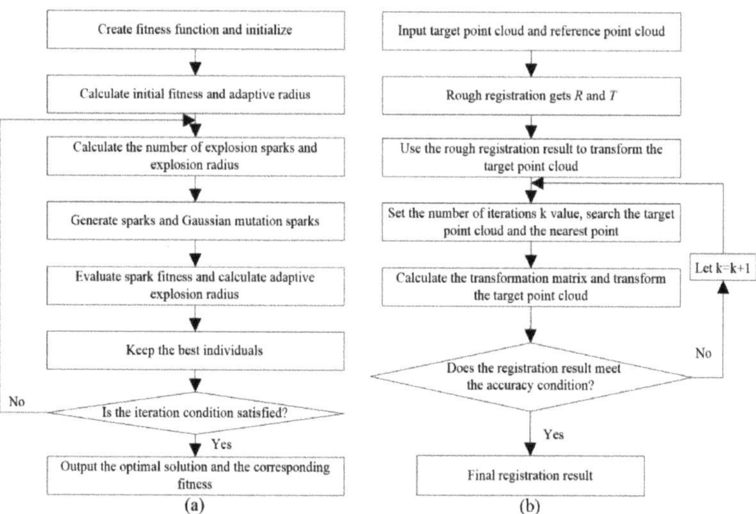

Figure 4. The registration method we designed, (**a**) applies AFWA for the rough registration process, (**b**) a two-step combined registration process of rough registration and fine registration.

4.4.2. Fine Registration Process

The main purpose of fine registration is to correct the previously obtained registration results and obtain more accurate registration results. Therefore, after the rough registration, we use the ICP based on KD-tree optimization to correct the result obtained by the rough registration. At the same time, the rotation variable R and translation variable T obtained in the rough registration process are extracted as the optimization parameters of the transformation operation, so that the p point set is transformed as follows

$$\mathbf{p}' = \mathbf{p} \times R + T \tag{2}$$

where $\mathbf{p}'$ is the new target point set after the transformation calculation. The fine registration steps for the proposed design using optimized ICP are:

(1) Input the calculated target point set $\mathbf{p}'$ and the original point set Q together. At this time, the KD-tree structure is used to store the point set Q. Then the focus is to search the closest neighbor point set $\mathbf{q}'$ of $\mathbf{p}'$, which is implemented by the nearest neighbor algorithm, and then set the iteration number k (the initial value is $k = 1$).

(2) Calculate the rotation variable R_k and translation variable T_k from $\mathbf{p}'$ to $\mathbf{q}'$. Here, the quaternion calculation method is used and the value of Equation (3) should be minimized.

$$\sum_{i=1}^{n} \left\| \mathbf{q}_i - \left(R_k \times \mathbf{p}_i' + T_k \right) \right\|^2 \tag{3}$$

Use the solved R_k and T_k to transform the $\mathbf{p}'$ to obtain a new target point set $\mathbf{p}''$, which is calculated as follows:

$$\mathbf{p}'' = R_k \times \mathbf{p}' + T_k. \tag{4}$$

(3) Calculate the average distance d_{k+1} between point set $\mathbf{p}''$ and point set $\mathbf{q}'$

$$d_{k+1} = \frac{1}{n} \sum_{i=1}^{n} \left\| \mathbf{q}_i' - \left(R_k \times \mathbf{p}_i'' + T_k \right) \right\|^2 . \tag{5}$$

Determine whether the convergence condition $\|d_{k+1} - d_k\| < \varepsilon$ is satisfied, where ε is the minimum iteration accuracy, and d_k is the average distance of the previous generation. If it is not satisfied, the point set p″ is used as the new initial target point set p′, and let $k = k + 1$, repeat steps 1–3 until the iteration condition is satisfied.

(4)　According to the obtained rotation variable R_k and translation variable T_k, the p point set is transformed, and finally, the final registration result is obtained together with the reference point cloud Q.

Figure 4b shows the process of the two-step combined registration method based on rough registration and fine registration.

5. Experiments and Validation

The comparative experiment we designed was implemented with Matlab. All the four experiments were completed on an Intel Core i5-4210m 2.6 GHz/8 GB and NVIDIA geforce GTX 850 computer. At present, in the research of 3D image registration methods, rough registration based on stochastic optimization algorithms is a popular method, and the genetic algorithm (GA) and particle swarm optimization (PSO) algorithms are mainly used [35]. We designed four experiments to verify the performance of our registration method, mainly from the aspects of registration accuracy and speed for comparison and verification. The first experiment is to use the registration method we introduced for registration. In the other two experiments, we used GA or PSO to replace the AFWA in the rough registration. The fine registration process remains unchanged, and the fine registration step based on optimized ICP is still used. In the last experiment, rough registration with AFWA and fine registration with ICP. In these four experiments, the input data are all the same. Figure 5 shows the point cloud registration results of the four experiments.

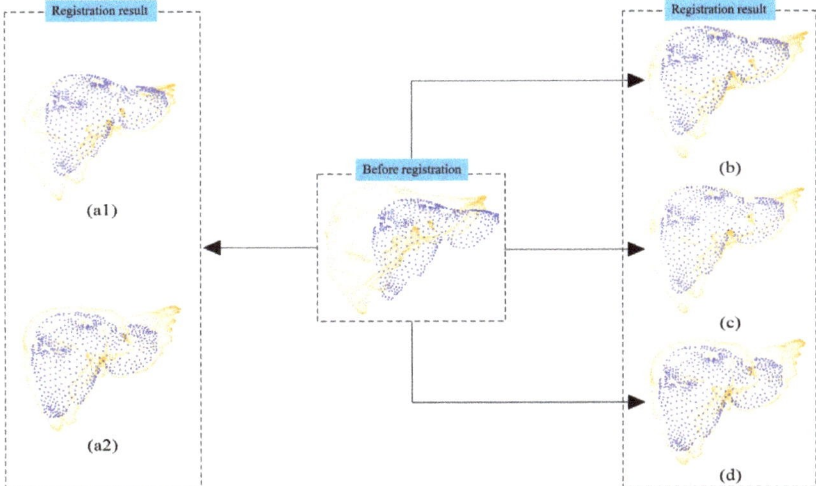

Figure 5. The results obtained by using four registration methods to perform registration, respectively, (**a1**) the rough registration result, rough registration with AFWA, (**a2**) the registration result, our registration method, (**b**) the registration result, rough registration with GA and fine registration with optimized ICP, (**c**) the registration result, rough registration with PSO and fine registration with optimized ICP, and (**d**) the registration result, rough registration with AFWA and fine registration with ICP.

In point cloud rough registration, Figure 6 shows the relationship between the number of iterations and the fitness of the three stochastic optimization algorithms.

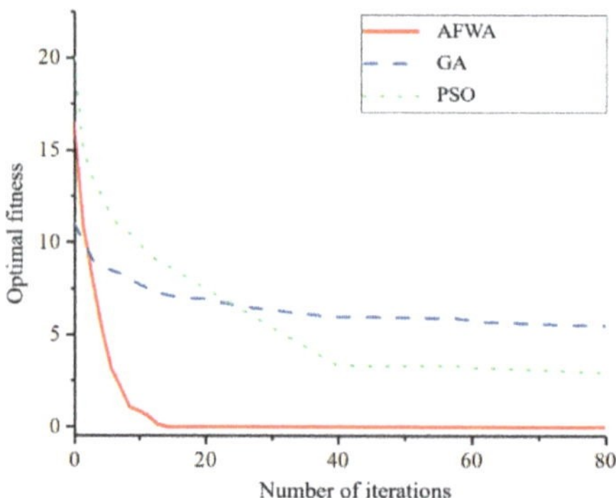

Figure 6. In the process of rough registration, the relationship between the number of iterations and fitness of the three stochastic optimization algorithms.

In the experiment, the four registration methods were tested. Table 1 shows the overall computation time and accuracy comparison of the four methods.

Table 1. In the experiment, the overall computation time and accuracy of the four registration methods are compared.

		GA + Improved ICP	PSO + Improved ICP	AFWA + ICP	Ours
Dataset 1	Registration time (s)	0.709	0.814	16.186	0.606
	Accuracy (mm)	0.0208	0.0019	0.0018	0.0018
Dataset 2	Registration time (s)	0.768	0.861	17.548	0.657
	Accuracy (mm)	0.0346	0.0027	0.0022	0.0022
Dataset 3	Registration time (s)	0.849	0.953	18.658	0.726
	Accuracy (mm)	0.0253	0.0023	0.0019	0.0019

6. Conclusions and Discussion

This study analyzes the 3D image registration technology in laparoscopic liver surgery navigation. The most important thing is to introduce a two-step combined registration method of rough registration and fine registration, which can quickly and accurately superimpose the preoperative liver model on the laparoscopic image. We segmented and reconstructed the preoperative CT images to obtain the 3D model and point cloud of the liver. During the surgery, we built the intraoperative 3D surface model point cloud, which was then registration. Using SLAM to track the camera motion can finally realize AR visualization. These works are applicable to laparoscopic liver tumor resection, while for open surgeries, there are easier and more accurate ways to reconstruct the 3D model of the liver, such as using an Intel RealSense RGB-D camera. Comparing our registration method with other registration methods based on a stochastic optimization algorithm, from the analysis in Figure 6, our method converges very fast and can converge in about 12 generations, which is obviously better than the other two registration methods based on the stochastic optimization algorithm. The registration accuracy of our proposed registration method in three tests was 0.0018 mm, 0.0022 mm, and 0.0019 mm, respectively, which also demonstrates the good robustness of the method. As can be seen from Table 1, our proposed registration method is clearly better than other registration methods in terms of computation time and registration accuracy. It is worth noting that our registration method has better performance for searching corresponding points, reduces a lot of iterative calculations in fine registration, and can also overcome the defect that ICP has the possibility of falling into a local optimum when it is applied. During laparoscopic liver tumor resection, the liver will be deformed to some extent due to the patient's breathing or the collision of the surgical instruments with the liver. If this deformation is to be compensated, one can consider building a deformable biomechanical model of the liver, deformation guidance of the model by the surface motion of the liver, and a non-uniform deformation field. However, our proposed registration method compensates for liver deformation by taking into account improved calculation time and registration accuracy for initial registration and multiple intraoperative updates of the registration. This approach relies on a hybrid operating room with a real-time instrument to provide real-time images intraoperatively. During the surgery, the first registration is performed first, followed by multiple intraoperative registration updates. The CT image provided intraoperatively is used as the reference image and the image provided by the laparoscopic camera is used as the target image, thus performing multiple registration updates. After the first registration, the liver is deformed to varying degrees compared to its initial state due to a number of effects. Intraoperative registration updates can compensate for this deformation, and it still works even if the liver has undergone a large deformation.

Our proposed 3D image registration method will have a beneficial effect on surgery navigation systems, especially it will improve the registration accuracy and speed of surgery navigation systems. It is predictable that the improved surgery navigation system will help doctors quickly locate the lesion while bringing a good user experience to the doctor, so as to perform more accurate and safer surgery. Before our proposed method is applied in a real surgery setting, consideration should also be given to how to eliminate the effects caused by electrocautery during the procedure, such as smoke and liver bleeding. In future research, the better real-time registration of 3D images should be achieved through the improvement of computer hardware technology and the search for higher performance and faster 3D image registration methods, so as to provide a strong technical foundation for precision medicine and clinical application.

Author Contributions: D.L.: Conceptualization, methodology, software, writing—original draft; M.W.: Investigation, supervision, validation, and writing—review. All authors have read and agreed to the published version of the manuscript.

Funding: This study was supported by the National Natural Science Foundation of China (No. 61972117) and the Natural Science Foundation of Heilongjiang Province of China (ZD2019E007).

Data Availability Statement: Publicly available datasets were analyzed in this study. These data can be found here: https://www.ircad.fr/research/3d-ircadb-01 (accessed on 3 January 2022).

Conflicts of Interest: The authors declare no conflict of interest.

References

1. Winne, C.; Khan, M.; Stopp, F.; Jank, E.; Keeve, E. Overlay visualization in endoscopic ENT surgery. *Int. J. Comput. Assist. Radiol. Surg.* **2011**, *6*, 401–406. [CrossRef] [PubMed]
2. Collins, T.; Pizarro, D.; Bartoli, A.; Canis, M.; Bourdel, N. Computer-assisted laparoscopic myomectomy by augmenting the uterus with pre-operative MRI data. In Proceedings of the 2014 IEEE International Symposium on Mixed and Augmented Reality (ISMAR), Munich, Germany, 10–12 September 2014; IEEE: Munich, Germany, 2014; pp. 243–248.
3. Pascau, J. Image-guided intraoperative radiation therapy: Current developments and future perspectives. *Expert Rev. Med. Devices* **2014**, *11*, 431–434. [CrossRef]
4. Roberts, D.W.; Strohbehn, J.W.; Hatch, J.F.; Murray, W.; Kettenberger, H. A frameless stereotaxic integration of computerized tomographic imaging and the operating microscope. *J. Neurosurg.* **1986**, *65*, 545–549. [CrossRef] [PubMed]
5. Kelly, P.J.; Kall, B.A.; Goerss, S.; Earnest, F. Computer-assisted stereotaxic laser resection of intra-axial brain neoplasms. *J. Neurosurg.* **1986**, *64*, 427–439. [CrossRef]
6. Wengert, C.; Cattin, P.C.; Duff, J.M.; Baur, C.; Székely, G. Markerless endoscopic registration and referencing. In Proceedings of the International Conference on Medical Image Computing and Computer-Assisted Intervention, Copenhagen, Denmark, 1–6 October 2006; Springer: Berlin/Heidelberg, Germany, 2006; pp. 816–823.
7. Okamoto, T.; Onda, S.; Matsumoto, M.; Gocho, T.; Futagawa, Y.; Fujioka, S.; Yanaga, K.; Suzuki, N.; Hattori, A. Utility of augmented reality system in hepatobiliary surgery. *J. Hepato-Biliary Pancreat. Sci.* **2013**, *20*, 249–253. [CrossRef] [PubMed]
8. Teber, D.; Guven, S.; Simpfendörfer, T.; Baumhauer, M.; Güven, E.O.; Yencilek, F.; Gözen, A.S.; Rassweiler, J. Augmented reality: A new tool to improve surgical accuracy during laparoscopic partial nephrectomy? Preliminary in vitro and in vivo results. *Eur. Urol.* **2009**, *56*, 332–338. [CrossRef]

9. Kenngott, H.G.; Neuhaus, J.; Müller-Stich, B.P.; Wolf, I.; Vetter, M.; Meinzer, H.P.; Köninger, J.; Büchler, M.W.; Gutt, C.N. Development of a navigation system for minimally invasive esophagectomy. *Surg. Endosc.* **2008**, *22*, 1858–1865. [CrossRef]

10. Kersten-Oertel, M.; Jannin, P.; Collins, D.L. The state of the art of visualization in mixed reality image guided surgery. *Comput. Med. Imaging Graph.* **2013**, *37*, 98–112. [CrossRef]

11. Schneider, C.; Nguan, C.; Longpre, M.; Rohling, R. Motion of the Kidney between Preoperative and Intraoperative Positioning. *IEEE Trans. Biomed. Eng.* **2013**, *60*, 1619–1627. [CrossRef]

12. Haouchine, N.; Cotin, S.; Peterlik, I.; Dequidt, J.; Sanz-Lopez, M.; Kerrien, E.; Berger, M.O. Impact of soft tissue heterogeneity on augmented reality for liver surgery. *IEEE Trans. Vis. Comput. Graph.* **2014**, *21*, 584–597. [CrossRef]

13. Fischer, J.; Eichler, M.; Bartz, D.; Strasser, W. A hybrid tracking method for surgical augmented reality. *Comput. Graph.* **2007**, *31*, 39–52. [CrossRef]

14. Totz, J.; Thompson, S.; Stoyanov, D.; Gurusamy, K.; Davidson, B.R.; Hawkes, D.J.; Clarkson, M.J. Fast semi-dense surface reconstruction from stereoscopic video in laparoscopic surgery. In Proceedings of the International Conference on Information Processing in Computer-Assisted Interventions, Fukuoka, Japan, 28 June 2014; Springer: Cham, Switzerland, 2014; pp. 206–215.

15. Haouchine, N.; Dequidt, J.; Peterlik, I.; Kerrien, E.; Berger, M.O.; Cotin, S. Image-guided simulation of heterogeneous tissue deformation for augmented reality during hepatic surgery. In Proceedings of the 2013 IEEE International Symposium on Mixed and Augmented Reality (ISMAR), Adelaide, Australia, 1–4 October 2013; IEEE: Adelaide, Australia, 2013; pp. 199–208.

16. Chang, P.L.; Handa, A.; Davison, A.J.; Stoyanov, D.; Edwards, P.E. Robust real-time visual odometry for stereo endoscopy using dense quadrifocal tracking. In Proceedings of the International Conference on Information Processing in Computer-Assisted Interventions, Fukuoka, Japan, 28 June 2014; Springer: Cham, Switzerland, 2014; pp. 11–20.

17. Krücker, J.; Viswanathan, A.; Borgert, J.; Glossop, N.; Yang, Y.; Wood, B.J. An electro-magnetically tracked laparoscopic ultrasound for multi-modality minimally invasive surgery. In Proceedings of the International Congress Series, Madrid, Spain, 18–20 April 2005; Elsevier: Amsterdam, The Netherlands, 2005; Volume 1281, pp. 746–751.

18. Martens, V.; Besirevic, A.; Shahin, O.; Schlaefer, A.; Kleemann, M. LapAssistent-computer assisted laparoscopic liver surgery. In Proceedings of the Biomedizinischen Technik (BMT) Conference, Rostock, Germany, 5–8 October 2010.

19. Hammill, C.W.; Clements, L.W.; Stefansic, J.D.; Wolf, R.F.; Hansen, P.D.; Gerber, D.A. Evaluation of a minimally invasive image-guided surgery system for hepatic ablation procedures. *Surg. Innov.* **2014**, *21*, 419–426. [CrossRef] [PubMed]

20. Feuerstein, M.; Mussack, T.; Heining, S.M.; Navab, N. Intraoperative laparoscope augmentation for port placement and resection planning in minimally invasive liver resection. *IEEE Trans. Med. Imaging* **2008**, *27*, 355–369. [CrossRef] [PubMed]

21. Rauth, T.P.; Bao, P.Q.; Galloway, R.L.; Bieszczad, J.; Friets, E.M.; Knaus, D.A.; Kynor, D.B.; Herline, A.J. Laparoscopic surface scanning and subsurface targeting: Implications for image-guided laparoscopic liver surgery. *Surgery* **2007**, *137*, 229. [CrossRef] [PubMed]

22. Shekhar, R.; Dandekar, O.; Bhat, V.; Philip, M.; Lei, P.; Godinez, C.; Sutton, E.; George, I.; Kavic, S.; Mezrich, R. Live augmented reality: A new visualization method for laparoscopic surgery using continuous volumetric computed tomography. *Surg. Endosc.* **2010**, *24*, 1976–1985. [CrossRef] [PubMed]

23. Fusaglia, M.; Tinguely, P.; Banz, V.; Weber, S.; Lu, H. A Novel Ultrasound-Based Registration for Image-Guided Laparoscopic Liver Ablation. *Surg. Innov.* **2016**, *23*, 397–406. [CrossRef]

24. Tam, G.K.L.; Cheng, Z.Q.; Lai, Y.K.; Langbein, F.C.; Liu, Y.; Marshall, D.; Martin, R.R.; Sun, X.F.; Rosin, P.L. Registration of 3D point clouds and meshes: A survey from rigid to nonrigid. *IEEE Trans. Vis. Comput. Graph.* **2012**, *19*, 1199–1217. [CrossRef]

25. Segal, A.; Hhnel, D.; Thrun, S. Generalized-ICP. In Proceedings of the Robotics: Science and Systems V, Seattle, DC, USA, 28 June–1 July 2009.

26. Bentley, J.L. Multidimensional binary search trees used for associative searching. *Commun. ACM* **1975**, *18*, 509–517. [CrossRef]

27. Liu, J.; Zhu, J.; Yang, J.; Meng, X.; Zhang, H. Three-dimensional point cloud registration based on ICP algorithm employing KD tree optimization. In Proceedings of the Eighth International Conference on Digital Image Processing (ICDIP 2016), Chengdu, China, 20–22 May 2016; International Society for Optics and Photonics: Bellingham, WA, USA, 2016; Volume 10033, p. 100334D.

28. Li, J.; Zheng, S.; Tan, Y. Adaptive fireworks algorithm. In Proceedings of the 2014 IEEE Congress on Evolutionary Computation (CEC), Beijing, China, 6–11 July 2014.

29. Shi, X.J.; Liu, T.; Han, X. Improved Iterative Closest Point (ICP) 3D point cloud registration algorithm based on point cloud filtering and adaptive fireworks for coarse registration. *Int. J. Remote Sens.* **2020**, *41*, 3197–3220. [CrossRef]

30. Chen, Y.L.; He, F.Z.; Zeng, X.T.; Li, H.R.; Liang, Y.Q. The explosion operation of fireworks algorithm boosts the coral reef optimization for multimodal medical image registration. *Eng. Appl. Artif. Intell.* **2021**, *102*, 104252. [CrossRef]

31. Zhang, W.Q.; Zhu, W.; Yang, J.; Xiang, N.; Zeng, N.; Hu, H.Y.; Jia, F.C.; Fang, C.H. Augmented reality navigation for stereoscopic laparoscopic anatomical hepatectomy of primary liver cancer: Preliminary experience. *Front. Oncol.* **2021**, *11*, 996. [CrossRef] [PubMed]

32. Pelanis, E.; Teatini, A.; Eigl, B.; Regensburger, A.; Alzaga, A.; Kumar, R.P.; Rudolph, T.; Aghayan, D.L.; Riediger, C.; Kvarnström, N.; et al. Evaluation of a novel navigation platform for laparoscopic liver surgery with organ deformation compensation using injected fiducials. *Med. Image Anal.* **2021**, *69*, 101946. [CrossRef] [PubMed]

33. Verhey, J.T.; Haglin, J.M.; Verhey, E.M.; Hartigan, D.E. Virtual, augmented, and mixed reality applications in orthopedic surgery. *Int. J. Med. Robot. Comput. Assist. Surg.* **2020**, *16*, e2067. [CrossRef] [PubMed]

34. Zheng, L.; Mai, C.; Liao, W.; Wen, Y.; Liu, G. 3D point cloud registration for apple tree based on Kinect camera. *Trans. Chin. Soc. Agric. Mach.* **2016**, *47*, 1–6.
35. Saiti, E.; Theoharis, T. An application independent review of multimodal 3D registration methods. *Comput. Graph.* **2020**, *91*, 153–178. [CrossRef]

Perspective

Virtual Reality for Safe Testing and Development in Collaborative Robotics: Challenges and Perspectives

Sergi Bermúdez i Badia [1], Paula Alexandra Silva [2], Diogo Branco [1], Ana Pinto [3], Carla Carvalho [4], Paulo Menezes [5], Jorge Almeida [4] and Artur Pilacinski [4,6,*]

1   FCEE, Nova Lincs, University of Madeira, 9020-105 Funchal, Portugal; sergi.bermudez@uma.pt (S.B.i.B.); diogo.branco@arditi.pt (D.B.)
2   DEI, CISUC, University of Coimbra, 3030-788 Coimbra, Portugal; paulasilva@dei.uc.pt
3   CeBER, University of Coimbra, 3030-788 Coimbra, Portugal; ana.pinto@dem.uc.pt
4   CINEICC, University of Coimbra, 3030-788 Coimbra, Portugal; ccarvalho@fpce.uc.pt (C.C.); jorgealmeida@fpce.uc.pt (J.A.)
5   ISR, DEEC, University of Coimbra, 3030-788 Coimbra, Portugal; paulomenezes@isr.uc.pt
6   Department of Neurosurgery, The Faculty of Medicine, University Hospital Knappschaftskrankenhaus Bochum GmbH, Ruhr-University Bochum, 44892 Bochum, Germany
*   Correspondence: ap@uc.pt

Abstract: Collaborative robots (cobots) could help humans in tasks that are mundane, dangerous or where direct human contact carries risk. Yet, the collaboration between humans and robots is severely limited by the aspects of the safety and comfort of human operators. In this paper, we outline the use of extended reality (XR) as a way to test and develop collaboration with robots. We focus on virtual reality (VR) in simulating collaboration scenarios and the use of cobot digital twins. This is specifically useful in situations that are difficult or even impossible to safely test in real life, such as dangerous scenarios. We describe using XR simulations as a means to evaluate collaboration with robots without putting humans at harm. We show how an XR setting enables combining human behavioral data, subjective self-reports, and biosignals signifying human comfort, stress and cognitive load during collaboration. Several works demonstrate XR can be used to train human operators and provide them with augmented reality (AR) interfaces to enhance their performance with robots. We also provide a first attempt at what could become the basis for a human–robot collaboration testing framework, specifically for designing and testing factors affecting human–robot collaboration. The use of XR has the potential to change the way we design and test cobots, and train cobot operators, in a range of applications: from industry, through healthcare, to space operations.

Keywords: collaborative robotics; acceptability; uncanny valley; user experience; augmented reality; virtual reality; extended reality

Citation: Badia, S.B.i.; Silva, P.A.; Branco, D.; Pinto, A.; Carvalho, C.; Menezes, P.; Almeida, J.; Pilacinski, A. Virtual Reality for Safe Testing and Development in Collaborative Robotics: Challenges and Perspectives. *Electronics* 2022, 11, 1726. https://doi.org/10.3390/electronics11111726

Academic Editor: Stefanos Kollias

Received: 29 April 2022
Accepted: 20 May 2022
Published: 29 May 2022

1. Introduction

Motor collaboration between humans is essential for activities ranging from working together at construction sites to performing complex surgeries. This is because the human ability to read the motor intentions of another human is unparalleled: a skilled technician does not need much instruction to hold up an element that the other one is welding; a nurse does not need much guidance when feeding her patient with a spoon. However, situations such as the COVID-19 pandemic reveal threats to this traditional model of collaboration. The contagion risk posed by human contact had a severe socio-economic impact imposing changes for industry across the board, from factories to hospitals and care homes [1]. While many institutions have rapidly switched to remote work and communication, many others could not do the same, as human contact is required in many industries. In situations of severe risk, such as a pandemic, human activities could be at least partially replaced by robots, thereby reducing contagion risk [2,3]. However, for industries requiring, at present, close human collaboration, this is less true. Even though the use of cobots could minimize the risk to humans, the collaboration between humans and robots is still far and away from nearing, let alone matching, the collaboration between humans [4,5].

Therefore, especially in the area where the interaction between humans and robots may represent a risk for the human, collaborative robots may become of vital help to their human operators. This paper reviews the current work, highlights existing issues and challenges and proposes novel approaches to the use of virtual reality (VR) and, more generally, extended reality (XR), as a tool for safe testing collaborative robotics. We used a narrative review, which is why we did not use explicit and systematic criteria for the search and critical analysis of the literature. It was not our intention to exhaust the sources of information; however, we tried to carry out a deep search that includes articles from 1997 to 2022. The selection of studies and their interpretation was performed to classify the main applications and critical factors involved in the use of XR in the broad domain of collaborative robotics, with special emphasis on the cobot, the users and the environment.

2. Human-Robot Collaboration, Safety and Acceptability

Human–robot collaboration (HRC) is a specific sub-domain of human–robot interaction (HRI), which studies a human operator and a robot working together on a common goal using physical manipulation [6]. The general idea of HRC is not new, and several companies have deployed collaborative robots capable of working along industry lines. Still, any progress in this domain is limited by the safety and acceptability of such collaboration [5,7]. Human safety is a critical factor: as industrial robots are often heavy and/or equipped with powerful effectors, they pose a physical danger. For this reason, most industrial robots are kept at a distance or inside safety cages (Figure 1). This solution is suboptimal for robots that are supposed to help humans perform their tasks since real cooperation assumes that both agents work simultaneously.

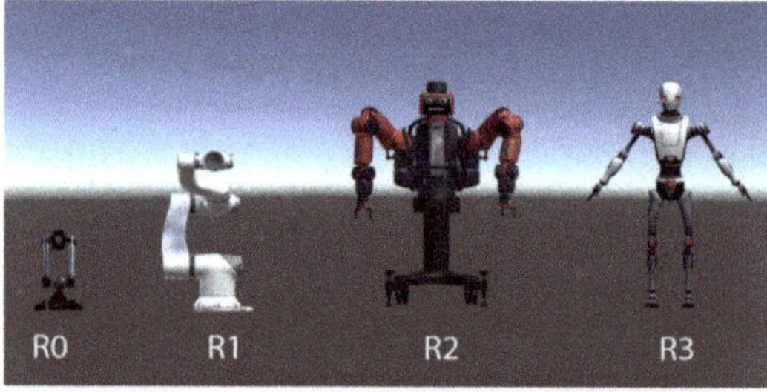

Figure 1. Top: Example VR robot models arranged according to their anthropomorphism. (R0) one arm basic; (R1) an articulated arm; (R2): two arms Baxter; (R3): a humanoid robot. Bottom: an example VR-collaboration scene used by the authors, developed with Unity Game Engine (Unity Technologies, San Francisco, CA, USA). The scene shows a basic tool-passing task in which subject kinematic and physiological data can be recorded in response to manipulated scene features (e.g., cobot appearance, speed, etc.).

Table 1 shows the different levels of collaboration with robots at present. Fenced robots are the non-collaborative, most popular ones. Then there are robots that allow for collaboration. Again, their use is usually limited to cases 2 and 3 due to safety. Finally, the last two columns denote actual dynamic collaboration.

Table 1. Types of collaboration with industrial robots at present. As the level of collaboration increases (left to right), so does the requirement for intrinsic safety features vs. external sensors. Source: IFR Position Paper [4], adapted from Bauer et al. [8].

Level of Collaboration	1 Cell	2 Coexistence	3 Sequential Collaboration	4 Cooperation	5 Responsive Collaboration
Requirement for intrinsic safety features vs. external sensors	Fenced robot	No fence but no shared workspace	Robot and worker both active in the workspace but movements are sequential	Robot and worker work on the same part at the same time, both in motion	Robot responds in real time to the movement of the operator.

When two agents are working together, they need to establish joint attention to form a joint intention and execute joint actions [9–11]. Mutual understanding of each other's actions and the acceptability of robotic actions to a human is therefore an important issue in the field of human–machine interaction (HMI) [12,13]. It is implicitly assumed that, in the robot–human diad, the human defines the intentions the robot has to adapt to [9]. However, unlike the presently available robots, the human brain comes equipped with specialized "computational machinery" for recognizing and predicting actions. The human brain is extremely efficient in recognizing other people's actions, for example, their action intentions or errors [14,15]. This recognition makes humans able to rapidly adapt to what the other human does, reacting accordingly. However, we do not know whether the human brain applies the same predictive processes to non-human agents as it does to humans [16]. For example, one could expect that, as collaborative robots become more human-like, the quality and efficiency of human interactions with them would steadily increase. This is not always true. In the domain of social HMI, it has been shown that if robots resemble humans too closely, they are perceived as strange and unpleasant to interact with [17]. This effect is called the "uncanny valley" and is not limited to humans: other social primates also show adverse behavior towards realistic avatars [18]. This suggests that the primate brain may have hardwired neural systems allowing for intuitive discriminating of "natural" behavior. While the "uncanny valley" has been described for social HMI [19], virtually nothing is known about its impact on collaborative motor performance. Likewise, although it was previously reported [20] that humans operating assistive robots perform better if these robots follow human-like movement patterns (e.g., the relationship between curvature and speed), it is not known whether the same applies to scenarios where humans and cobots work autonomously (such as while cooperating).

Human actions are predictable in the sense that arm/joint configurations define the degrees of freedom of movement, allowing the brain to construct models of the other person's actions based on natural motor repertoire [21]. For observing robot actions, this is less obvious, as robotic arms do not have the default biomechanical design constraints the human arm has and can execute much more complex movements (such as 360-degree rotations). Yet, the correct prediction of the other agent's movements is needed for adapting one's own actions and, as such, efficient cooperation. The intuitiveness of the other agent's actions is of vital importance in situations where human cognitive effort has to be minimal, such as when under threat, stress, fatigue or heightened cognitive load. That is why it is important to understand how different robot designs (more or less human-like in terms of appearance and motion) might impact how humans perceive them and how this perception impacts manual collaboration.

Using VR allows testing human interactions with diverse virtual models (digital twins) of real cobots, including those popular in industry. Several cobot models, such as Baxter or Kinova, already have their digital twins extensively developed and implemented in different VR platforms, such as Unity 3D (Unity Technologies), including advanced motion planners and the physics of their virtual robotics limbs. The use of such digital twins allows, likewise, using the same robot-control-system (e.g., ROS) framework for controlling both the virtual and real industrial robots. Moreover, VR allows for the development of cobot models beyond the existing robot designs. This allows the testing of solutions not limited by the readily available technology and different, even hypothetical, robot models of different appearance or action patterns, such as in the study by Weistroffer et al. [22]. While these authors report a complex relationship between robot appearance, motion patterns, human performance, self-reports and physiological signals, it is important to emphasize that they did not measure more detailed performance indicators, such as human-motion-patterns (speed and accuracy) or eye-gaze data. Therefore, it remains to be further uncovered how robot anthropomorphism affects more subtle user performance.

Figure 1 shows VR robot models of increased anthropomorphism, similar to those used by Weistroffer et al. [22]. The bottom of Figure 1 shows an example VR collaboration scene with Baxter, in which the cobot passes a tool to the user, mimicking real interaction. Note that the robot has a face, a feature found on the real Baxter. Robot anthropomorphism, apart from possibly affecting human motor performance, can also affect higher-order cognitive aspects such as the feeling of presence (see, e.g., Dubosc et al. [23]), or attributing blame in the case of error (see, e.g., Furlough et al. [24]). The use of VR allows flexible manipulation of robot and scene designs to capture these cognitive aspects.

3. The Use of Virtual Reality for Testing Human–Robot Collaboration

Human–robot collaboration carries a physical risk to humans. For example, the robot arm can strike the operator or otherwise harm them. Therefore, operators can be stressed while collaborating with robots, which may result in their abnormal behavior, such as increased cognitive load or reduced motor performance [25]. Such dangerous scenarios are difficult to test in the natural world, implying significant limitations to user-experience testing of cobots.

In recent years, a feasible solution to test cobots while maintaining human safety is to use immersive VR environments [22,26,27]. Oyekan et al. [28] describe that, to design a collaborative environment to understand human reactions to both predictable and unpredictable robot motions, a virtual reality digital twin of a physical layout can be used. Dombrowski et al. [29] give us an interactive simulation of HRC—a technique that uses real-time physics simulation to immerse the design engineer or production planner inside a responsive virtual model of the factory—to optimize and validate manufacturing processes to achieve a better understanding of the risks and complexity of the assembly processes. Taken together, these studies demonstrate diverse approaches to testing different types of interaction scenarios and virtual cobots. This virtual testing can also be conducted for dangerous scenarios, but without putting humans at risk. For example, VR allows constructing scenes where the user is within the reach of a robot arm, thus collecting the user's psychophysiological measures and movement patterns in those simulated dangerous scenes, but without an actual risk to the user.

4. A Framework for Extended Reality in Testing Human–Robot Collaboration

In the field of software engineering and human–computer interaction (HCI), specific methodologies exist that guide the design cycles of novel solutions and products in those areas [30] (see, e.g., Sommerville [31]). In the area of HRC, those systematic approaches are scarce to non-existent. While, at present, such agile approaches exist in robot design [32,33], these assume a given robot type and rely primarily on user feedback, such as self-reports and other subjective measures of user experience. The use of such subjective measures is, however, not without problems, as we will discuss later.

An XR framework for designing and testing HRC allows for an iterative development process, arguably at a reduced cost, since different iterations could be developed and tested before real-world deployment. In tandem, the study of human comfort with the robot could become a central part of the design. A feedback loop between the development team and users can be implemented more easily when using virtual, as compared to real, cobot designs, leading to more agile cycles of design and redesign. This is particularly important, as it allows for the design, implementation and testing of collaboration models at higher levels of abstraction, without requiring to deal with low-level motor control and perceptual issues. The VR scenarios themselves may simulate a range of scenes, from those taking place in a factory to those of an assistive robot in a care facility. Such virtual scenarios can mimic realistic environments (such as a specific factory line) or hypothetical ones (such as a space station).

Given the flexibility of VR, in experiments, one can manipulate several variables relating to different aspects of collaboration scenes. Based on the literature reviewed here, spanning across years from 1997 to 2022 and queried through major scientific article databases, we identify features implemented or that are possible to implement in such scenes, and we classify them as variables about the cobot, the user and the environment. We summarize those variables in Table 2.

Table 2. Critical variables for cobot testing using virtual/extended reality are defined based on the literature reviewed here. We divide manipulated (independent) variables as those about either the cobot, the context, or the user. In each box, examples of each manipulated/measured variable are provided.

Critical Variables for HRC Experiments				
Manipulated (Independent)			Measured (Dependent)	
Cobot	Environment	User	Subjective	Objective
• Anthropomorphism • Presence of gaze • Speed • Accuracy • Fluidity • Proximity • Size	• Auditory noise • Scene type (e.g., factory) • Lighting	• Demographics (gender, age) • Cognitive load • Experience with technology	• Acceptability/trust ratings • Attributing blame • Sense of presence (realism)	• Physiological responses • Motor efficiency • Pupillometry

It is known that robot anthropomorphism influences the human's emotional and social perception of the robot, for example, the willingness to sacrifice it [34,35]. This is an important factor to consider when deciding on cobot design ergonomy, as different human emotional attitudes to the robot may influence collaboration efficiency in different situations (such as rescue operations). Onnasch and Roessler [35] provide a compelling taxonomy of different aspects of cobot design and their impact on human behavior in different interaction contexts.

The presence of cobot gaze is especially interesting, as it has been included on some cobots. For example, the company Rethink Robotics included gaze as a feature in their Baxter and Sawyer cobots to putatively increase cobot acceptability [36] (Kessler, 2017). This is because eye gaze is a critical element for human social life, as it communicates intentions [28], allowing the interaction partner to act accordingly on these intentions. Eye gaze predictively guides human hand actions [37] and is attracted to object affordances [38,39]. Gaze is also crucial for reading other agents' intentions [10]. Despite how important gaze is for collaboration between humans, to our knowledge, the question of how the presence of the cobot gaze impacts human movement parameters has not yet been investigated. In this way, whether the human brain relies on gaze in perceiving actions and intentions of non-human agents and whether this similarly informs human actions as other humans' gaze does, remains to be determined.

5. VR in Testing Cognitive and Social Aspects of Collaboration

Richards [40] proposes that the best way of achieving a higher level of collaboration between a human and robot is for the robot to mimic (to some extent or another) the behaviors of its human counterparts. In maintaining interaction efficiency, we need to understand the boundaries between human–robot capabilities, beliefs, intent and control. More specifically, we need to know how designers need to consider cognitive and social processes (e.g., trust, acceptability and attribution of blame) in an HRC for designing better cobots and collaboration conditions.

Trust is one of the requisites for building a successful human–robot collaboration [41]. It is the attitude that an agent will help achieve an individual's goals in a situation characterized by uncertainty and vulnerability [42]. Trust also represents a calculative orientation toward risk [43]; by trusting, we assume a potential gain, while by distrusting, we are avoiding a possible loss [44]. Research on trusting robots shows that the relationship between trust and joint physical coordination is critical when human workers interact with robots in a collaborative task [45]. In HRC contexts, "affective" trust better predicts the willingness to use a robot by human workers, and both types of trust—cognitive (e.g., reliability and predictability of the robot and robot attributes) and affective (e.g., proximity and personality)—are ensured by the statements of apology and competence that the robots manifest [46]. The technology (cobot) acceptance by humans in a collaborative workplace is a predictive factor of the success of the human–robot interaction [7]. The real-time trust that results from the study of Desai et al. [47] confirms traditional post-run survey approaches for human–robot trust can be masked by primacy–recency bias and demonstrate that early drops in reliability negatively impact real-time trust differently than middle or late drops. In agreement with the same authors, robot trust feedback can improve autonomy control allocation during low reliability without altering real-time trust levels. It should be noted that feedback interface designs using semantic symbols lead to more abrupt real-time trust changes than non-semantic symbols. The research of Oyekan et al. [28] suggests that greater autonomy for the robot will result in greater attribution of blame in work tasks. In general, the order of amount of blame was humans, robots, and environmental factors. If the scenario described the robot as nonautonomous, the participants attributed almost as little blame to them as to the environmental factors; in contrast, if the scenario described the robot as autonomous, the participants attributed almost as much blame to them as to the human.

The studies that aim to analyze the cognitive and social processes in technology demonstrated the importance of these topics to the correct implementation and actual usage of the same. For that reason, it is fundamental to understand which factors could increase the trust, acceptability, etc., of users in HRC. The use of VR allows for the manipulation of robot characteristics that might affect cognitive and social processes (c.f. Weistroffer et al. [22]). Moreover, it allows combining behavioral and subjective measures of cognitive, emotional and social human factors with their physiological markers to yield a full picture of human factors in HRC [22,48,49]. Table 3 summarizes identified issues in VR experiments on HRC and proposed remedies.

Table 3. Summary of identified issues in VR experiments on HRC and proposed remedies.

Issues in HRC Studies and Proposed Remedies for VR Experiments	
Problem	**Suggested Remedy**
1. Testing using real robots is limited to current designs only [22] 2. Low statistical power/small sample sizes [22,50] 3. Self-reports/physiological markers alone are not sensitive enough for assessing UX [50] 4. Results on HRC difficult to generalize across populations 5. Limited feeling of presence	1. Use VR models of hypothetical cobot designs to manipulate more variables 2. Increase samples (use statistical power calculators to determine sample size); increase within-subject repetitions 3. Combine self-reports with psychophysiological markers (heart rate, pupillometry, etc.); use standardized tools for measuring trust/acceptability; use time-resolved measures of stress, as provided by physiological markers 4. Test subject populations of diverse demographic characteristics (gender, age, experience with robots, etc.) 5. Increase immersion by using higher fidelity of stimuli, sound and haptic information

6. Combined Use of User Experience Questionnaires and Objective Measures in HRC

VR simulations of HRC tasks can be very complex and require subjects to grab objects handed to them by the robot, hand an object to the robot, or simultaneously/jointly with a cobot to reach for a target object [22]. However, the validity of VR simulations for HRC relies on three intertwined concepts: immersion, presence, and embodiment. Immersion is modulated by the quality of the sensory information given by VR systems and the amount to which their interaction can support users' sensorimotor contingencies (SCs) [51]. The better the immersion of a system, the higher the precision of the presentation of sensory stimuli (such as display resolution and field of view, sound and haptic information) and the more SCs supported (such as head, hand, arm, or full-body tracking). Immersion, in turn, has an impact on the experience of being there, on presence. Despite the lack of a consensual definition, presence might be defined as the psychological state in which a person reacts to a VE as they would in the physical world [52]. Presence is regarded to be the primary process that makes VR operate. However, there is no direct link between immersion and a sensation of being present. There is, however, widespread agreement that presence is a multi-component concept [53]. The sensation of presence, according to Slater, is based on the place illusion (PI)—the illusion of being there—and the plausibility illusion (PSI)—the believability of what is going on [51]. PSI is heavily reliant on the implemented VEs.

Hence, a VR system that ensures the necessary conditions for presence [54]—whereas PI is more directly related to the immersive features of a VR system with adequate immersive properties, embodiment and plausible and believable scenarios—can elicit behavioral and psychophysiological responses [55–57] consistent with real-world counterparts. Modern VR setups allow for relatively precise recordings of human-hand motion capture [58] with the feedback of the user's hand enabling a more or less embodied experience. The possibility of the inclusion of virtual models of anthropomorphic hands mimicking a user's own, as well as a variety of other end effectors (including different tools), allows for testing different levels of embodiment and their impact on collaborative situations, not limited to user's own body, like in real-life testing. Similarly, users' hand movements can indicate the levels of acceptability of motor cooperation with different cobot types. Natural hand velocity profiles for object-oriented movements are single-peak [59] and the presence of multiple peaks indicates a change in plan, such as that of adapting to cobot movement (e.g., Flash and Henis [60]). Analysis of velocity profiles is routinely used in motor neuroscience for assessing hand trajectory programming. Hand trajectories—in combination with hand speed, movement duration and precision—can be a good, objective indicator of human motor performance in collaborating with different types of cobots.

In addition, users can be wearing a haptic glove providing tactile sensation, to increase immersion and effectiveness. Human hand actions critically depend on the presence of haptic feedback [61]. In joint actions, forces applied to the object by each partner provide an important cue about their intentions, and the current state of the action and help coordination [62]. As pointed out by Bauer et al. [9], robot touch may also serve other communication purposes, important for establishing communication (such as a handshake); therefore, including it in VR scenes with cobots seems to be an important issue to be solved. While the use of haptic technologies significantly improves the embodiment of virtual scenes [58], haptics is not currently widespread due to the limited number of commercially naturalistic haptic interfaces.

Finally, the VR setting allows the recording of biological signals. These can be skin conductance (e.g., Weistroffer et al. [22]), heart rate (Etzi et al.; Weistroffer et al. [22,50]) and muscle activity, using respective sensors. The inclusion of these sensors allows for obtaining the objective metrics of user stress, independent of their self-reports (e.g., Etzi et al. [50]). Physiological responses recorded online, such as skin conductance level and heart rate variability, can be used to further detect stress levels, such as the activation of the fight or flight mechanism [63] in dangerous situations, e.g., when the virtual cobot hits the subject. This, together with participants' self-reports, provides a more in-depth perspective of cobot acceptability than questionnaires alone. For example, a situation where participants' positive self-reports are combined with physiological markers indicating stress would surface a more complex emotional state that could then be further disentangled [50].

Combining different psychophysiological and behavioral signals can be exploited by using machine learning tools to analyze them. This approach can help pinpoint subtle effects that user experience (UX) questionnaires would not be sensitive enough to measure. For example, it is possible to use a questionnaire to ask users about their level of stress/comfort with alternative cobots scenarios, after they have completed a series of tasks. However, the results of those questionnaires would not answer more important and interesting questions, such as: When did stress kick in? When were users most stressed? Were users stressed on the same task for each scenario or did some cause more/less stress? The retrospective nature of questionnaires means that the results that can be collected through them are too coarse-grained to accurately address more precise questions [64]. A portfolio of psychophysiological measures affords us the possibility of using more concrete measurements of the state of the human body to accompany post-fact questionnaires. This is particularly relevant in situations that could potentially involve risk and safety issues. Weistroffer et al. [22] demonstrated the feasibility of combining user questionnaires with physiological measures during human collaboration with virtual cobots of different levels of anthropomorphism and human-like vs. non-human-like effector velocity profiles. Their research showed that, while anthropomorphic robots gathered more user attention on their appearance, physiological signals did not reflect this effect. More recently, Etzi et al. [50] demonstrated that, while subjects physiological responses in a collaborative task did not indicate discomfort with changing robot velocity, their subjective self-reports did. Taking this integrative feedback approach and correlating psychophysiological measures with subjective questionnaires would provide us with a fuller and richer picture of what an ideal cobot scenario would be for a human, than that we would obtain from only the task performance data and subjective post-experience responses. Moreover, it allows avoiding subject responses to questionnaires to be driven mainly by their guessing of experimental demands, i.e., the demand characteristics of the VR scenario (c.f., McCambridge et al. [65]).

It is important to note that these above-mentioned studies integrating self-reports with physiological measures, employed subject samples smaller than in typical psychophysiological studies using similar methods. That is, Weistroffer et al. [22] used a sample of 13 subjects while Etzi et al. [50] had a sample size of 10. Using sample sizes this small has been repeatedly discussed in the relevant literature as one of the main reasons for low statistical power and difficulty in replicating findings (e.g., Button et al. [66]). For this reason, the failure to find physiological markers of stress when self-reports indicate it might result from low statistical power. Future studies using physiological markers should therefore employ higher sample sizes, e.g., determined by a-priori power analysis to warrant generalizability of their findings.

7. Telepresence and Teleoperation Scenarios

The use of VR provides an unprecedented opportunity to test cobots in hypothetical telepresence/teleoperation scenarios. Examples of use cases of teleoperation include factories, atomic power plants, assembly operations in space or the sea, and search and rescue operations [67,68]. Simulating remote presence based on HRIHRI is especially useful if the operating environment is hazardous and, therefore, placing a human operator at the site is not safe.

These scenarios may require shifting the user's point of view from their first-person perspective to a third-person perspective, or birds' eye view, which is common for teleoperation. It is important to consider how this shifting of perspective might affect the performance of, and/or cognitive load on, the human operator. Several authors have pinpointed the issues with embodiment in teleoperation when the operator directly controls the robot [69,70]. However, it remains to be determined how perspective and embodiment factors influence situations where the operator interacts with an otherwise autonomous robot.

Critically for teleoperation scenarios, VR allows using simulated delays and noise corruption and the responsive visual feedback being delayed or noisy in a way that emulates real-world teleoperation-related noise and streaming issues.

8. The Use of Augmented Reality

While virtual reality is based on creating an immersive digital environment, augmented reality (AR) provides an additional overlay enhancing the real world [71]. This usually has a form of an animated overlay over the visual scene, providing the user with additional information such as visual cues to the task, instrument parameters, etc. Such overlay can, for example, provide the operator with cues helping to establish joint attention (e.g., Marques et al. [11]). Such use of AR for cobot technology has been demonstrated in several studies. For example, Liu and Wang [72] explored the potential of AR as a worker support system in manufacturing tasks. They designed a system for assembly training and monitoring using AR. Above each assembly part and tool, a 3D text was displayed, providing assembly instructions to assemble objects in a specified sequence, together with a robot. A somewhat-similar concept was provided by Hietanen et al. [73], showing that AR overlays can be used to enhance user interaction with the production system, albeit with some limitations, demonstrating that currently available head-mounted displays might not be suitable for use in industry lines. On the more cognitive side of HRC, Palmirini et al. [74] developed an AR interface positively affecting human trust in cobots, as measured by psychometric methods. Michalos et al. [75] proposed in their study that, to improve operator's safety and acceptance in hybrid assembly environments, a tool using the immersion capabilities of AR technology must be applied.

Although the use of AR allows for enhancing HRC through the addition of virtual interfaces, cues, etc., its efficacy first needs to be tested. Such testing of AR interface can be easier to do in VR, where the virtual interfaces can be emulated in a range of scenarios as described above, and thereby would be tried against several options. This, in turn, results in an agile development of solutions that are not limited to a specific laboratory/experimental context.

9. Considering Operator Gender and Age in Cobot Testing

The use of VR has potential beyond the variety of simulated scenarios. The relatively flexible setting up and easy-to-use equipment allow testing a variety of subjects, including those from outside the pool of current cobot operators. This flexibility provides several opportunities in assessing how personal characteristics interact with different characteristics of cobots, yet research on personal attributes moderating human collaboration with robots is lacking, which seems an important gap to be filled.

One conceivable factor is the gender of the operator. Although evidence for a substantial influence of gender on motor actions and especially collaborative manual behavior is scarce, men and women differ in their upper arm and hand biomechanics, which translates to some visuomotor skills critical for collaboration, such as visuomotor coordination while using the upper arm [12,13]. As noted before, collaborative robots may have different anthropomorphic features. Yet, previous studies have shown that males were sensitive to the differences between robotic and anthropomorphic movements, while women largely ignored those differences [76,77]. For this reason, gender seems to potentially affect the measured motor efficiency of collaborating with cobots. We expect gender to further impact acceptability, stress and trust in at least some collaboration scenarios.

Similar to gender, age might play an important role in cobot acceptability, due to factors such as experience with technology, visuomotor abilities, etc. Cobot acceptability seems especially important in the context of assistive robots aimed at the older population, as this group of users seems to value the physical attractiveness and social likeability of robots more than their younger counterparts [78]. Furthermore, analysis of gaze behavior has shown that, while younger people pay attention to several body parts, older adults focus significantly more on the robot face [78]. In this way, it is possible that the use of eye gaze might increase the cobots' perceived friendliness and, likewise, the acceptance in a specific age or gender group. With the increasing presence of robots in areas that range from industrial plants to care homes, it becomes crucial to develop and fine–tune how human–robot collaboration takes place. To accommodate well for the personal characteristics of the individuals involved in this collaboration can be key not only to the collaboration effectiveness and efficiency but also to the quality of the interaction and experience between humans and robots.

10. Conclusions and Future Directions

Based on the current literature we can delineate several opportunities that the use of XR provides in advancing cobot research, development and deployment. We believe that, in the domain of development, the use of simulations and digital twins results in more agile development cycles for cobot solutions and flexibility in tested cobot designs. Such development would benefit from a general framework, highlighting important variables to consider in developing such simulations. In this paper, we propose what could be the backdrop for such a framework. We summarize critical variables in HRC VR experiments in Table 2 and provide a list of common issues and their proposed remedies in Table 3.

First of all, simulating diverse cobot designs, including hypothetical ones, allows for assessing cobot characteristics on operator efficiency and comfort without the restrictions posed by testing operators at the workplace with actual robots. Simulating diverse scenes and environments allows for assessing workplace and collaboration features, but foremost allows for safely testing dangerous scenarios, leveraging on immersive VR.

Testing operator performance can itself be performed using different measures, such as hand and eye motion tracking and physiological signals to yield objective and controlled performance measures. These measures can be combined with more traditional data, such as user self-reports and questionnaires, to construct a full picture of actual human interactions with collaborative robots. Future work should consider bigger sample sizes than those used to date, especially when measuring physiological signals.

Augmented reality has been used to enhance user performance and training in collaborating with cobots. The additional interface offered by AR can cue the user, e.g., about action sequences they are supposed to perform, but many more applications of the technology are conceivable in both training and generally improving human performance. The use of AR is very likely to increase as more cobots are deployed and, as such, research in this direction seems to have large potential.

Flexibility in designing scenes offered by VR can be also used to emulate remote operation, by introducing noise and delays typical for teleoperation scenarios. This gives an opportunity to expose/train operators in situations beyond cooperating with a robot in the same physical space. To date, studies on HRC and telepresence seem somewhat missing and we believe this direction has the potential to be explored further.

The use of XR opens up a whole array of possibilities to safely and quickly test cobot designs and collaboration scenarios without putting humans at the risk of harm. Modern XR technologies allow the integration of a wide variety of sensory modalities to create aware and immersive scenes. This way, testing cobots can be taken beyond the physical constraints of currently available cobot models and real-world settings. Furthermore, the development process can become more efficient by considering human reactions (i.e., psychological and physiological), leading to a more human-centered, holistic and efficient approach in human–robot collaboration.

Author Contributions: Conceptualization, A.P. (Artur Pilacinski), S.B.i.B., P.A.S., A.P. (Ana Pinto), J.A., P.M. and D.B.; writing—original draft preparation, A.P. (Artur Pilacinski), S.B.i.B., P.A.S., D.B.; writing—review and editing, A.P. (Artur Pilacinski), S.B.i.B., P.A.S., A.P. (Ana Pinto), C.C., D.B.; visualization, D.B., A.P. (Artur Pilacinski); supervision, A.P. (Artur Pilacinski), P.A.S., S.B.i.B.; project administration, A.P. (Artur Pilacinski), S.B.i.B.; funding acquisition, A.P. (Artur Pilacinski), S.B.i.B., J.A., D.B. All authors have read and agreed to the published version of the manuscript.

Funding: University of Coimbra: Sementes 2020 (A.P. (Artur Pilacinski)); Fundação para a Ciência e Tecnologia: PTDC/MHC-PCN/6805/2014 (A.P. (Artur Pilacinski)); NOVA-LINCS [PEest/UID/CEC/04516/2019] (S.B.i.B.); Fundação para a Ciência e Tecnologia: 2021.05646.BD (D.B.); FCT—Fundação para a Ciência e a Tecnologia UIDB/05037/2020 (A.P. (Ana Pinto)). European Research Council (ERC) under the European Union's Horizon 2020 research and innovation programme (Grant agreement No. 802553—"ContentMAP") (J.A.). APC were covered by the University of Coimbra.

Acknowledgments: The authors want to thank Ioannis Iossifidis and members of Proaction and Neurorehabilitation Labs for their valuable input.

Conflicts of Interest: The authors declare no conflict of interest.

References

1. Contreras, F.; Baykal, E.; Abid, G. E-Leadership and Teleworking in Times of COVID-19 and Beyond: What We Know and Where Do We Go. *Front. Psychol.* **2020**, *11*, 590271. [CrossRef]
2. Caselli, M.; Fracasso, A.; Traverso, S. Robots and Risk of COVID-19 Workplace Contagion: Evidence from Italy. *Technol. Forecast. Soc. Chang.* **2021**, *173*, 121097. [CrossRef]
3. Guizzo, E.; Klett, R. How Robots Became Essential Workers in the COVID-19 Response. Available online: https://spectrum.ieee.org/how-robots-became-essential-workers-in-the-covid19-response (accessed on 13 January 2022).
4. *IFR Position Paper Demystifying Collaborative Industrial Robots*; International Federation of Robotics: Frankfurt, Germany, 2018.
5. Towers-Clark, C. Keep The Robot In The Cage—How Effective (And Safe) Are Co-Bots? Available online: https://www.forbes.com/sites/charlestowersclark/2019/09/11/keep-the-robot-in-the-cagehow-effective--safe-are-co-bots/ (accessed on 13 January 2022).
6. Grosz, B.J. Collaborative Systems (AAAI-94 Presidential Address). *AI Mag.* **1996**, *17*, 67.
7. Bröhl, C.; Nelles, J.; Brandl, C.; Mertens, A.; Schlick, C.M. TAM Reloaded: A Technology Acceptance Model for Human-Robot Cooperation in Production Systems. In *Proceedings of the HCI International 2016 – Posters' Extended Abstracts*; Communications in Computer and Information Science, 617; Stephanidis, C., Ed.; Springer International Publishing: Cham, Switzerland, 2016; pp. 97–103.
8. Bauer, W.; Bender, M.; Braun, M.; Rally, P.; Scholtz, O. Lightweight Robots in Manual Assembly–Best to Start Simply. In *Examining Companies Initial Experiences with Lightweight Robots*; Frauenhofer-Institut für Arbeitswirtschaft und Organisation IAO: Stuttgart, Germany, 2016; pp. 1–32.
9. Bauer, A.; Wollherr, D.; Buss, M. Human–robot collaboration: A survey. *Int. J. Humanoid Robot.* **2008**, *05*, 47–66. [CrossRef]
10. Zuberbühler, K. Gaze Following. *Curr. Biol. CB* **2008**, *18*, R453–R455. [CrossRef]
11. Marques, B.; Silva, S.S.; Alves, J.; Araujo, T.; Dias, P.M.; Sousa Santos, B. A Conceptual Model and Taxonomy for Collaborative Augmented Reality. *IEEE Trans. Vis. Comput. Graph.* **2021**, 1–21. [CrossRef]
12. Gromeier, M.; Koester, D.; Schack, T. Gender Differences in Motor Skills of the Overarm Throw. *Front. Psychol.* **2017**, *8*, 212. [CrossRef]
13. Moreno-Briseño, P.; Díaz, R.; Campos-Romo, A.; Fernandez-Ruiz, J. Sex-Related Differences in Motor Learning and Performance. *Behav. Brain Funct.* **2010**, *6*, 74. [CrossRef]
14. Blakemore, S.-J.; Decety, J. From the Perception of Action to the Understanding of Intention. *Nat. Rev. Neurosci.* **2001**, *2*, 561–567. [CrossRef]
15. Eaves, D.L.; Riach, M.; Holmes, P.S.; Wright, D.J. Motor Imagery during Action Observation: A Brief Review of Evidence, Theory and Future Research Opportunities. *Front. Neurosci.* **2016**, *10*, 514. [CrossRef]
16. Martin, A.; Weisberg, J. Neural foundations for understanding social and mechanical concepts. *Cogn. Neuropsychol.* **2003**, *20*, 575–587. [CrossRef]
17. MacDorman, K.F.; Green, R.D.; Ho, C.-C.; Koch, C.T. Too Real for Comfort? Uncanny Responses to Computer Generated Faces. *Comput. Hum. Behav.* **2009**, *25*, 695–710. [CrossRef]
18. Steckenfinger, S.A.; Ghazanfar, A.A. Monkey Visual Behavior Falls into the Uncanny Valley. *Proc. Natl. Acad. Sci. USA* **2009**, *106*, 18362–18366. [CrossRef]
19. Kahn, P.H.; Ishiguro, H.; Friedman, B.; Kanda, T.; Freier, N.G.; Severson, R.L.; Miller, J. What Is a Human?: Toward Psychological Benchmarks in the Field of Human–Robot Interaction. *Interact. Stud. Soc. Behav. Commun. Biol. Artif. Syst.* **2007**, *8*, 363–390. [CrossRef]
20. Maurice, P.; Huber, M.E.; Hogan, N.; Sternad, D. Velocity-Curvature Patterns Limit Human–Robot Physical Interaction. *IEEE Robot. Autom. Lett.* **2018**, *3*, 249–256. [CrossRef]
21. Spüler, M.; Niethammer, C. Error-Related Potentials during Continuous Feedback: Using EEG to Detect Errors of Different Type and Severity. *Front. Hum. Neurosci.* **2015**, *9*, 155. [CrossRef]
22. Weistroffer, V.; Paljic, A.; Callebert, L.; Fuchs, P. A Methodology to Assess the Acceptability of Human-Robot Collaboration Using Virtual Reality. In *Proceedings of the the 19th ACM Symposium on Virtual Reality Software and Technology, Singapore, 6–9 October 2013*; ACM Press: New York, NY, USA, 2013; pp. 39–48.

23. Dubosc, C.; Gorisse, G.; Christmann, O.; Fleury, S.; Poinsot, K.; Richir, S. Impact of Avatar Facial Anthropomorphism on Body Ownership, Attractiveness and Social Presence in Collaborative Tasks in Immersive Virtual Environments. *Comput. Graph.* **2021**, *101*, 82–92. [CrossRef]

24. Furlough, C.; Stokes, T.; Gillan, D.J. Attributing Blame to Robots: I. The Influence of Robot Autonomy. *Hum. Factors J. Hum. Factors Ergon. Soc.* **2021**, *63*, 592–602. [CrossRef]

25. Rabby, K.M.; Khan, M.; Karimoddini, A.; Jiang, S.X. An Effective Model for Human Cognitive Performance within a Human-Robot Collaboration Framework. In Proceedings of the 2019 IEEE International Conference on Systems, Man and Cybernetics (SMC), Bari, Italy, 6–9 October 2019; pp. 3872–3877. [CrossRef]

26. Dianatfar, M.; Latokartano, J.; Lanz, M. Review on Existing VR/AR Solutions in Human–Robot Collaboration. *Procedia CIRP* **2021**, *97*, 407–411. [CrossRef]

27. Duguleana, M.; Barbuceanu, F.G.; Mogan, G. Evaluating Human-Robot Interaction during a Manipulation Experiment Conducted in Immersive Virtual Reality. In Proceedings of the International Conference on Virtual and Mixed Reality, Orlando, FL, USA, 9–14 July 2011; pp. 164–173.

28. Oyekan, J.O.; Hutabarat, W.; Tiwari, A.; Grech, R.; Aung, M.H.; Mariani, M.P.; López-Dávalos, L.; Ricaud, T.; Singh, S.; Dupuis, C. The Effectiveness of Virtual Environments in Developing Collaborative Strategies between Industrial Robots and Humans. *Robot. Comput.-Integr. Manuf.* **2019**, *55*, 41–54. [CrossRef]

29. Dombrowski, U.; Stefanak, T.; Perret, J. Interactive Simulation of Human-Robot Collaboration Using a Force Feedback Device. *Procedia Manuf.* **2017**, *11*, 124–131. [CrossRef]

30. Holtzblatt, K.; Beyer, H. *Contextual Design*, 2nd ed.; Morgan Kaufmann: Burlington, MA, USA, 2016; ISBN 978-0-12-801136-2.

31. Sommerville, I. *Software Engineering*, 10th ed.; Pearson: Boston, MA, USA, 2016; ISBN 978-0-13-394303-0.

32. Tonkin, M.; Vitale, J.; Herse, S.; Williams, M.A.; Judge, W.; Wang, X. *Design Methodology for the UX of HRI: A Field Study of a Commercial Social Robot at an Airport*; ACM Press: New York, NY, USA, 2018; ISBN 978-1-4503-4953-6.

33. Zhong, V.J.; Schmiedel, T. A User-Centered Agile Approach to the Development of a Real-World Social Robot Application for Reception Areas. In Proceedings of the Companion of the 2021 ACM/IEEE International Conference on Human-Robot Interaction, Boulder, CO, USA, 8–11 March 2021; pp. 76–80.

34. Onnasch, L.; Roesler, E. Anthropomorphizing Robots: The Effect of Framing in Human-Robot Collaboration. *Proc. Hum. Factors Ergon. Soc. Annu. Meet.* **2019**, *63*, 1311–1315. [CrossRef]

35. Onnasch, L.; Roesler, E. A Taxonomy to Structure and Analyze Human–Robot Interaction. *Int. J. Soc. Robot.* **2021**, *13*, 833–849. [CrossRef]

36. Kessler, S. This Industrial Robot Has Eyes Because They Make Human Workers Feel More Comfortable. Available online: https://qz.com/958335/why-do-rethink-robotics-robots-have-eyes/ (accessed on 13 January 2022).

37. Johansson, R.S.; Westling, G.; Bäckström, A.; Flanagan, J.R. Eye–Hand Coordination in Object Manipulation. *J. Neurosci.* **2001**, *21*, 6917–6932. [CrossRef]

38. Osiurak, F.; Rossetti, Y.; Badets, A. What Is an Affordance? 40 Years Later. *Neurosci. Biobehav. Rev.* **2017**, *77*, 403–417. [CrossRef]

39. Pilacinski, A.; De Haan, S.; Donato, R.; Almeida, J. Tool Heads Prime Saccades. *Sci. Rep.* **2021**, *11*, 11954. [CrossRef]

40. Richards, D. Escape from the Factory of the Robot Monsters: Agents of Change. *Team Perform. Manag. Int. J.* **2017**, *23*, 96–108. [CrossRef]

41. Salem, M.; Lakatos, G.; Amirabdollahian, F.; Dautenhahn, K. Would You Trust a (Faulty) Robot? Effects of Error, Task Type and Personality on Human-Robot Cooperation and Trust. In In Proceedings of the 2015 10th ACM/IEEE International Conference on Human-Robot Interaction (HRI), Portland, OR, USA, 2–5 March 2015; pp. 141–148. [CrossRef]

42. Lee, J.D.; See, K.A. Trust in Automation: Designing for Appropriate Reliance. *Hum. Factors J. Hum. Factors Ergon. Soc.* **2004**, *46*, 50–80. [CrossRef]

43. Bacharach, M.; Guerra, G.; Zizzo, D.J. The Self-Fulfilling Property of Trust: An Experimental Study. *Theory Decis.* **2007**, *63*, 349–388. [CrossRef]

44. Gambetta, D. Can We Trust? In *Trust: Making and Breaking Cooperative. Relations, Electronic Edition*; Department of Sociology, University of Oxford: Oxford, UK, 2000; Volume 13.

45. Wang, Y.; Lematta, G.J.; Hsiung, C.-P.; Rahm, K.A.; Chiou, E.K.; Zhang, W. Quantitative Modeling and Analysis of Reliance in Physical Human–Machine Coordination. *J. Mech. Robot.* **2019**, *11*, 060901. [CrossRef]

46. Cameron, D.; Collins, E.; Cheung, H.; Chua, A.; Aitken, J.M.; Law, J. Don't Worry, We'll Get There: Developing Robot Personalities to Maintain User Interaction after Robot Error. In *Conference on Biomimetic and Biohybrid Systems*; Springer: Cham, Switzerland, 2016; pp. 409–412. [CrossRef]

47. Desai, M.; Kaniarasu, P.; Medvedev, M.; Steinfeld, A.; Yanco, H. Impact of Robot Failures and Feedback on Real-Time Trust. In Proceedings of the 2013 8th ACM/IEEE International Conference on Human-Robot Interaction (HRI), Tokyo, Japan, 3–6 March 2013; pp. 251–258. [CrossRef]

48. Gupta, K.; Hajika, R.; Pai, Y.S.; Duenser, A.; Lochner, M.; Billinghurst, M. In AI We Trust: Investigating the Relationship between Biosignals, Trust and Cognitive Load in VR. In Proceedings of the VRST '19: 25th ACM Symposium on Virtual Reality Software and Technology, Parramatta, Australia, 12–15 November 2019; pp. 1–10.

49. Gupta, K.; Hajika, R.; Pai, Y.S.; Duenser, A.; Lochner, M.; Billinghurst, M. Measuring Human Trust in a Virtual Assistant Using Physiological Sensing in Virtual Reality. In Proceedings of the 2020 IEEE Conference on Virtual Reality and 3D User Interfaces, VR, Atlanta, GA, USA, 22–26 March 2020; pp. 756–765. [CrossRef]

50. Etzi, R.; Huang, S.; Scurati, G.W.; Lyu, S.; Ferrise, F.; Gallace, A.; Gaggioli, A.; Chirico, A.; Carulli, M.; Bordegoni, M. Using Virtual Reality to Test Human-Robot Interaction During a Collaborative Task. In Proceedings of the International Design Engineering Technical Conferences and Computers and Information in Engineering Conference, Anaheim, CA, USA, 18–21 August 2019; p. V001T02A080.

51. Slater, M. Place Illusion and Plausibility Can Lead to Realistic Behaviour in Immersive Virtual Environments. *Philos. Trans. R. Soc. B Biol. Sci.* **2009**, *364*, 3549–3557. [CrossRef] [PubMed]

52. Slater, M.; Wilbur, S. A Framework for Immersive Virtual Environments (FIVE): Speculations on the Role of Presence in Virtual Environments. *Presence Teleoperators Virtual Environ.* **1997**, *6*, 603–616. [CrossRef]

53. Baños, R.M.; Botella, C.; Alcañiz, M.; Liaño, V.; Guerrero, B.; Rey, B. Immersion and Emotion: Their Impact on the Sense of Presence. *Cyberpsychol. Behav.* **2004**, *7*, 734–741. [CrossRef] [PubMed]

54. Slater, M.; Lotto, B.; Arnold, M.M.; Sánchez-Vives, M.V. How We Experience Immersive Virtual Environments: The Concept of Presence and Its Measurement. *Anu Psicol* **2009**, *40*, 193–210.

55. Slater, M.; Antley, A.; Davison, A.; Swapp, D.; Guger, C.; Barker, C.; Pistrang, N.; Sanchez-Vives, M.V. A Virtual Reprise of the Stanley Milgram Obedience Experiments. *PLoS ONE* **2006**, *1*, e39. [CrossRef] [PubMed]

56. Slater, M.; Rovira, A.; Southern, R.; Swapp, D.; Zhang, J.J.; Campbell, C.; Levine, M. Bystander Responses to a Violent Incident in an Immersive Virtual Environment. *PLoS ONE* **2013**, *8*, e52766. [CrossRef] [PubMed]

57. Martens, M.A.; Antley, A.; Freeman, D.; Slater, M.; Harrison, P.J.; Tunbridge, E.M. It Feels Real: Physiological Responses to a Stressful Virtual Reality Environment and Its Impact on Working Memory. *J. Psychopharmacol.* **2019**, *33*, 1264–1273. [CrossRef] [PubMed]

58. Buckingham, G. Hand Tracking for Immersive Virtual Reality: Opportunities and Challenges. *Front. Virtual Real.* **2021**. [CrossRef]

59. Morasso, P. Spatial Control of Arm Movements. *Exp. Brain Res.* **1981**, *42*, 223–227. [CrossRef]

60. Flash, T.; Henis, E. Arm Trajectory Modifications During Reaching Towards Visual Targets. *J. Cogn. Neurosci.* **1991**, *3*, 220–230. [CrossRef]

61. Rao, A.K.; Gordon, A.M. Contribution of Tactile Information to Accuracy in Pointing Movements. *Exp. Brain Res.* **2001**, *138*, 438–445. [CrossRef]

62. Kosuge, K.; Kazamura, N. Control of a Robot Handling an Object in Cooperation with a Human. In Proceedings of the 6th IEEE International Workshop on Robot and Human Communication. RO-MAN'97 SENDAI, Sendai, Japan, 29 September–1 October 1997; pp. 142–147.

63. Bradley, M.M.; Codispoti, M.; Cuthbert, B.N.; Lang, P.J. Emotion and Motivation I: Defensive and Appetitive Reactions in Picture Processing. *Emot. Wash. DC* **2001**, *1*, 276–298. [CrossRef]

64. Lazar, J.; Feng, J.H.; Hochheiser, H. *Research Methods in Human-Computer Interaction*, 2nd ed.; Morgan Kaufmann Publishers: Burlington, MA, USA, 2017; ISBN 978-0-12-809343-6.

65. McCambridge, J.; de Bruin, M.; Witton, J. The Effects of Demand Characteristics on Research Participant Behaviours in Non-Laboratory Settings: A Systematic Review. *PLoS ONE* **2012**, *7*, e39116. [CrossRef] [PubMed]

66. Button, K.S.; Ioannidis, J.P.A.; Mokrysz, C.; Nosek, B.A.; Flint, J.; Robinson, E.S.J.; Munafò, M.R. Power Failure: Why Small Sample Size Undermines the Reliability of Neuroscience. *Nat. Rev. Neurosci.* **2013**, *14*, 365–376. [CrossRef] [PubMed]

67. Tachi, S. From 3D to VR and Further to Telexistence. In Proceedings of the 2013 23rd International Conference on Artificial Reality and Telexistence (ICAT), Tokyo, Japan, 11–13 December 2013; pp. 1–10.

68. Kim, J.-H.; Starr, J.W.; Lattimer, B.Y. Firefighting Robot Stereo Infrared Vision and Radar Sensor Fusion for Imaging through Smoke. *Fire Technol.* **2015**, *51*, 823–845. [CrossRef]

69. Ewerton, M.; Arenz, O.; Peters, J. Assisted Teleoperation in Changing Environments with a Mixture of Virtual Guides. *Adv. Robot.* **2020**, *34*, 1157–1170. [CrossRef]

70. Toet, A.; Kuling, I.A.; Krom, B.N.; van Erp, J.B.F. Toward Enhanced Teleoperation Through Embodiment. *Front. Robot. AI* **2020**, *7*, 14. [CrossRef]

71. Cipresso, P.; Giglioli, I.A.C.; Raya, M.A.; Riva, G. The Past, Present, and Future of Virtual and Augmented Reality Research: A Network and Cluster Analysis of the Literature. *Front. Psychol.* **2018**, *9*, 2086. [CrossRef]

72. Liu, H.; Wang, L. An AR-Based Worker Support System for Human-Robot Collaboration. *Procedia Manuf.* **2017**, *11*, 22–30. [CrossRef]

73. Hietanen, A.; Pieters, R.; Lanz, M.; Latokartano, J.; Kämäräinen, J.-K. AR-Based Interaction for Human-Robot Collaborative Manufacturing. *Robot. Comput.-Integr. Manuf.* **2020**, *63*, 101891. [CrossRef]

74. Palmarini, R.; del Amo, I.F.; Bertolino, G.; Dini, G.; Erkoyuncu, J.A.; Roy, R.; Farnsworth, M. Designing an AR Interface to Improve Trust in Human-Robots Collaboration. *Procedia CIRP* **2018**, *70*, 350–355. [CrossRef]

75. Michalos, G.; Karagiannis, P.; Makris, S.; Tokçalar, Ö.; Chryssolouris, G. Augmented Reality (AR) Applications for Supporting Human-Robot Interactive Cooperation. *Procedia CIRP* **2016**, *41*, 370–375. [CrossRef]

76. Abel, M.; Kuz, S.; Patel, H.J.; Petruck, H.; Schlick, C.M.; Pellicano, A.; Binkofski, F.C. Gender Effects in Observation of Robotic and Humanoid Actions. *Front. Psychol.* **2020**, *11*, 797. [CrossRef] [PubMed]
77. Nomura, T. Robots and Gender. *Gend. Genome* **2017**, *1*, 18–26. [CrossRef]
78. Oh, Y.H.; Ju, D.Y. Age-Related Differences in Fixation Pattern on a Companion Robot. *Sensors* **2020**, *20*, 3807. [CrossRef]

Article

Gaze-Based Interaction Intention Recognition in Virtual Reality

Xiao-Lin Chen [1,2] and Wen-Jun Hou [1,3,*]

1   Beijing Key Laboratory of Network Systems and Network Culture, Beijing University of Posts and Telecommunications, Beijing 100876, China; cxl95@163.com
2   School of Automation, Beijing University of Posts and Telecommunications, Beijing 100876, China
3   School of Digital Media and Design Arts, Beijing University of Posts and Telecommunications, Beijing 100876, China
*   Correspondence: hou1505@163.com

Abstract: With the increasing need for eye tracking in head-mounted virtual reality displays, the gaze-based modality has the potential to predict user intention and unlock intuitive new interaction schemes. In the present work, we explore whether gaze-based data and hand-eye coordination data can predict a user's interaction intention with the digital world, which could be used to develop predictive interfaces. We validate it on the eye-tracking data collected from 10 participants in item selection and teleporting tasks in virtual reality. We demonstrate successful prediction of the onset of item selection and teleporting with an 0.943 F_1-Score using a Gradient Boosting Decision Tree, which is the best among the four classifiers compared, while the model size of the Support Vector Machine is the smallest. It is also proven that hand-eye-coordination-related features can improve interaction intention recognition in virtual reality environments.

Keywords: intention prediction; virtual reality; gaze-based interaction

Citation: Chen, X.-L.; Hou, W.-J. Gaze-Based Interaction Intention Recognition in Virtual Reality. *Electronics* **2022**, *11*, 1647. https://doi.org/10.3390/electronics11101647

Academic Editors: Jorge C. S. Cardoso, André Perrotta, Paula Alexandra Silva and Pedro Martins

Received: 15 April 2022
Accepted: 18 May 2022
Published: 21 May 2022

Publisher's Note: MDPI stays neutral with regard to jurisdictional claims in published maps and institutional affiliations.

1. Introduction

The Metaverse has recently attracted a great deal of attention in industry and academia, especially after Facebook changed its name to Meta. If the Metaverse is realized in the future, extended reality technology, including virtual reality technology, will be one of its essential supporting technologies. Biocca and Delaney [1] define virtual reality (VR) as "the sum of the hardware and software systems that seek to perfect an all-inclusive, sensory illusion of being present in another environment". The core characteristics of VR are immersion, interaction and imagination [2]. Immersion and interaction mean higher requirements for human–computer interaction in VR systems. Interaction should be more natural and intuitive. The first step is to identify and understand the interaction that the user wants to perform so that the system can provide appropriate help in time. Interaction intent recognition enables the system to provide shortcuts to the user by predicting the intended interaction, facilitating the interaction, and reducing the operational load of the user. For example, if the system knows what object the user would like to interact with within the virtual environment, it can connect a certain input command to the inferred interaction target and allow the user to complete the entire interaction without manual pointing, which can greatly reduce the physical and cognitive load of the user. Especially under the concept of the Metaverse, 24/7-wearable AR and VR devices for production and work are facing the problem that prolonged usage can exacerbate fatigue, so adaptive interaction interface that can accurately predict the interaction intention of the user has the potential to reinvent human–computer interaction under extended reality.

Research on the application of eye tracking in VR and human–computer interaction began early [3], but has not been widely used due to the cost and accuracy of the eye-tracking equipment. In 2017, we witnessed the acquisitions of companies that can provide eye-tracking technology by well-known companies in VR and augmented reality, highlighting the importance of eye tracking in this field. In 2019, companies such as FOVE Inc.,

Microsoft and HTC had already provided systems with built-in eye-tracking for professional and consumer markets. The applications of eye movements in VR fall into four main categories [4]: diagnostic (eye-movement behavior analysis), active (as a human–computer interface), passive (gaze-contingent rendering), and expressive (synthesizing eye movements of virtual avatars). This research mainly focuses on active applications; that is, eye movement as a human–computer interface.

A drawback in gaze-based interfaces is the Midas touch problem, i.e., unintentionally activated commands while the user is looking at an interactive element [5]. Fixation or dwell time is an indicator of an intention of the user to select an object through eye gaze alone [6–9]. However, this time threshold can negatively impact the user experience. For example, when the required dwell time is too short, it puts pressure on the user to look away and avoid unwanted selection. On the contrary, it may result in a longer wait time if it is too long [10]. If the interaction intention of the user can be recognized through natural eye-movement behavior rather than intentional, the mental and operational load of the user can be greatly reduced. Another common way to avoid the Midas touch problem is using a physical trigger as a confirmation mechanism, such as a hand controller or keyboard [6,8,11–13]. In such a case, it also makes sense to recognize the interaction intent to simplify physical buttons' operation or give more information as visual feedback based on the recognition result.

The eye has been said to be a mirror to the soul or window into the brain. This may be the first reason eye movements have attracted researchers' interest. There are many studies related to eye movements in the field of attention [14–18]. Eye movements can indicate areas of interest (active or passive attraction) and quantify the changes in human attention. Therefore, they are widely used in visual attention modeling. Eye movements can also reflect human perception [19], cognitive state [20,21], decision-making processes [22,23], and working memory [15]. Eye movements have also been used in studies of human activity classification [24–27], especially in human–computer interaction [24,27–32].

These studies have demonstrated that human eye-movement behavior can be significantly different across activities. All of the above studies focus on understanding human behavior and thinking through eye movements, which is a prerequisite and basis for the application of eye movements in intention recognition. Gaze behavior reflects cognitive processes and can give hints of our thinking and intentions.

An intention is an idea or plan of what you will do. A great deal of existing gaze-based intention recognition research aims to recognize the intention of daily human behavior [25,26,33–35] or higher-level intention involving game strategy [36]. The interaction intention in this study is the way that the user wants to interact with the computer system, i.e., to identify the interaction intention of the user before he/she performs the actual interaction. However, the interaction intent we want to identify here is low-level intent; more specifically, the intent to perform an interaction without involving complex contextual relationships and specific interaction environments. Similar to the task-independent interaction intent prediction studied by Brendan et al. [37], the application context of our study is in VR.

Our approach tracks the eye movements of the user in controller-based interaction in VR and fuses the eye movements and hand-eye coordination information collected via gaze and controller to predict the current intention of the user. Briefly, our research is conducted as follows. Initially, we collect controller and gaze data in two controller-based interaction tasks in VR (selection and teleporting) and build a multimodality database. We then extract gaze-based features from this database and train intention recognition models using supervised machine-learning techniques. Finally, we use a separate dataset to verify the accuracy of our models. The main contributions of this paper are as follows:

- We introduce a new dataset of human interaction intentions behind human gaze and hand behaviors. It contains gaze-and controller-related data of selection and teleporting in VR from multiple participants.

- We propose a gaze-controller-based feature-set representation based on human vision and behavioral studies to predict user intention through the gaze. These features are neither subject nor interface specific.
- We train four classifiers with supervised machine-learning and evaluate them in several aspects, including F_1-Score and model size. In addition, we perform feature selection to assess the relevance and redundancy of feature representations. The experimental results show that for behaviors from different people, the Gradient Boosting Decision Tree (GBDT) approach achieves F_1-Score of 0.924 for binary classification and 0.953 for three-class classification. Such results offer the possibility of a more natural implementation of the interaction interface paradigm, i.e., more intelligent delivery of low-cost interaction patterns by providing the right interventions at the right time.

Section 2 gives an outline of state-of-the-art gaze-based intention recognition studies. Our approach consists of three major parts: data collection, feature extraction, and intention recognition. They are detailed in Section 3. Section 4 compares and analyzes different classifiers' classification performance and feature importance. Section 5 includes a discussion of our work and a summary of future directions. Section 6 concludes our work.

2. Related Work

The term intent has different definitions in different fields. To avoid ambiguity, the term interaction intention in this study needs to be clarified. In human–computer interaction, the intent is either explicit or implicit. An explicit intent is directly input into the system through the interaction interface. Implicit intent involves the internal activities of users. It requires the system to infer the intentions based on some hints such as natural facial expressions, behaviors, and eye movements. This is a key feature of intelligent interactive interfaces, i.e., understanding the current state of users and predicting the following action. The ultimate goal of our research is to enable computer systems, like humans, to understand and predict users' behavior and purpose for intuitive and safe interaction. Van-Horenbeke and Peer [38] explore human behavior, planning, and goal (intent) recognition as a holistic problem. They argue that behaviors and goals are incremental in granularity (i.e., a series of behaviors constitute intentions) and in time (i.e., behavior recognition focuses more on actions that occur simultaneously, while intention recognition focuses on upcoming actions). On the other hand, planning is more complex, focusing more on the relationship between a series of behaviors or intentions and the specific meaning in the semantic context in the interaction. In our study, interaction intention recognition is the least fine-grained intention recognition. Let us consider the action of pressing a button. The expected interaction result behind the series of actions, including finding a specific location and pressing it, is the "interaction intent" in this study, i.e., selection or teleporting. We do not consider the deeper intent of winning a game or switching to a better visual perspective, i.e., the interaction intent is relatively weakly linked to the semantic context of the interaction.

Eye movements are a common source of information in intention or behavior recognition. Table 1 summarizes the research on using eye-related data to classify daily behaviors and intention classification in computer environments. According to the table, the most commonly used classification algorithms include Support Vector Machine (SVM), Logistic Regression (LR), and Random Forest (RF). Our study also chooses to perform a cross-sectional comparison of these classification algorithms. These studies are also aimed at different environments. The application environments of the above studies are mainly personal computers or tablets, and there are relatively few studies in VR. Our study is to recognize interaction intention of the user in controller-based interaction in VR based on eye-movement data.

Table 1. Task, activity, and intention classification studies using eye movement data.

Reference	Year	Platform	Scope	Classifier	Performance	Tasks/Activities/Intentions
[39]	2014	PC	Intention recognition	Nearest Neighborhood (NN) Support Vector Machine (SVM)	Average accuracy: 79.81 ± 4.93 Average accuracy: 85.26 ± 0.70	Navigational intent Informational intent
[40]	2014	PC	Intention recognition	Support Vector Machine (SVM)	Average accuracy: 90%	Navigational intent Informational intent
[41]	2017	PC	Intention recognition	Nearest Neighborhood (NN) Support Vector Machine (SVM)	Average accuracy: 85%	Unintentional intention Purposeful intention
[42]	2012	PC	Intention prediction	Support Vector Machine (SVM)	ROC-AUC: 0.807 Accuracy: 76%	Issue a command or not
[43]	2018	PC	Intention prediction	Support Vector Machine (SVM)	Accuracy: 77.2%	Monitoring Tracking Decision Burst Off loop
[36]	2013	PC	Cognitive states prediction	Support Vector Machine (SVM)	Best accuracy: 32 %	8-tiles puzzle game: Cognitions Evaluations Plans Intentions Current move
[28]	2004	PC	Activity recognition	-	-	Reading comprehension Mathematical reasoning Searching Object manipulation
[29]	2011	PC	Activity recognition	LHMM	Accuracy: 51.3% Accuracy: 89.1%	Evaluate website traffic task E-Learning quiz task
[30]	2013	PC	Activity recognition	Logistic Regression	Average accuracy: 53.18%	Retrieve values Filter Compute derived value Find extremum Sort
[24]	2018	PC	Activity recognition	Support Vector Machine (SVM) K-Nearest Neighbour (K-NN) Random Forest	F1 score: SVM 0.71 K-NN 0.61 Random Forest 0.73	Read Watch Browse Play Search Interpret Debug Write
[44]	2015	Reality	Intention prediction	Support Vector Machine (SVM)	Accuracy: 76%	Making sandwich
[33]	2009	Reality	Activity recognition	Support Vector Machine (SVM)	Average precision: 76.1% Average recall: 70.5%	Copy Read Write Video Browse Null
[34]	2011	Reality	Activity recognition	Support Vector Machine (SVM)	Average accuracy: 80.2% Average precision: 76.1% Average recall: 70.%	Reading or not reading Copy, read, write video, browse, null Visual memory (familiar/unfamiliar images)
[35]	2012	Reality	Activity recognition	Support Vector Machine (SVM)	Mean average precision 57%	Copy Read Write Video Browse Null
[25]	2019	Reality	Activity recognition	Random Forest	Average accuracy 67%	Common navigation tasks: Self-positioning and orientation. Local environment target search Map target search Route memorization Walking to the destination
[26]	2020	Reality	Activity recognition	CNN	Average Precision: 40.41%	26 common action classes
[45]	2015	Tabletop	Intention prediction	Support Vector Machine (SVM)	88% success rate	Drag Maximize Minimize Scroll Free-form drawing

Table 1. *Cont.*

Reference	Year	Platform	Scope	Classifier	Performance	Tasks/Activities/Intentions
[39]	2014	PC	Intention recognition	Nearest Neighborhood (NN) Support Vector Machine (SVM)	Average accuracy: 79.81 ± 4.93 Average accuracy: 85.26 ± 0.70	Navigational intent Informational intent
[46]	2019	VR	Intention prediction	Long Short-Term Memory (LSTM) Topology	Accuracy 99.94% Precision 99.92% Recall 99.96% F_1-Score 99.94%	Navigation: Needing navigation aid No need for navigation aid
[37]	2021	VR	Intention prediction	Logistic Regression	Average PR-AUC = 0.12 Average ROC-AUC = 0.77	Issue a command or not
[27]	2020	VR	Activity recognition	Support Vector Machine (SVM) Logistic Regression Random Forest	Prediction accuracy: SVM: 80.23% Logistic Regression: 74.74% Random Forest: 79.50%	Shopping Goal-directed search Exploratory search

Alghofaili et al. [46] classify whether users need navigation assistance in VR environments through Long Short-Term Memory (LSTM) topology. It determines whether the user loses his/her way by analyzing the eye-movement behavior of the user in VR roaming scenarios. Pfeiffer et al. [27] classify the type of search (goal or exploration based) when shopping in cave-based VR. Their study also relies mainly on eye-movement data for training and evaluating three classifiers: SVM, LR, and RF, where SVM has the highest accuracy of 80.2%.

The most similar work to our study is the work of Brendan et al. [37]. Their study predicts whether a user will make a selection interaction or not in VR. In their study, a separate LR classifier is trained for each participant, but the overall results are not very satisfactory, with an average PR-AUC of 0.12. However, in their study, they also find that the classifiers for participants are very similar in terms of feature selection, which to the extent indicates that the interaction intention of the user is common in eye movement-based features. There is some commonality in the eye movement-based features. Therefore, the training dataset in our study is composed of eye-movement data and controller data generated by multiple users during the two interaction tasks of selection and telepoting. We want the trained models to determine whether the user wants to interact or not and the interaction type (selection or telepoting).

The superiority of our work over the existing works that aim to classify user interaction intention in VR is twofold. First, many studies are content-related, since they focus on highly specific application scenarios such as VR navigation [46] and shopping [27]. Our work can be applied in all areas that utilize basic interaction tasks such as selecting and teleporting. Application areas can range from simple scene-roaming to more complicated game interactions. Second, our recognition model is more accurate than some existing works [27,37], making it a better candidate for practical use.

3. Materials and Methods

3.1. Data Collecting

3.1.1. Participants

Ten participants (five female and five male) volunteered for this experiment. Their ages ranged between 22 and 27. All participants had normal or corrected-to-normal vision by using glasses or lenses during the experiment. Most participants were either undergraduate or graduate students. All participants had used VR Head Mounted Display (HMD) before. A pretest was conducted before the formal experiment to help the participants prepare.

3.1.2. Physical Setup

The virtual environment was displayed through an HTC VIVE Pro Eye integrated with an eye tracker. The screen had a 1440 × 1600 pixels/eye resolution with a 110° field of view. The HMD's highest refresh rate was 90 Hz. The refresh rate of the built-in eye tracker was 120 Hz, which offered tracking precision of 0.5–1.1°. The experiment was conducted on a PC with an Intel Core i7-9700 CPU, an NVIDIA GeForce GTX 1070 8G GPU, and 16G

DDR4 2666 Hz RAM. The experimental platform was developed using Unity 2019.4 and C#.

3.1.3. Experiment Design

We designed two basic VR interactive tasks for experiments. One used ray casting to select the target sphere (Figure 1). The other was teleporting to the target location (Figure 2).There were two reasons for choosing these two tasks: first, these two primary tasks are relatively simple, but they are very similar in interaction behavior; second, they are often used in actual VR applications. The most complex interaction in the current VR application scenario was the game. For example, in the game "Half-life: Alyx" released in 2020, selecting an item from a distance and teleporting are the basic interaction tasks. Other, more straightforward scenes, such as the Home scenario of SteamVR, also included these two tasks. They are also used as experimental tasks in many studies [37,47].

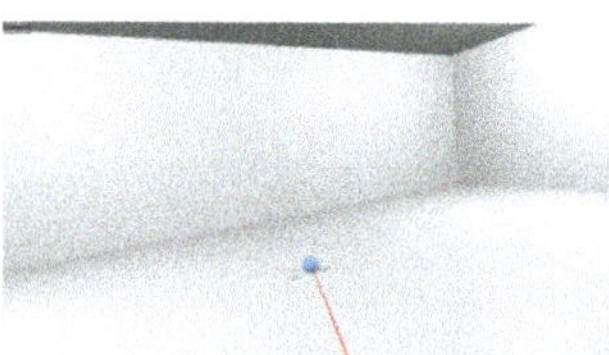

Figure 1. Using controllers to select the target sphere.

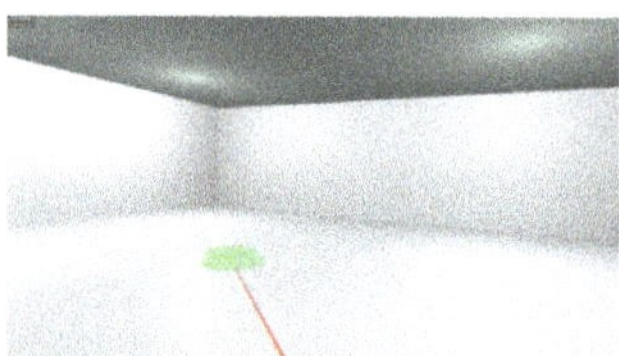

Figure 2. Using controllers to teleport to the target position.

The virtual environment was an empty room with the participant in the center. Participants were asked to repeat one of the two tasks 20 times in each session. The position of each target sphere or each target position was random. Each task was conducted in five sessions; that is, a total of 10 sessions for each participant.

3.1.4. Data Set

The raw data collected from the experiment consisted of gaze-related data, controller-related data, helmet-position coordinates, timestamps, and task types. Gaze-related data include the combined gaze-origin position, combined normalized gaze-direction vector, the corresponding timestamp and pupil diameter, and eye openness for either eye (Figure 3). In addition, we also acquired 3D gaze points in real-time with the help of a ray-based method [48]. The gaze direction vector and the corresponding gaze original position were used to find the intersection with the reconstructed 3D scene, representing the 3D gaze-points. The handle-related data were mainly the coordinates of the intersection points of the handle rays with the environment. One hundred tests were performed on ten subjects. After removing invalid data, 98 sets of valid data were obtained, i.e., a total of 250,380 raw data.

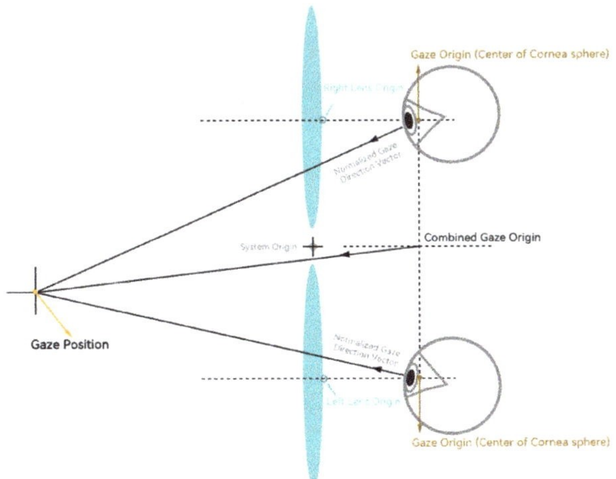

Figure 3. Eye tracker output data description.

One thing to note is that although the data collection frequency of the eye-tracking device was 120 Hz, our experimental platform was developed on Unity, so the actual data-collection frequency depended on the refresh frequency of the Update function. However, the increasing demand for GPU graphics rendering or the saturation of computing power led to a temporary decrease in the data collection frequency. The sampling frequency in this experiment fluctuates between 60 Hz and 40 Hz, with an average of 46 Hz. This will be taken into account in the subsequent feature extraction.

3.2. Proposed Method

3.2.1. Data Pre-Processing

Our processing pipeline is visualized in Figure 4. The first step filled the missing data mainly caused by blinking. The last valid data were directly filled in the blanks. There were 9552 blank data points, accounting for about 3.8%. The next step converted right-handed coordinates to left-handed. The eye-related data were obtained using the SDK (SRanpial) through a Unity script. According to the document of SRanpial, Gaze Original is the point in the eye from which the gaze ray originates, and Gaze Direction Normalized is the normalized gaze direction of the eye. They are both based on a right-handed coordinate system. However, Unity is based on a left-handed coordinate system. Therefore, we needed to multiply their X coordinates by -1 to convert the right-handed coordinate system to left-handed. Then, we transformed the Gaze Original vectors from the eye-in-head frame to the eye-in-world frame by adding the coordinates of the main camera to the Gaze Original vectors.

3.2.2. Ground Truth

We used the trigger/pad events from the hand controller to mark the ground truth of input datasets. It was uncertain how far in advance the intention could be predicted. We also needed to ensure sufficient training samples, so we chose two time thresholds to divide the data. The 20 or 40 sets of samples preceding a click were considered as positive samples; that is, the sampled data within 400 milliseconds as ground truth generation (GTG) type1 or 800 milliseconds as GTG type2 before the interaction occurred. In addition, we also tried to train two types of interaction-intention prediction models. One was a binary classifier, to predict whether users want to issue a command or not. The other was a three-class classifier which predicts whether users want to select, teleport, or execute no command at all. Positive samples needed to be further divided into two types according to interaction tasks: selection or teleporting.

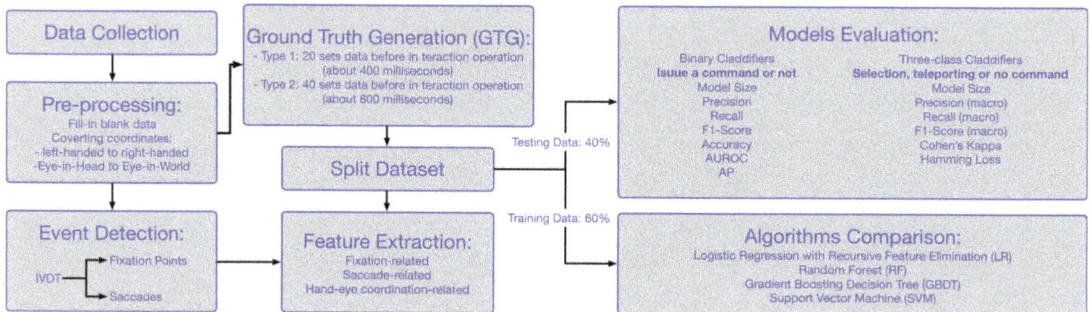

Figure 4. The pipeline to detect eye events, extract features, and train and evaluate models.

3.2.3. Eye Event Detection and Feature Extraction

Many previous studies selected eye-based features to capture spatiotemporal characteristics based on two fundamental eye movements—fixation points and saccades. Our method utilizes four types of features for interaction-intention prediction: fixation, saccade, pupil, and hand-eye coordination. We extracted them from each fixation and saccade. We summarize these features in Table 2. Therefore, eye event detection is required before feature extraction to classify these two types of eye movements.

Komogortsev and Karpov [49] proposed a ternary classification algorithm called velocity and dispersion threshold identification (I-VDT). We chose it to classify the two types of eye movements. It first identifies saccades by the velocity threshold. Subsequently, it identifies smooth pursuits from fixation by a modified dispersion threshold and duration. The original algorithm needs an initial time window to carry out. However, in a VR environment, due to increasing graphic rendering requirements or the limited computing power of GPUs, the data collection frequency is unstable and often reduced. Since the raw data is obtained using the SDK (SRanpial) through a Unity script, the data-collection frequency depends on the graphic engine's processing rate. To solve this problem, we adjusted the algorithm. Instead of setting an initial window, we checked whether it met the minimum fixation duration after determining a group of fixation points. In addition, we also checked the dispersion distance between the centroids of two adjacent fixation groups. They merged if they were too close (below the dispersion threshold). Moreover, the smooth pursuit was not one of our classification categories, so we modified the algorithm.

The I-VDT algorithm in this paper employs three velocity, dispersion, and minimum fixation-duration thresholds. The specific values of these three parameters are determined by previous research [50]. The velocity threshold is 140 degrees per second. The minimum fixation duration is 110 milliseconds. The maximum dispersion angle is 5.75 degrees. I-VDT begins by calculating point-to-point velocities for each eye-data sample. Then, I-VDT classifies (Algorithm A1) each point as a fixation or saccade point based on a simple velocity threshold: if the point's velocity is below the threshold, it is a fixation point; otherwise, it is a saccade point. Then, we check whether each fixation group meets the minimum fixation duration and whether the dispersion distance between adjacent fixation groups meets the maximum dispersion distance. If both are met, it is regarded as a fixation at centroid (x, y, z) of the fixation group points with the first point's timestamp as fixation start timestamp and the duration of the points as the fixation duration.

Each gaze sample should belong to fixation or saccade after classification by I-VDT. So, to represent all these features as a continuous-time series, we set the value for each gaze sample as the feature value from the most recent fixation or saccade event, i.e., each was carried forward in time until the next detected event. Pupil-related and hand-eye-coordination-related features were all calculated based on the fixation or scanning data group to which the sample belonged. As for hand-eye coordination, related features were based on the distance between points of gaze and controller-ray intersection with the virtual environment at the same time. Specifically, let $G_t < x,y,z >$ be the positions of gaze in the virtual environment at time t during the execution of a particular task; let $C_t < x,y,z >$ represent the position of the intersection point of the controller ray with the virtual environment at time t. We argue that the distance between these points $D_t = |G_t - C_t|$ strongly correlates with whether the user executes interaction. Çığ, Ç and Sezgin [45] confirmed that the distance between strokes and gaze in pen-based touchscreen interaction is related to task types, and different task types have completely different rise/fall characteristics. We assume the same in VR controller interaction, so we choose this feature type. See Table 2 for specific features.

Table 2. Features derived from fixation, saccade, pupillary responses, and hand-eye coordination.

Types	Features
Fixation Related	Fixation detection: Sample-level boolean indicating whether a sample was part of a fixation or not Fixation duration Standard deviation of gaze position on x-axis, y-axis, and z-axis during fixation Skewness of gaze position on x-axis, y-axis, and z-axis during fixation Kurtosis of gaze position on x-axis, y-axis, and z-axis during fixation Average velocity of gaze samples during fixation Path length of gaze samples during fixation Dispersion of gaze samples during fixation
Saccade Related	Saccade duration Standard deviation of gaze position on x-axis, y-axis, and z-axis M3S2K of gaze velocity during saccade Saccadic ratio: peak velocity/saccade duration Saccade amplitude
Pupil Related	M3S2K of left-eye pupil during a fixation or a saccade M3S2K of right eye pupil during a fixation or a saccade
Hand-Eye-Coordination-related	M3S2K of the distance between gaze position and the hit point of the controller ray during a fixation or saccade

Note: M3S2K refers to the computation of mean, median, maximum, standard deviation, skewness, and Kurtosis values.

3.2.4. Metrics

We chose accuracy, precision, recall, F_1-Score, and model size to evaluate binary classifiers.

Accuracy is the ratio of correct predictions. If $\hat{y}_i$ is the predicted value of the i-th sample and y_i is the corresponding true value, then the ratio of correct predictions over $n_{samples}$ samples is defined as

$$\text{Accuracy}(y,\hat{y}) = \frac{\sum_{i=0}^{n_{samples}-1} 1(\hat{y}_i = y_i)}{n_{samples}} \tag{1}$$

where $1(x)$ is an indicator function.

Precision is the ability of the classifier not to label negative samples as positive, and recall is the ability of the classifier to find all positive samples. The calculation formulas are as follows:

$$\text{Precision} = \frac{TP}{TP + FP} \tag{2}$$

$$\text{Recall} = \frac{TP}{TP + FN} \tag{3}$$

where TP, FP, and FN are the numbers of true positives, false positives, and false negatives, respectively.

F_1-Score is the weighted harmonic mean of precision and recall with equal importance. The F_1-Score is defined as

$$F_1 = \frac{2 * (\text{Precision} \times \text{Recall})}{\text{Precision} + \text{Recall}} \tag{4}$$

In addition to the above metrics, for binary classification, we also use average precision (AP) and AUROC (the area under the receiver operating characteristic curve) to evaluate binary classifiers.

The value of AP is between 0 and 1 and higher is better. AP is defined as

$$AP = \sum_n (R_n - R_{n-1})P_n \tag{5}$$

where P_n and R_n are the precision and recall at the n-th threshold. With random predictions, the AP is the ratio of positive samples.

A receiver operating characteristic (ROC), or ROC curve, is a graphical plot that illustrates the performance of a binary classifier as its discrimination threshold varies. It is created by plotting the ratio of true positives to all positives (TPR = true positive rate) versus the ratio of false positives to all negatives (FPR = false positive rate), at various threshold settings. By computing the area under the ROC curve (AUROC), the curve information is summarized in one number. The closer to 1, the better.

As for three-class classifiers, we chose Hamming loss, Cohen's kappa, model size, and the macro average of precision, recall, and F_1-Score.

Let n_{labels} be the number of classes or labels, the Hamming loss $L_{Hamming}$ is defined as:

$$L_{Hamming}(y, \hat{y}) = \frac{\sum_{i=0}^{n_{labels}-1} 1(\hat{y}_i \neq y_i)}{n_{labels}} \tag{6}$$

The closer to zero, the better.

The calculation formulas of macro average metrics are as follows:

$$Precision_{macro} = \frac{\sum_{l \in L} P(y_l, \hat{y}_l)}{|L|} \tag{7}$$

$$Recall_{macro} = \frac{\sum_{l \in L} R(y_l, \hat{y}_l)}{|L|} \tag{8}$$

$$F1_{macro} = \frac{\sum_{l \in L} F_1(y_L, \hat{y}_l)}{|L|} \tag{9}$$

where L is the set of labels, and $P(y_l, \hat{y}_l), R(y_l, \hat{y}_l), F_1(y_l, \hat{y}_l)$ are the Precision, Recall, F_1-Score of class or label l, respectively.

A kappa score is a number between -1 and 1. Scores above 0.8 are generally considered good agreement; zero or lower means no agreement (practically random labels).

3.2.5. Classifiers

We used the features described in the previous sections to build models that automatically classify observations as positive (interaction intention) or negative. There are plenty of candidate classification algorithms. We explored LR models, RF, GBDT, and SVM (which are commonly used for gaze data (Table 1)) to predict interaction intention in VR. All the above algorithms are implemented by Scikit-learn (https://github.com/scikit-learn/scikit-learn, accessed on 1 April 2022) [51], an open source machine-learning library in Python. We performed parameter tuning to find the optimal parameters for each classifier with F_1-Score. The optimal parameters for each classifier are given in Appendix B Table A1.

4. Results

All evaluations were performed using Scikit-learn. The evaluations were measured in line with the standard three-step machine-learning pipeline, where we first extracted features from the dataset and split the data into training and test datasets, then trained classifier models using training data, and finally measured all metrics using test data. We evaluated the hyper-parameters of each model using a grid search with two-fold cross-validation based on F_1-Score.

4.1. Performance of Binary Classifiers

Table 3 presents an overview of the main results of the best classification performance for each combination of algorithms and GTG methods for binary classification.

We compare the performance across all combinations of four classifiers, two GTG methods, and two feature sets. Table 3 shows the performance using LR, SVM, RF, and GBDT . The LR classifier performed poorly for both feature sets. As our dataset is highly complex and multi-dimensional, the LR classifier proved unsuitable for our purpose. The F_1-Scores of the other three classifiers are higher than 86%, which is worthy of further analysis.

We can see an improvement in the F1-Score when hand-eye-coordination-related features were used. The F1-Scores of the other three classifiers were improved by 1–3% by incorporating hand-eye-coordination-related features. Table 3 also shows that the GTG methods influenced the classifiers' performance for the Whole Feature Set. When using the Whole Feature Set, the GBDT classifier achieved a maximum $F1$-Score of 92.4% using 20 sets of data before interaction operation (400 milliseconds, GTG type1) as positive samples and 87.3% with 40 sets of data before interaction operation (800 milliseconds, GTG type2) as positive samples. However, the difference between the two GTG methods was less significant when using the Eye-Only Feature Set. One possible explanation can be related to the fact that hand-eye-coordination-related features are more sensitive to time. In other words, the relevant features have substantial differences only when they are very close to the time of interaction.

Table 3. Binary classification results for the combinations of four classifiers (RF, GBDT, LR, and SVM), two feature sets, and two GTG methods.

GTG		20 Sets Data before Interaction Operation (400 Milliseconds)				40 Sets Data before Interaction Operation (800 Milliseconds)			
Algorithm		RF	GBDT	RFE + LR	SVM	RF	GBDT	RFE + LR	SVM
Whole Feature Set	Accuracy	0.976	0.976	0.838	0.964	0.949	0.947	0.801	0.928
	Precision	0.954	0.947	0.476	0.894	0.962	0.947	0.679	0.908
	Recall	0.890	0.902	0.172	0.880	0.874	0.882	0.709	0.859
	F_1-Score	0.921	0.924	0.253	0.887	0.916	0.914	0.693	0.883
	AUROC	0.994	0.993	0.875	0.980	0.987	0.981	0.850	0.962
	AP	0.980	0.970	0.460	0.940	0.980	0.970	0.660	0.950
	Size	83 MB	54.5 MB	37KB	10.3 MB	139.2 MB	32.9 MB	19 KB	21.8 MB
Eye-Only Feature Set (No Hand-Eye Coordination-related Feature)	Accuracy	0.972	0.974	0.836	0.959	0.945	0.943	0.758	0.927
	Precision	0.958	0.949	0.465	0.881	0.966	0.947	0.653	0.899
	Recall	0.862	0.884	0.190	0.854	0.857	0.868	0.508	0.868
	F_1-Score	0.908	0.916	0.270	0.868	0.908	0.906	0.571	0.883
	AUROC	0.993	0.990	0.853	0.973	0.985	0.977	0.817	0.960
	AP	0.980	0.950	0.430	0.920	0.980	0.970	0.600	0.940
	Size	100.1 MB	66.1 MB	37KB	8.4 MB	156.4 MB	41.2 MB	39 KB	18.2 MB

In addition to standard evaluation metrics in machine learning, we also chose the model size as a reference because the ultimate goal of our research is to achieve real-time classification, so the smaller the model, the better. RF and GBDT had similar classification performances, but the GBDT model was relatively small. RF and GBDT are ensemble classifiers, which means the final models contain many decision trees. The SVM classifier only needed to record the final classification hyperplane so that the model was smaller than the other two.

Table 4 lists the top-ten features according to RF and GBDT importance scores when predicting whether users want to issue a command or not with the Eye-only Feature Set or Whole Feature Set.

Table 4. Top-ten importance features based on RF and GBDT feature importance scores of binary classifiers.

Algorithm		RF		GBDT	
GTG	Feature Set	Features	Importance	Features	Importance
20 sets data before interaction operation (400 milliseconds)	Whole Feature Set	[C] Min of distance	0.060	[C] Median of distance	0.120
		[F] Fixation duration	0.051	[S] Saccade duration	0.064
		[C] Median of distance	0.023	[C] Skewness of distance	0.055
		[C] Mean of distance	0.013	[F] Fixation duration	0.040
		[F] Average velocity of gaze samples during fixation	0.007	[S] Average velocity of gaze samples during saccades	0.040
		[C] Max of distance	0.006	[C] Max of distance	0.036
		[F] Dispersion of gaze samples during fixation	0.006	[C] Min of Distance	0.034
		[F] Fixation detection	0.005	[C] Standard deviation of distance	0.026
		[S] Average velocity of gaze samples during saccades	0.004	[C] Mean of distance	0.026
		[S] Max velocity of gaze samples During saccades	0.003	[F] Dispersion of gaze samples during fixation	0.024
	Eye-only Feature Set (No Hand-Eye Coordination-related Feature)	[F] Fixation duration	0.071	[F] Fixation duration	0.089
		[F] Average velocity of gaze samples during fixation	0.020	[F] Average velocity of gaze samples during fixation	0.081
		[F] Dispersion of gaze samples during fixation	0.020	[P] Kurtosis of right-eye pupil diameter	0.051
		[F] Fixation detection	0.011	[F] Path length of gaze samples during fixation	0.050
		[S] Max velocity of gaze samples during saccades	0.010	[F] Dispersion of gaze samples during fixation	0.037
		[S] Average velocity of gaze samples during saccades	0.009	[S] Average velocity of gaze samples during saccades	0.036
		[F] Path length of gaze samples during fixation	0.006	[P] Mean of left-eye pupil diameter	0.031
		[F] Standard deviation of z-axis coordinate of the gaze position during fixation	0.006	[P] Standard deviation of right-eye pupil diameter	0.028
		[S] Median velocity of gaze samples during saccades	0.006	[F] Standard deviation of z-axis coordinate of the gaze position during fixation	0.028
		[F] Standard deviation of x-axis coordinate of the gaze position during fixation	0.005	[P] Mean of right-eye pupil diameter	0.026
40 sets data before interaction operation (800 milliseconds)	Whole Feature Set	[C] Min of distance	0.043	[C] Median of distance	0.074
		[F] Fixation duration	0.026	[S] Saccade duration	0.064
		[F] Fixation detection	0.012	[C] Min of distance	0.046
		[C] Median of distance	0.010	[C] Standard deviation of distance	0.043
		[C] Mean of distance	0.010	[F] Kurtosis of y-axis coordinate of the gaze position during fixation	0.037
		[S] Average velocity of gaze samples during saccades	0.008	[F] Average velocity of gaze samples during fixation	0.034
		[S] Max velocity of gaze samples during saccades	0.007	[C] Mean of distance	0.031
		[C] Max of distance	0.007	[P] Kurtosis of right-eye pupil diameter	0.030

Table 4. *Cont.*

Algorithm Feature Set	RF Features	Importance	GBDT Features	Importance
GTG				
	[F] Dispersion of gaze samples during fixation	0.007	[F] Skewness of y-axis coordinate the gaze position during fixation	0.030
	[F] Average velocity of gaze samples during fixation	0.006	[S] Max Velocity of gaze samples during saccades	0.029
Eye-Only Feature Set (No Hand-Eye Coordination-related Feature)	[F] Fixation Duration	0.064	[F] Fixation duration	0.136
	[S] Average velocity of gaze samples during saccades	0.017	[S] Standard deviation of y-axis coordinate of the gaze position during saccade	0.064
	[F] Fixation detection	0.013	[F] dispersion of gaze samples during fixation	0.055
	[S] Max velocity of gaze samples during saccades	0.013	[S] Min velocity of gaze samples during saccades	0.047
	[S] Median velocity of gaze samples during saccades	0.010	[P] Skewness of left-eye pupil diameter	0.038
	[S] Min velocity of gaze samples during saccades	0.009	[F] Skewness of x-axis coordinate of the gaze position during fixation	0.031
	[S] Saccade amplitude	0.009	[P] Mean of left-eye pupil diameter	0.028
	[F] Average velocity of gaze samples during fixation	0.009	[F] Average velocity of gaze samples during fixation	0.027
	[F] Dispersion of gaze samples during fixation	0.009	[S] Average velocity of gaze samples during saccades	0.027
	[S] Saccadic ratio	0.009	[S] Median velocity of gaze samples during saccades	0.023

Note: [F] stands for fixation-related feature; [S] stands for saccade-related feature; [P] stands for pupil-related feature; [C] stands for hand-eye-coordination-related feature.

For the Whole Feature Set, taking the example of the GBDT classifier with the highest $F1$-Score using GTG type1, the top-10 important features consisted of six hand-eye-coordination-related features, two fixation-related features, and one saccade-related feature. The top-10 features of other classifiers were highly consistent with this one. The four hand-eye-coordination-related features—min, max, median, and mean of distance—received high importance. As for eye-only features, three features about the velocity of gaze samples, such as the average velocity of gaze samples during fixation or saccade and the maximum velocity of gaze samples during saccade, also scored high in importance, the same as fixation-related features—fixation duration and dispersion of gaze samples during fixation.

For the Eye-Only Feature Set, taking the example of the GBDT classifier with the highest F_1-Score using GTG type1, the top-10 important features consisted of five fixation-related features, four pupil-related features, and one saccade-related feature. Overall, the important eye-only features were the same as the classifiers that used the Whole Feature Set.

4.2. Performance of Three-Class Classifiers

For three-class classifiers, except LR, the F1-Scores of the other three algorithms are above 0.9. The GBDT is still the best classification algorithm, followed by RF and SVM. Table 5 shows an overview of the main results for three-class classifiers.

Table 5. Three-class classification results for the combinations of four classifiers (RF, GBDT, LR, and SVM), two feature sets, and two GTG methods.

GTG Algorithm		20 Sets Data before Interaction Operation (400 Milliseconds)				40 Sets Data before Interaction Operation (800 Milliseconds)			
		RF	GBDT	RFE + LR	SVM	RF	GBDT	RFE + LR	SVM
Whole Feature Set	Precision (macro)	0.963	0.964	0.280	0.936	0.956	0.969	0.633	0.922
	Recall (macro)	0.916	0.923	0.333	0.930	0.893	0.939	0.557	0.909
	F_1-Score (macro)	0.939	0.943	0.305	0.933	0.921	0.953	0.582	0.915
	Cohen's kappa	0.906	0.912	0.000	0.866	0.879	0.927	0.395	0.830
	Hamming loss	0.026	0.024	0.159	0.036	0.056	0.034	0.263	0.072
	Size	107.5 MB	86.2 MB	4 KB	10.3 MB	180 MB	148.4 MB	5 KB	21.8 MB
Eye-Only Feature Set (No Hand-Eye Coordination-related Feature)	Precision (macro)	0.964	0.964	0.280	0.953	0.957	0.964	0.582	0.907
	Recall (macro)	0.899	0.907	0.333	0.898	0.885	0.915	0.476	0.895
	F_1-Score (macro)	0.929	0.934	0.304	0.923	0.917	0.938	0.492	0.901
	Cohen's kappa	0.891	0.899	0.000	0.846	0.871	0.902	0.274	0.801
	Hamming loss	0.029	0.027	0.159	0.039	0.059	0.046	0.290	0.085
	Size	118.9 MB	87.9 MB	3 KB	34.6 MB	175.6 MB	144.6 MB	4 KB	15.9 MB

In terms of GTG, for the GBDT algorithm, the two GTGs had little difference in classification performance, while for RF and SVM, the result of GTG type1 was better than that of type2. For the feature sets, as we estimated, the classification performance of the Eye-Only Feature Set was worse than the Whole Feature Set by 0.006–0.016 (F_1-Score). As for the model size, the GBDT had a better classification performance with a smaller model size than the RF. SVM was the smallest model, the same as binary classifiers.

Table 6 lists the top ten features of three-class classifiers using RF and GBDT. The features related to hand-eye coordination are still of high importance. However, some new features, especially those related to the y-axis distribution of fixation points, have a significant difference between the two interactive tasks of selection and blinking. However, it may also indicate that these indicators may be related to the design of the interactive interface.

Table 6. Top-ten importance features based on Random Forest and GBDT feature importance scores of three-class classifiers.

Algorithm GTG	Feature Set	RF Features	Importance	GBDT Features	Importance
20 sets data before interaction operation (400 milliseconds)	Whole feature set	[F] Fixation duration	0.062	[F] Fixation duration	0.078
		[C] Min of distance	0.058	[C] Min of distance	0.058
		[C] Median of distance	0.046	[C] Median of distance	0.054
		[C] Mean of distance	0.043	[F] Standard deviation of y-axis coordinate of the gaze position during fixation	0.051
		[F] Dispersion of gaze samples during fixation	0.042	[F] Dispersion of gaze samples during fixation	0.048
		[F] Standard deviation of y-axis coordinate of the gaze position during fixation	0.041	[C] Mean of distance	0.045
		[F] Path length of gaze samples during fixation	0.037	[F] Kurtosis of y-axis coordinate of the gaze position during fixation	0.032
		[F] Average velocity of gaze samples during fixation	0.032	[F] Average velocity of gaze samples during fixation	0.031
		[F] Kurtosis of y-axis coordinate of the gaze position during fixation	0.027	[F] path length of gaze samples during fixation	0.027
		[F] Skewness of y-axis coordinate of the gaze position during fixation	0.024	[F] Skewness of y-axis coordinate of the gaze position during fixation	0.026
	Eye-Only Feature Set (No Hand-Eye Coordination-related Feature)	[F] Fixation duration	0.084	[F] Fixation duration	0.082
		[F] Dispersion of gaze samples during fixation	0.058	[F] Path length of gaze samples during fixation	0.062
		[F] Standard deviation of y-axis coordinate of the gaze position during fixation	0.054	[F] Standard deviation of y-axis coordinate of the gaze position during fixation	0.053
		[F] Path length of gaze samples during fixation	0.046	[F] Dispersion of gaze samples during fixation	0.045
		[F] Average velocity of gaze samples during fixation	0.039	[F] Average velocity of gaze samples during fixation	0.043
		[F] Skewness of y-axis coordinate of the gaze position during fixation	0.039	[F] Skewness of y-axis coordinate of the gaze position during fixation	0.037
		[F] Kurtosis of y-axis coordinate of the gaze position during fixation	0.036	[F] Kurtosis of y-axis coordinate of the gaze position during fixation	0.033
		[S] Average velocity of gaze samples during saccades	0.030	[S] Max Velocity of Gaze samples during saccades	0.030
		[S] Max velocity of gaze samples during saccades	0.028	[S] Average velocity of gaze samples during saccades	0.029
		[F] Standard deviation of z-axis coordinate of the gaze position during fixation	0.025	[F] Standard deviation of z-axis coordinate of the gaze position during fixation	0.026
		[C] Min of distance	0.067	[C] Min of distance	0.173
		[F] Fixation duration	0.054	[F] Standard deviation of y-axis coordinate of the gaze position during fixation	0.116
		[F] Standard deviation of y-axis coordinate of the gaze position during fixation	0.048	[F] Fixation duration	0.086

Table 6. *Cont.*

Algorithm		RF		GBDT	
GTG	Feature Set	Features	Importance	Features	Importance
40 sets data before interaction operation (800 milliseconds)	Whole Feature Set	[C] Median of distance	0.043	[C] Max of distance	0.027
		[F] Dispersion of gaze samples during fixation	0.043	[C] Mean of distance	0.027
		[C] Mean of distance	0.041	[S] Average velocity of gaze samples during saccades	0.027
		[F] Path Length of gaze samples during fixation	0.036	[S] Max velocity of gaze samples during saccades	0.025
		[F] Kurtosis of y-axis coordinate of the gaze position during fixation	0.028	[F] Fixation detection	0.025
		[F] Average velocity of gaze samples during fixation	0.028	[P] Max of right-eye pupil diameter	0.023
		[C] Max of distance	0.026	[S] saccadic ratio	0.022
	Eye-Only Feature Set (No Hand-Eye Coordination-related Feature)	[F] Fixation duration	0.075	[F] Fixation duration	0.090
		[F] Standard deviation of y-axis coordinate of the gaze position during fixation	0.055	[F] Standard deviation of y-axis coordinate of the gaze position during fixation	0.047
		[F] Dispersion of gaze samples during fixation	0.044	[F] Skewness of y-axis coordinate of the gaze position during fixation	0.043
		[F] Path length of gaze samples during fixation	0.039	[F] Dispersion of gaze samples during fixation	0.042
		[S] Average velocity of gaze samples during saccades	0.035	[F] Path length of gaze samples during fixation	0.042
		[S] Max velocity of gaze samples during saccades	0.033	[S] Max velocity of gaze samples during saccades	0.038
		[F] Average velocity of gaze samples during fixation	0.032	[F] Kurtosis of y-axis coordinate of the gaze position during fixation	0.037
		[F] Kurtosis of y-axis coordinate of the gaze Position during fixation	0.030	[F] Average velocity of gaze samples during fixation	0.033
		[F] Skewness of y-axis coordinate of the gaze position during fixation	0.030	[S] Average velocity of gaze samples during saccades	0.032
		[S] Median velocity of gaze samples during saccades	0.027	[S] Median velocity of gaze samples during saccades	0.029

Note: [F] stands for fixation-related feature, [S] stands for saccade-related feature, [P] stands for pupil-related feature, [C] stands for Hand-Eye-Coordination-Related feature.

5. Discussion

The research of binary classifiers mainly explores which features can separate intentional behavior from unintentional behavior. The research of classifiers is to explore which features may be particularly relevant to the two tasks in our experiment. It can be said that binary classifiers can play a comparative role to three-class classifiers. In general, the features in binary classifiers are independent of the coordinate axis. The y-axis—that is, the vertical gaze coordinate distribution in three-class classification—plays a vital role in distinguishing the two types of tasks. It should be noted that when we select features at the beginning, we avoid features related to absolute coordinates and retain features related to the distribution law of coordinates. The above phenomenon may be because the selection task requires the user to keep staring at the target until visual feedback indicates that the interaction is completed. However, the teleporting task only requires clarification of the destination, so there is no need to keep staring at destination but to prepare for the change of perspective after teleporting. This phenomenon needs to be further explored in later research.

In the selection of features, we used two feature sets. The major difference was whether to include the hand-eye-coordination-related features. On the one hand, we wanted to verify whether the features of hand-eye coordination can improve the accuracy of interaction intention recognition in a multimodal interaction system, including controller and eye movement. The results show that the hand-eye-coordination index is important in predicting interaction intention. On the other hand, we should also consider whether the interaction intention of users can be effectively predicted with only eye-movement data and without controller-related data. Our study shows that only the features related to eye movement can be used to classify the interaction intention and the classification performance is also acceptable.

We used two kinds of methods to generate datasets. The main difference was how many groups of sampled data were included before the interaction occurred. We expected the system to deduce the interaction intention in advance. We selected 400 milliseconds and 800 milliseconds for comparative analysis. The classification result of the 800-millisecond classifier was slightly inferior to that of the 400-millisecond classifier, which is understandable. The generation time of real interaction intention was short, especially for our experiment's simple interaction tasks. If a long period is selected for data generation, the difference of features under different categories will not be significant, and the classification performance will naturally decline. However, it is not always good to use a shorter period. The shorter the period is, the fewer data we can generate in the dataset. In that way, the robustness of the trained model may decline. The choice of this time length needs to be determined through further experimental research and combined with the user's expectation of the intention prediction system.

As for the selection of algorithms, GBDT had the best performance. Its classification performance was not inferior to RF, and its model size was smaller than RF's. When we transformed the model into a real-time classifier, it was more likely to reduce latency. The model size of the SVM was small enough, but the overall classification performance still lagged behind the other two algorithms. In addition, SVM is more dependent on hyperparameters and takes the longest time to train.

We declare several limitations of our work, despite our best efforts to minimize them. First, the dataset is not entirely naturalistic. The number of participants was limited, so it was necessary to use data from new participants to verify the performance of the models. The experimental environment was also relatively simple. Whether more complex interaction scenarios will impact the classification performance still needs to be verified by follow-up research.

In the light of promising findings reported in this paper, we envision several immediate follow-ups to our work, as well as long-term research directions to explore. An immediate extension might involve conducting experiments to see if our classification models apply to other more complex interaction environments rather than a concise experiment environment only. We want to explore two factors. One is whether the targets of different dimensions will affect the prediction results of the classifier (the selection target in this experiment is a sphere if it is replaced by a plane). The other is whether the interface complexity will affect the prediction results of the classifier (if there are multiple targets or locations in the environment at the same time). We also want to build an online prediction system to verify the performance of classifiers. Further experiments would evaluate the usability aspects of this setup and compare it to state-of-the-art online interaction intention prediction mechanisms in the literature. Another possible direction might involve conducting experiments to see if our prediction system can successfully recognize other interaction tasks.

6. Conclusions

This paper explored hand-eye-coordination-related features to improve interaction intention recognition in a VR environment. We collected a dataset of eye-movement data and controller-related data from 10 participants as they performed two basic interaction tasks: selection and teleporting. We extracted a Whole Feature Set, including fixation-related, saccade-related, pupil-related, and hand-eye-coordination-related features, and an Eye-Only Feature Set without hand-eye-coordination-related features. We obtained a high binary classification performance score (F_1-Score = 0.924) using the combination of the Whole Feature Set, GTG method type1, and the GBDT classifier, as well as a high three-class classification performance score (F_1-Score = 0.953) using the combination of the Whole Feature Set, GTG method type2, and the GBDT classifier. The results show that hand-eye-coordination-related features improve interaction intention recognition in VR environments. The GBDT had the best classification performance among the four classifiers, and its model size was smaller than the RF's. Generally, this work provides the groundwork for its exploration and towards building a robust and generalizable model for eye-based interaction-intention recognition in VR. We believe that predicting the interaction intention will eventually enable us to build systems that save users the trouble of switching during basic interaction tasks.

Author Contributions: Conceptualization, X.-L.C. and W.-J.H.; methodology, X.-L.C. and W.-J.H.; software, X.-L.C.; validation, X.-L.C. and W.-J.H.; formal analysis, X.-L.C.; data curation, X.-L.C.; writing—original draft preparation, X.-l.C.; writing—review and editing, W.-J.H.; supervision, W.-J.H.; funding acquisition, X.-L.C. All authors have read and agreed to the published version of the manuscript.

Funding: This work was supported by BUPT Excellent Ph.D. Students Foundation (Grant No.: CX2019112). This work was also funded by Beijing University of Posts and Telecommunications-China Mobile Research Institute Joint Center (Research Title: Research on 6G new business models and business communication quality indicators).

Institutional Review Board Statement: Not applicable.

Informed Consent Statement: Informed consent was obtained from all subjects involved in the study.

Data Availability Statement: Publicly available datasets were analyzed in this study. This data can be found here: https://www.scidb.cn/s/nQN7fm, accessed on 1 April 2022.

Acknowledgments: Thanks to all participants.

Conflicts of Interest: The authors declare no conflict of interest. The funders had no role in the design of the study; in the collection, analyses, or interpretation of data; in the writing of the manuscript, or in the decision to publish the results.

Abbreviations

The following abbreviations are used in this manuscript:

VR	Virtal Reality
HMD	Head Mounted Display
LR	Logistic Regression
SVM	Support Vector Machine
RF	Random Forest
GBDT	Gradient Boosting Decision Tree
GTG	Ground Truth Generation

Appendix A

The I-VDT algorithm in this paper employs three thresholds of velocity, dispersion, and minimum fixation duration. The specific values of these three parameters are determined by previous research. The velocity threshold is 140 degrees per second. The minimum fixation during is 110 milliseconds. The maximum dispersion angle is 5.75 degrees. See Algorithm A1 below for details.

Algorithm A1 Velocity and Dispersion-Threshold Identification

---

Require: p_i:3D gaze position with timestamps, (x, y, z, t); V_i:normalized gaze direction vector with timestamps, Vel:velocity threshold; DD_{max}: maximum fixation dispersion distance threshold; $Duration_{min}$: minimum fixation duration threshold;
Ensure: f_i:representative coordinates corresponding to fixations groups, and the the starting time and duration of these fixations groups, $(x_f, y_f, z_f, t_{start}, d)$
     // calculate the instantaneous visual angle
     for $i = 0 \rightarrow n - 1$ **do**
3:      $v_i = \dfrac{\arccos \frac{V_i \cdot V_{i+1}}{\|V_i\|\|V_{i+1}\|}}{|t_{i+1} - t_i|} \times 5.73 \times 10^4$
     end for
     Initialize Previous fixation group PFG and current fixation group CFG
6:      save p_0 into PFG
     save p_1 into CFG
     for $i = 2 \rightarrow n - 1$ **do**
9:      Calculate the CFG centroid coordinates (x, y, z)
     Calculate the dispersion distance (DD) between CFG centroid coordinates and p_i coordinates
     if $v_i < Vel$ **then**
12:      save p_i into CFG
     else
     if CFG is not empty **then**
15:      Calculate the duration d of the points in CFG
     if $d > Duration_{min}$ **then**
     Calculate the dispersion distance (DD) between the first point in CFG and the last point in PFG
18:      **if** $DD < DD_{max}$ **then**
     Merge CFG into PFG
     else
21:      Calculate the PFG centroid coordinates (x_f, y_f, z_f)
     Save the timestamp t of the first point in PFG as t_{start}
     Calculate the duration d of points in PFG
24:      **Initialize** PFG
     Merge CFG into PFG
     Initialize CFG
27:      save p_i into CFG
     end if
     else
30:      **Initialize** CFG
     save p_i into CFG
     end if
33:      **end if**
     end if
     end for

Appendix B

Table A1 shows the optimal parameters for each classifier discussed in this paper.

Table A1. The optimal parameters of each classifier are selected by grid search.

	Ground Truth Generation	Algrithm	Random Forest	Gradient Boosting Decision Tree	Logistic Regression with Recursive Feature Elimination	Support Vector Machine
2-class	20 sets data before interaction operation (400 milliseconds)	Whole Feature Set	max_depth: 29 max_features: 0.1 min_samples_leaf: 1 min_samples_split: 2 n_estimators: 100 criterion: entropy	max_depth: 24 max_features: 0.1 min_samples_leaf: 2 min_samples_split: 7 n_estimators: 100 learning_rate: 1.0	Optimal number of features: 38	C: 100.0 gamma: 0.1 kenel:RBF
		Eye-Only Feature Set (No Hand-Eye Coordination-related Feature)	max_depth: 30 max_features: 0.1 min_samples_leaf: 1 min_samples_split: 2 n_estimators: 100 criterion: entropy	max_depth: 25 max_features: 0.1 min_samples_leaf: 3 min_samples_split: 2 n_estimators: 100 learning_rate: 1.0	Optimal number of features: 39	C: 100.0 gamma: 0.1 kenel:RBF
	40 sets data before interaction operation (800 milliseconds)	Whole Feature Set	max_depth: 35 max_features: 0.1 min_samples_leaf: 1 min_samples_split: 2 n_estimators: 100 criterion: entropy	max_depth: 17 max_features: 1 min_samples_leaf: 19 min_samples_split: 8 n_estimators: 100 learning_rate: 1.0	Optimal number of features: 16	C: 10.0 gamma: 0.1 kenel:RBF
		Eye-Only Feature Set (No Hand-Eye Coordination-related Feature)	max_depth: 30 max_features: 0.1 min_samples_leaf: 1 min_samples_split: 2 n_estimators: 100 criterion: entropy	max_depth: 17 max_features: 1.0 min_samples_leaf: 14 min_samples_split: 2 n_estimators: 100 learning_rate 1.0	Optimal number of features: 41	C: 100.0 gamma: 0.1 kenel:RBF
3-class	20 sets data before interaction operation (400 milliseconds)	Whole Feature Set	max_depth: 27 max_features: 0.1 min_samples_leaf: 1 min_samples_split: 2 n_estimators: 100 criterion: gini	max_depth: 22 max_features: 0.1 min_samples_leaf: 12 min_samples_split: 2 n_estimators: 100 learning_rate: 0.1	Optimal number of features: 1	C: 100.0 gamma: 0.1 kenel:RBF

Table A1. *Cont.*

Ground Truth Generation	Algrithm	Random Forest	Gradient Boosting Decision Tree	Logistic Regression with Recursive Feature Elimination	Support Vector Machine
	Eye-Only Feature Set (No Hand-Eye Coordination-related Feature)	max_depth: 29 max_features: 0.1 min_samples_leaf: 1 min_samples_split: 2 n_estimators: 99 criterion: entropy	max_depth: 12 max_features: 0.1 min_samples_leaf: 1 min_samples_split: 3 n_estimators: 100 learning_rate: 0.1	Optimal number of features: 1	C: 10.0 gamma: 1.0 kenel:RBF
40 sets data before interaction operation (800 milliseconds)	Whole Feature Set	max_depth: 29 max_features: 0.1 min_samples_leaf: 1 min_samples_split: 2 n_estimators: 98 criterion: gini	max_depth: 18 max_features: 0.9 min_samples_leaf: 12 min_samples_split: 9 n_estimators: 99 learning_rate 0.3	Optimal number of features: 48	C: 10.0 gamma: 0.1 kenel:RBF
	Eye-Only Feature Set (No Hand-Eye Coordination-related Feature)	max_depth: 29 max_features: 0.1 min_samples_leaf: 1 min_samples_split: 2 n_estimators: 88 criterion: gini	max_depth: 17 max_features: 0.1 min_samples_leaf: 14 min_samples_split: 2 n_estimators: 100 learning_rate: 0.1	Optimal number of features: 40	C: 10.0 gamma: 0.1 kenel:RBF

References

1. Biocca, F.; Delaney, B. Immersive Virtual Reality Technology. In *Communication in the Age of Virtual Reality*; L. Erlbaum Associates Inc.: Mahwah, NJ, USA, 1995; pp. 57–124.
2. Burdea, G.C.; Coiffet, P. *Virtual Reality Technology*; John Wiley & Sons: Hoboken, NJ, USA, 2003.
3. Duchowski, A.T. A breadth-first survey of eye-tracking applications. *Behav. Res. Methods Instrum. Comput.* **2002**, *34*, 455–470. [CrossRef] [PubMed]
4. Duchowski, T. Gaze-based interaction: A 30 year retrospective. *Comput. Graph.* **2018**, *73*, 59–69. [CrossRef]
5. Jacob, R., Eye Tracking in Advanced Interface Design. In *Virtual Environments and Advanced Interface Design*; Oxford University Press, Inc.: Oxford, MS, USA, 1995; pp. 258–288.
6. Hansen, J.; Rajanna, V.; MacKenzie, I.; Bækgaard, P. A Fitts' Law Study of Click and Dwell Interaction by Gaze, Head and Mouse with a Head-Mounted Display. In Proceedings of the Workshop on Communication by Gaze Interaction (COGAIN '18), Warsaw, Poland, 14–17 June 2018; Association for Computing Machinery: New York, NY, USA, 2018. [CrossRef]
7. Blattgerste, J.; Renner, P.; Pfeiffer, T. Advantages of Eye-Gaze over Head-Gaze-Based Selection in Virtual and Augmented Reality under Varying Field of Views. In *Proceedings of the Symposium on Communication by Gaze Interaction*; ACM: New York, NY, USA, 2018.
8. Rajanna, V.; Hansen, J. Gaze Typing in Virtual Reality: Impact of Keyboard Design, Selection Method, and Motion. In Proceedings of the 2018 ACM Symposium on Eye Tracking Research & Applications (ETRA '18), Warsaw, Poland, 14–17 June 2018; Association for Computing Machinery: New York, NY, USA, 2018. [CrossRef]
9. Pai, Y.; Dingler, T.; Kunze, K. Assessing hands-free interactions for VR using eye gaze and electromyography. *Virtual Real.* **2019**, *23*, 119–131. [CrossRef]
10. Piumsomboon, T.; Lee, G.; Lindeman, R.; Billinghurst, M. Exploring natural eye-gaze-based interaction for immersive virtual reality. In Proceedings of the 2017 IEEE Symposium on 3D User Interfaces (3DUI), Los Angeles, CA, USA, 18–19 March 2017; pp. 36–39. [CrossRef]
11. Qian, Y.; Teather, R. The Eyes Don't Have It: An Empirical Comparison of Head-Based and Eye-Based Selection in Virtual Reality. In Proceedings of the 5th Symposium on Spatial User Interaction (SUI '17), Brighton, UK, 16–17 October 2017; Association for Computing Machinery: New York, NY, USA, 2017; pp. 91–98. [CrossRef]
12. Kytö, M.; Ens, B.; Piumsomboon, T.; Lee, G.; Billinghurst, M., Pinpointing: Precise Head- and Eye-Based Target Selection for Augmented Reality. In Proceedings of the 2018 CHI Conference on Human Factors in Computing Systems, Montreal, QC, Canada, 21–27 April 2018; Association for Computing Machinery: New York, NY, USA, 2018; pp. 1–14.
13. Luro, F.; Sundstedt, V. A Comparative Study of Eye Tracking and Hand Controller for Aiming Tasks in Virtual Reality. In Proceedings of the 11th ACM Symposium on Eye Tracking Research & Applications (ETRA '19), Denver, CO, USA, 25–28 June 2019; Association for Computing Machinery: New York, NY, USA, 2019. [CrossRef]
14. Scott, N.; Zhang, R.; Le, D.; Moyle, B. A review of eye-tracking research in tourism. *Curr. Issues Tour.* **2019**, *22*, 1244–1261. [CrossRef]
15. Kim, S.J.; Laine, T.H.; Suk, H.J. Presence Effects in Virtual Reality Based on User Characteristics: Attention, Enjoyment, and Memory. *Electronics* **2021**, *10*, 1051. [CrossRef]
16. Wolfe, J.M.; Horowitz, T.S. Five factors that guide attention in visual search. *Nat. Hum. Behav.* **2017**, *1*, 0058. [CrossRef]
17. Wolfe, J.M. Guided Search 6.0: An updated model of visual search. *Psychon. Bull. Rev.* **2021**, *28*, 1060–1092. [CrossRef]
18. McNally, R.J. Attentional bias for threat: Crisis or opportunity? *Clin. Psychol. Rev.* **2019**, *69*, 4–13. [CrossRef]
19. Anobile, G.; Arrighi, R.; Castaldi, E.; Burr, D.C. A Sensorimotor Numerosity System. *Trends Cogn. Sci.* **2021**, *25*, 24–36. [CrossRef]
20. Liu, X.; Chen, T.; Xie, G.; Liu, G. Contact-Free Cognitive Load Recognition Based on Eye Movement. *J. Electr. Comput. Eng.* **2016**, *2016*, 1–8. [CrossRef]
21. Kamińska, D.; Smółka, K.; Zwoliński, G. Detection of Mental Stress through EEG Signal in Virtual Reality Environment. *Electronics* **2021**, *10*, 2840. [CrossRef]
22. Al-Moteri, M.O.; Symmons, M.; Plummer, V.; Cooper, S. Eye tracking to investigate cue processing in medical decision-making: A scoping review. *Comput. Hum. Behav.* **2017**, *66*, 52–66. [CrossRef]
23. Brunyé, T.T.; Gardony, A.L. Eye tracking measures of uncertainty during perceptual decision making. *Int. J. Psychophysiol.* **2017**, *120*, 60–68. [CrossRef] [PubMed]
24. Srivastava, N.; Newn, J.; Velloso, E. Combining Low and Mid-Level Gaze Features for Desktop Activity Recognition. *Proc. ACM Interact. Mob. Wearable Ubiquitous Technol.* **2018**, *2*, 1–27. [CrossRef]
25. Liao, H.; Dong, W.; Huang, H.; Gartner, G.; Liu, H. Inferring user tasks in pedestrian navigation from eye movement data in real-world environments. *Int. J. Geogr. Inf. Sci.* **2019**, *33*, 739–763. [CrossRef]
26. Xu, B.; Li, J.; Wong, Y.; Zhao, Q.; Kankanhalli, M.S. Interact as You Intend: Intention-Driven Human-Object Interaction Detection. *IEEE Trans. Multimed.* **2020**, *22*, 1423–1432. [CrossRef]
27. Pfeiffer, J.; Pfeiffer, T.; Meißner, M.; Weiß, E. Eye-Tracking-Based Classification of Information Search Behavior Using Machine Learning: Evidence from Experiments in Physical Shops and Virtual Reality Shopping Environments. *Inf. Syst. Res.* **2020**, *31*, 675–691. [CrossRef]
28. Iqbal, S.T.; Bailey, B.P. Using Eye Gaze Patterns to Identify User Tasks. *He Grace Hopper Celebr. Women Comput.* **2004**, *6*, 2004.

29. Courtemanche, F.; Aïmeur, E.; Dufresne, A.; Najjar, M.; Mpondo, F. Activity recognition using eye-gaze movements and traditional interactions. *Interact. Comput.* **2011**, *23*, 202–213. [CrossRef]

30. Steichen, B.; Carenini, G.; Conati, C. User-adaptive information visualization: Using eye gaze data to infer visualization tasks and user cognitive abilities. In Proceedings of the 2013 international conference on Intelligent user interfaces-IUI '13, Santa Monica, CA, USA, 19–22 March 2013; ACM Press: New York, NY, USA, 2013. [CrossRef]

31. Yang, J.J.; Gang, G.W.; Kim, T.S. Development of EOG-Based Human Computer Interface (HCI) System Using Piecewise Linear Approximation (PLA) and Support Vector Regression (SVR). *Electronics* **2018**, *7*, 38. [CrossRef]

32. Paing, M.P.; Juhong, A.; Pintavirooj, C. Design and Development of an Assistive System Based on Eye Tracking. *Electronics* **2022**, *11*, 535. [CrossRef]

33. Bulling, A.; Ward, J.A.; Gellersen, H.; Tröster, G. Eye movement analysis for activity recognition. In Proceedings of the 11th International Conference on Ubiquitous Computing, Orlando, FL, USA, 30 September–3 October 2009; ACM: New York, NY, USA, 2009. [CrossRef]

34. Bulling, A.; Roggen, D.; Tröster, G. What's in the Eyes for Context-Awareness? *IEEE Pervasive Comput.* **2011**, *10*, 48–57. [CrossRef]

35. Ogaki, K.; Kitani, K.M.; Sugano, Y.; Sato, Y. Coupling eye-motion and ego-motion features for first-person activity recognition. In Proceedings of the 2012 IEEE Computer Society Conference on Computer Vision and Pattern Recognition Workshops, Providence, RI, USA, 16–21 June 2012. [CrossRef]

36. Bednarik, R.; Eivazi, S.; Vrzakova, H. A Computational Approach for Prediction of Problem-Solving Behavior Using Support Vector Machines and Eye-Tracking Data. In *Eye Gaze in Intelligent User Interfaces*; Springer: London, UK, 2013; pp. 111–134. [CrossRef]

37. Brendan, D.; Peacock, C.; Zhang, T.; Murdison, T.S.; Benko, H.; Jonker, T.R. Towards Gaze-Based Prediction of the Intent to Interact in Virtual Reality. In Proceedings of the ACM Symposium on Eye Tracking Research and Applications (ETRA '21 Short Papers), Stuttgart, Germany, 25–29 May 2021; Association for Computing Machinery: New York, NY, USA, 2021. [CrossRef]

38. Van-Horenbeke, F.A.; Peer, A. Activity, Plan, and Goal Recognition: A Review. *Front. Robot. AI* **2021**, *8*, 106. [CrossRef]

39. Jang, Y.M.; Mallipeddi, R.; Lee, S.; Kwak, H.W.; Lee, M. Human intention recognition based on eyeball movement pattern and pupil size variation. *Neurocomputing* **2014**, *128*, 421–432. [CrossRef]

40. Jang, Y.M.; Mallipeddi, R.; Lee, M. Identification of human implicit visual search intention based on eye movement and pupillary analysis. *User Model. User-Adapt. Interact.* **2014**, *24*, 315–344. [CrossRef]

41. Lisha, M.A.; Jian, L.V.; Pan, W.; Shan, J.; Ping, Z. Research on Implicit Intention Recognition and Classification Based on Eye Movement Pattern. *J. Graph.* **2017**, *38*, 332.

42. Bednarik, R.; Vrzakova, H.; Hradis, M. What do you want to do next: A novel approach for intent prediction in gaze-based interaction. In Proceedings of the Symposium on Eye Tracking Research and Applications, Santa Barbara, CA, USA, 28–30 March 2012; ACM Press: New York, NY, USA, 2012. [CrossRef]

43. Liang, Y.; Wang, W.; Qu, J.; Yang, J. Application of Eye Tracking in Intelligent User Interface. In Proceedings of the 2018 3rd International Conference on Communication, Image and Signal Processing, Sanya, China, 16–18 November 2018; pp. 333–340.

44. Huang, C.; Andrist, S.; Sauppé, A.; Mutlu, B. Using gaze patterns to predict task intent in collaboration. *Front. Psychol.* **2015**, *6*, 1049. [CrossRef]

45. Çığ, Ç.; Sezgin, T.M. Gaze-based prediction of pen-based virtual interaction tasks. *Int. J. Hum.-Comput. Stud.* **2015**, *73*, 91–106. [CrossRef]

46. Alghofaili, R.; Sawahata, Y.; Huang, H.; Wang, H.; Shiratori, T.; Yu, L. Lost in Style: Gaze-Driven Adaptive Aid for VR Navigation. In Proceedings of the 2019 CHI Conference on Human Factors in Computing Systems, Glasgow, UK, 4–9 May 2019; Association for Computing Machinery: New York, NY, USA, 2019; pp. 1–12.

47. Zagata, K.; Gulij, J.; Halik, Ł.; Medyńska-Gulij, B. Mini-Map for Gamers Who Walk and Teleport in a Virtual Stronghold. *ISPRS Int. J. Geo-Inf.* **2021**, *10*, 96. [CrossRef]

48. Mansouryar, M.; Steil, J.; Sugano, Y.; Bulling, A. 3D Gaze Estimation from 2D Pupil Positions on Monocular Head-Mounted Eye Trackers. In Proceedings of the Ninth Biennial ACM Symposium on Eye Tracking Research & Applications (ETRA '16), Charleston, SC, USA, 14–17 March 2016; Association for Computing Machinery: New York, NY, USA, 2016; pp. 197–200. [CrossRef]

49. Komogortsev, O.V.; Karpov, A. Automated classification and scoring of smooth pursuit eye movements in the presence of fixations and saccades. *Behav. Res. Methods* **2013**, *45*, 203–215. [CrossRef]

50. Chen, X.; Hou, W. Identifying Fixation and Saccades in Virtual Reality. *arXiv* **2002**, arXiv:2205.04121.

51. Pedregosa, F.; Varoquaux, G.; Gramfort, A.; Michel, V.; Thirion, B.; Grisel, O.; Blondel, M.; Prettenhofer, P.; Weiss, R.; Dubourg, V.; et al. Scikit-learn: Machine Learning in Python. *J. Mach. Learn. Res.* **2011**, *12*, 2825–2830.

Article

Personalized Virtual Reality Environments for Intervention with People with Disability

Manuel Lagos Rodríguez [1], Ángel Gómez García [2], Javier Pereira Loureiro [1] and Thais Pousada García [1,*]

[1] TALIONIS Research Group, CITIC, Universidade da Coruña, 15071 A Coruña, Spain; m.lagos@udc.es (M.L.R.); javier.pereira@udc.es (J.P.L.)
[2] LIA2 Research Group, CITIC, Universidade da Coruña, 15071 A Coruña, Spain; angel.gomez@udc.es
* Correspondence: thais.pousada.garcia@udc.es

Abstract: Background: Virtual reality (VR) is a technological resource that allows the generation of an environment of great realism while achieving user immersion. The purpose of this project is to use VR as a complementary tool in the rehabilitation process of people with physical and cognitive disabilities. An approach based on performing activities of daily living is proposed. Methods: Through joint work between health and IT professionals, the VR scenarios and skills to be trained are defined. We organized discussion groups in which health professionals and users with spinal injury, stroke, or cognitive impairment participated. A testing phase was carried out, followed by a qualitative perspective. As materials, Unity was used as a development platform, HTC VIVE as a VR system, and Leap Motion as a hand tracking device and as a means of interacting with the scenarios. Results: A VR application was developed, consisting of four scenarios that allow for practicing different activities of daily living. Three scenarios are focused on hand mobility rehabilitation, while the remaining scenario is intended to work on a cognitive skill related to the identification of elements to perform a task. Conclusions: Performing activities of daily living using VR environments provides an enjoyable, motivating, and safe means of rehabilitation in the daily living process of people with disabilities and is a valuable source of information for healthcare professionals to assess a patient's evolution.

Keywords: immersive environment; cognitive impairment; physical disability; unity; rehabilitation

Citation: Lagos Rodríguez, M.; García, Á.G.; Loureiro, J.P.; García, T.P. Personalized Virtual Reality Environments for Intervention with People with Disability. *Electronics* **2022**, *11*, 1586. https://doi.org/10.3390/electronics11101586

Academic Editors: Jorge C. S. Cardoso, André Perrotta, Paula Alexandra Silva and Pedro Martins

Received: 14 April 2022
Accepted: 15 May 2022
Published: 16 May 2022

Publisher's Note: MDPI stays neutral with regard to jurisdictional claims in published maps and institutional affiliations.

1. Introduction

During the rehabilitation process of people with disabilities carried out in therapeutic centers, appropriate work environments are generally reproduced in order to improve different physical and cognitive skills. Activities of daily living are frequently practiced in these sessions to increase the patient's functional capacity as well as consequently his or her independence. The configuration of an appropriate workspace to practice such activities can become very complex. This may give rise to physical or economic barriers that hinder access to such environments by healthcare institutions and entities.

Thus, in order to practice different household tasks such as cooking, setting the table, doing the laundry, or gardening, it may be necessary to have certain rooms in a house as well as the necessary equipment. In addition to household chores, other examples of daily activities that are often the subject of therapeutic intervention include shopping in a supermarket, driving a car, moving around a city independently, and activities related to personal care.

It is unfeasible for a healthcare center to have effective environments for the development of all of the above activities. In addition, the execution of certain activities in real spaces without adequate training may pose a risk to the user and his or her environment, including such activities as driving vehicles, handling certain household products, and using kitchen utensils.

As a solution to the previous problem, the use of virtual reality (VR) has been proposed to generate virtual spaces for the practice of daily life activities. VR is a technological resource that allows the generation of environments with a real appearance and interaction with the different elements of scenarios in a similar way as would happen in the real world [1].

The use of virtual environments represents an interesting possibility to reduce the physical and economic limitations in existing configurations, and can generate a safe way of working with those tasks that can represent a danger for the user until he or she has the appropriate skills. In addition, the use of VR can offer better results in certain exercises compared to conventional rehabilitation. This has been pointed out in a study carried out with people with Parkinson's in which the use of VR and conventional rehabilitation was compared [2].

In that case, a VR application was created consisting of four scenarios = representing activities of daily life configured in such a way as to train various physical and cognitive abilities. Three of the scenarios aimed at improving the grip and manipulation of items with the hands and practicing wrist movement. The remaining scenario was intended for practicing the identification of items in the performance of an everyday task.

Unlike other studies [3–5], games or commercial applications of general scope have not been used. These are personalized environments adapted to people who require training for some physical or cognitive pathology. Other noteworthy cases include the use of the Leap Motion device, which allows users to interact with their hands in VR scenarios without the need for controls.

2. Related Work

After a bibliographic analysis of the use of VR equipment in intervention processes for people with disabilities, several studies with satisfactory results stand out and invite further research along these lines. An example of this is a case study on elderly people, including two people with Parkinson's disease [6]. In this study, two tests were performed in virtual environments: a purely observational test used as the first contact with VR, and another that involved the movement of different parts of the body and coordination movements. As equipment, the authors used HTC VIVE VR glasses, two controllers, and sensors to delimit the workspace, and used TheBlue [7] and NVIDIA VR FunHouse [8] games on the software side. TheBlue is a relaxing and mainly observational experience, and was used to introduce users to VR technology. At the end of its run, users were asked about the presence of cybersickness. In the NVIDIA VR FunHouse game, the participants had to perform different tasks such as picking up and throwing objects or popping balloons. Through these tasks, they worked on body movement and coordination. As a result, the authors noted that all participants completed the tests successfully and without adverse effects. In addition, they pointed out that one of the participants used a wheelchair and that because of his pathology he had dorsal kyphosis (curvature of the spine in the cervical area), which made it difficult for him to fully visualize the virtual environment. In order to solve this problem, they tilted his chair during the activity. All users were satisfied with the tests and were willing to repeat them, even recommending their use. As for improvement, one of the participants mentioned possible improvements to the ergonomics of the controls, and one of the users reported slight fatigue at the end of the second test as well as an increase in saliva.

Several studies about the use of VR in the rehabilitation of people following strokes have considered this technology a promising technique in the rehabilitation process [9,10]. As concerns our proposal, we find very interesting the work carried out by researchers at the Polytechnic University of Barcelona in collaboration with the Functional Diversity Association of Osona (ADFO) [11]. In this study, two VR applications were developed oriented to the treatment of physical and cognitive disabilities, mainly aimed at people who had suffered a stroke. The physical rehabilitation application consisted of an avatar to guide the user in the movement of different parts of the body. The cognitive rehabilitation application uses an environment that simulates a supermarket, in which the user has to do their shopping. The objective is to improve memory, coordination, and attention. The hardware used once again consists of HTC VIVE VR glasses, two controllers, and sensors that delimit the workspace. On the software side, it is worth noting that the VR scenarios were developed by the project's researchers. This allowed them to customize the work environments and direct them to the treatment of a series of specific physical and cognitive abilities, a very positive and differentiating aspect compared to other studies.

It is necessary to mention the WalkinVR application [12], which aims to make VR accessible to people with disabilities. This tool can adapt any game or VR environment from the Steam VR catalog. In addition, it is compatible with multiple VR devices, such as HTC VIVE and Oculus Rift. Its main functions are:

- Virtual Movement and Rotation, a function aimed at users with mobility problems, such as those who use wheelchairs or are bedridden;
- Assisted Play, a function designed for people with mobility disorders that make it difficult to use controllers to move around in a game;
- Controller Position Adjustment, which allows solving certain movement restrictions, such as having to place the game controllers at a certain height;
- Hand Tracking. a function aimed at people who, due to their pathology, cannot hold game controllers.

Despite not being a tool aimed at the rehabilitation of people with disabilities, this is a very interesting option in many ways, as it facilitates access to VR for people with certain physical or cognitive conditions. In addition, various games can be used for the rehabilitation of certain pathologies, which provides a safe, enjoyable, and motivating medium for intervention sessions for people with disabilities. However, by using commercial games there is no customization of activities to the pathologies of the users, something that is considered in our proposal.

In summary, the studies analyzed above suggest that VR can be a very interesting tool in rehabilitation sessions for people with disabilities. The use of VR environments is an enjoyable, motivating, and safe way to perform physical or cognitive exercises. Our proposal is not limited to the rehabilitation of physical or cognitive skills; rather, it has exercises for working on both types of skills. Particularly noteworthy is the customization of scenarios and the use of the hands as a means of interaction, which eliminates the need for controls. The non-use of controls is noteworthy because it significantly increases immersion in VR environments and provides greater realism in the performance of activities. In addition, in the context of rehabilitation it is even more important, as some users with physical pathologies involving their hands may have difficulties in holding and handling controllers. Interaction through the hands for a long time can be more fatiguing than other means of interaction [13]. Another differentiating aspect of our proposal is the special care involved in the design of virtual environments. Numerous 3D models were configured in order to be as close to reality as possible. Likewise, deep work was performed in the development and implementation of the functions in order to achieve a realistic interaction with the objects of the scene. As a result, great immersion of the user in the VR scenarios was achieved along with an interaction system that allows movements very close to reality.

Although VR has been presented as a promising technology in the rehabilitation process of people with disabilities, there are limitations to be aware of. The use of VR scenarios by blind or deaf people can offer certain barriers without adequate help. However, there are studies that deal with this problem and that provide hopeful results [14,15]. In addition, an interesting study [16] reflects the importance of analyzing factors that influence the perceived stigma associated with the use of assistive technologies. Greater care in development could reduce technology abandonment and perceived stigmas. Possible difficulties in implementing VR systems in rehabilitation centers should be taken into account. Healthcare professionals do not usually have close contact with ICTs, and care must be taken in designing applications in order to make them simple and intuitive.

3. Materials and Methods

Meetings were scheduled with different entities and organizations in the health field, particularly those working with people with disabilities. The objective of these meetings was to obtain information about the work sessions implemented by professionals for the users of these centers. Different needs and improvements in the therapeutic sessions that could be solved through the use of VR scenarios were registered. Thus, the pathologies to be treated, the virtual scenarios, and the actions that users should perform in these environments were defined. Four scenarios were agreed upon: the first scenario represented a living room and was intended to train movement with the hands (displacement, grip, and rotation). Users had to place different pieces of fruit in a fruit bowl, which involves practicing different grip modes. The second scenario simulated a kitchen, and aimed to improve difficulties with wrist movement. Users had to turn on a faucet, fill a jug with water, and turn the faucet off again. The third scenario again represented a living room; its purpose was to train the handling of cutlery. Users had to hold a fork and knife and cut a piece of meat. The fourth scenario represented a kitchen, and its objective was to practice the identification of elements required to carry out a task. Users had to select the foods needed to prepare breakfast and the appropriate cutlery to eat it.

Once the scenarios were defined, the appropriate components for the development of the application were selected and configured. Unity was used as the development platform for the VR scenarios, in combination with different plugins and libraries. To generate the VR experience, HTC VIVE glasses and a Leap Motion device were used. Sections 3.1 and 3.2 describe in detail the software and hardware used.

After the development of the application, three discussion groups were organized with the centers of people with disabilities. Four users and four health professionals participated in each session. All participants had a spinal injury, stroke, or cognitive impairment. Each user tried only those VR environments suitable for their pathology. The tests were developed in the controlled environment of the research group's lab. A qualitative perspective was followed in which the perceptions and opinions of users and health professionals were recorded. In order to do this, several questions were defined:

- How people felt about the use of VR equipment and scenarios;
- Whether VR environments were considered useful in the rehabilitation process;
- What components would participants remove;
- What elements could be incorporated into the design of the scenarios;
- How the users felt during and after finishing the activity.

3.1. Hardware

- HTC VIVE is VR equipment created by HTC and Valve [17]. It performs the visual representation of VR applications and allows the user to interact with the scenarios. The main components of this equipment are glasses, controls, and position sensors.

Additionally, it is necessary to have computer equipment that meets the minimum requirements specified by the manufacturer and to prepare the physical space where the users will move. An area of a minimum of 3 m^2 and a maximum of 15 m^2 is required for the position sensors to be placed. After all of the components have been installed, we simply need to run a game on the computer, put on the glasses, grab the controls and step into the play area.

- Leap Motion: This is an optical system capable of tracking the hands and fingers, allowing interaction with digital content such as games or applications of different kinds [18]. This device integrates perfectly with the HTC VIVE glasses (Figure 1). Considering the objectives of this project, the use of Leap Motion provides great benefits in the interaction with the elements of the different scenarios. In addition to providing much more realistic environments, the amount of movements that can be performed with the hands is not comparable to that which a remote control allows.

For these reasons, it was decided to make use of this device in all of the scenarios we developed.

Figure 1. HTC VIVE glasses with Leap Motion.

3.2. Software

- Unity [19]: This is a real-time 3D development platform which allows the creation of interactive environments for multiple platforms, including PC, consoles, or VR equipment [20]. It is available with Windows, Mac OS, or GNU/Linux operating systems and offers various types of licenses classified according to whether it is to be used in a personal or business environment.
- OpenVR [21]: This is an API created by Valve that allows access to the hardware of different VR equipment such as HTC VIVE or Oculus Rift. It is necessary to add OpenVR in Unity to correctly compile the environment and obtain a functional program for the HTC VIVE device.
- SteamVR [22]: This runtime environment allows the use of applications developed with the OpenVR API.
- Unity Leap Motion Modules: The manufacturer of Leap Motion provides various modules [23] that must be imported into Unity to access the API of the device, as well as different resources that will facilitate the development of the different scenarios. Specifically, in this project, the Core and Interaction Engine modules were used.

4. VR Application

The solution proposed in this project consists of a VR application that has different environments simulating situations of daily life, configured in such a way as to allow training of various physical and cognitive abilities. The use of spaces that represent situations of real life is intended to allow the user to train that physical and cognitive problem naturally through action or challenge that he/she is likely to have to perform daily.

As the usual Unity programming methodology was followed, first, the different 3D objects that form the VR scenarios were created and imported. Then, the programming process was carried out, consisting of the creation of a set of scripts to provide functionality to the objects and allow user interaction.

Four scenarios were developed to represent environments in which activities related to food and cooking take place. Three of the scenarios aimed at treating mobility problems in the hands, such as the grasping of objects or the rotation and prono-supination movement of the wrist. The fourth scenario is intended to improve problems of a cognitive nature, in particular, the identification of elements to perform a common task of daily living, such as preparing breakfast.

These scenarios were modeled and implemented in the form of a test, with specific goals that the user must achieve. In this way, the person who participates in the test can feel more motivated to reach the established objectives in comparison with traditional rehabilitation sessions. During the test with VR scenarios the user trains and progressively improves their abilities, reducing their functional limitations. In the execution of each environment, the user knows their progress in the pursuit of the objectives to be achieved through the use of visual or sound signals. After finishing the objective, a report is generated with data of interest on the specific skill being worked on that (Table 1) together with a live visualization of the test, allowing healthcare professionals to assess and check the user's evolution.

Table 1. Example of the data recorded in the Fruit Bowl scenario.

User	Test Time (Min.)	Left Hand						Right Hand					
		ABD	ADD	PF	WE	PE	PI	ABD	ADD	PF	WE	PI	PE
User 1	1.02	0°	76.52°	34.25°	54.79°	102.51°	52.40°	31.37°	36.64°	50.58°	38.05°	39.00°	131.54°
User 2	2.01	44.77°	48.33°	22.37°	71.48°	179.92°	18.19°	63.53°	80.06°	37.53°	71.83°	31.51°	179.87°
User 3	0.51	9.45°	57.74°	0°	45.477°	33.14°	23.59°	0°	0.24°	0°	5.51°	9.36°	0°
User 4	2.18	79.48°	81.29°	80.17°	45.88°	123.9°	45.16°	0°	76.52°	52.30°	11.59°	22.45°	52.86°
User 5	0.46	25.13°	48.05°	53.14°	54.83°	47.27°	36.45°	92.46°	35.22°	36.95°	74.72°	28.21°	105.20°
User 6	0.58	77.09°	53.85°	45.46°	40.11°	138.64°	14.96°	0°	5.57°	29.53°	0°	3.76°	2.12°
User 8	3.25	63.23°	85.38°	64.09°	41.87°	178.62°	36.01°	53.84°	95.71°	30.29°	69.38°	0°	178.92°
Mean	1.25	37.39°	56.40°	37.44°	44.30°	100.5°	28.35°	30.15°	41.25°	29.65°	33.89°	16.79°	81.31°

(ABD) Abduction, (ADD) Adduction, (PF) Palmar Flexion, (WE) Wrist Extension, (PE) Prono-supination External, (PI) Prono-supination Internal.

When starting the application, a home screen is displayed in which the user's first and last name must be entered. This information is later used as a means of identification in the activity report. After the user's data has been inserted, the VR application starts and the user sees an initial scenario through which he or she can select each work environment or end the application.

4.1. Description of Environments

- *Fruit bowl* scenario: Simulates the living room of a house; the mission proposed to the user is to place a series of pieces of fruit in a fruit bowl (Figure 2). The therapeutic objective of this scenario is to improve the grip and manipulation of the elements (which have different forms) with the hands. Thus, the use of different pieces of fruit allows the training of different grip modes, and the action of placing them in a fruit bowl allows the practice of hand movement combined with the grasping of elements. The movement that the user has to perform is spherical power grasping, implicating the long finger flexors to grasp and long finger extensors to release each fruit.

Figure 2. Fruit bowl scenario.

- *Tap* scenario: Simulates a kitchen; the objective proposed to the user is to fill a jug with water (Figure 3). To meet the goal, the user must open the screw tap using a rotational movement, fill the jug, and close it again. The aim is to improve difficulties present in the movement of the wrist. The action of turning the screw tap to open and close it, which allows training the rotation of the wrist in both directions of rotation while at the same time training a common action of daily life. The user has to perform a precision disc grasp. In addition, with this dam the user has to rotate the wrist to the right or left depending on whether they want to open or close the tap.

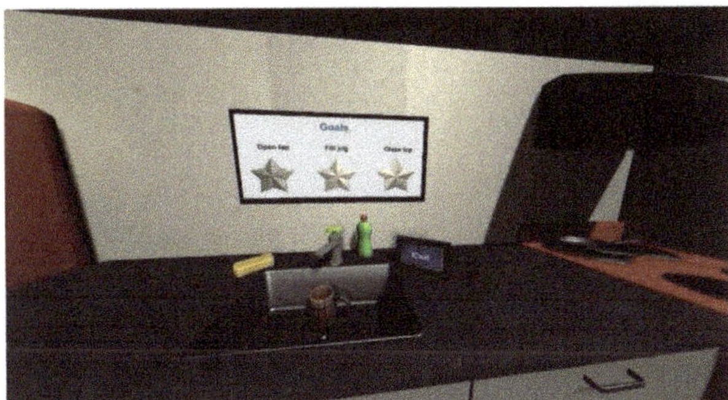

Figure 3. Tap scenario.

- *Cutlery* scenario: Again, the scenario represents a living room of a house; the action proposed to the user is to grab a knife and a fork and cut a piece of meat (Figure 4). The purpose of this scenario is to improve the grip and handling of cutlery. Users may have problems with the handling of cutlery at different meals of the day, which can limit their independence. The use of VR provides a safe, realistic, and motivating means of training. In this case, the movement and grasp implicate more manipulative dexterity. The user has to apply precision grasping with three fingers or with a thumb and finger. The movement implicates the flexo-extension of the elbow and rotation of the shoulders.

Figure 4. Cutlery scenario.

- *Breakfast* scenario: This environment, unlike the previous ones, aims to improve a problem of a cognitive nature related to the identification of elements to carry out an activity of daily life (Figure 5). Its execution implies working on the movement of the hands. Eating and cooking are common tasks in our daily lives, and are an ideal way to train those difficulties in identifying elements to perform certain tasks. In this scenario, the user must find several types of cutlery and different foods commonly present at breakfast, such as milk or cereals, and others that have nothing to do with it, such as a lemon or chicken breasts. In this way, the user has to select the food needed to prepare breakfast and the appropriate cutlery to take it. In this case, the grasp again implicates the whole hand, both spherical and cylindrical power grasping, involving the extensor and flexors muscles of the fingers. In addition, the user has to move over the space in order to transport objects from the worktop to the table and turn him or herself.

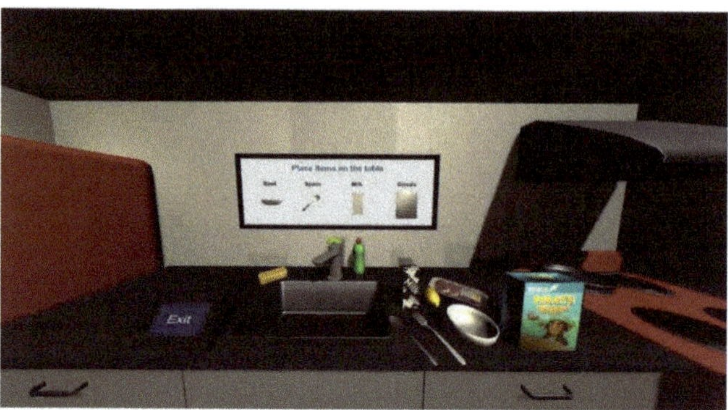

Figure 5. Breakfast scenario.

4.2. Visual and Audio Signals

To make the different environments more realistic, audio has been added to the different actions that take place during the execution of each environment. In addition, visual and audio cues have been added in order to provide feedback to the user on their progress in achieving the objectives of each test.

In the Fruit Bowl scenario, audio is played to simulate the real sound of hands touching the fruit. Regarding the tracking of the objectives, 3D models of the different pieces of fruit are used to create a panel where each fruit is displayed in black and white. Each fruit on the panel is associated with the real object that the user grabs with their hand; when a piece of fruit is placed in the fruit bowl, its representation on the panel changes from black and white to full color, and a sound is emitted each time one of the fruits is placed in the fruit bowl. In addition, after all the fruits have been placed in the fruit bowl a special sound is emitted to indicate the end of the game.

In the Tap scenario, when the screw tap is moved audio is played to simulate the sound it would make in reality, and the audio simulates the water flow. A panel with three objects representing a star was added, with each star associated with a game goal; the achievement of each goal illuminates its corresponding star and emits a sound. In addition, when all three objectives are achieved a special sound is emitted to indicate the end of the game.

In the Cutlery scenario, a sound is played simulating the cutting of the meat. The progress in the activity can be checked through a light bar and a percentage indicator, which vary as cuts are made. When the bar reaches 100%, a sound is played to indicate the end of the activity.

Finally, the implementation of the Breakfast scenario is carried out in the same way as the Fruit Bowl scenario; that each time one of the breakfast elements is placed on the indicated table, its representation on the panel changes from black and white to color, and a sound is played.

4.3. Data Export

One of the objectives of the project is the export of the data generated during the activity in the different scenarios for subsequent evaluation by a therapist. Thus, after the activity is finished a report is generated with different data of interest for the evaluation of the user. On the one hand, the maximum value (in degrees) that each hand is able to move in the X, Y, and Z axes is saved; thus, the lateral, vertical, and rotational movement of each hand can be evaluated. In addition, the name of the activity, the time of duration, and the date of completion are recorded. Table 1 shows an example of the data recorded by the Fruit Bowl scenario from the test sessions with entities and organizations.

To make it easier to keep track of different users, the application generates a separate file for each user and saves their activity logs in an orderly manner, taking into account the user's data indicated on the main screen of the application. If the user later uses the application again, the application will check to see whether a file already exists for that user. If it exists, it will continue to save the user's data. Otherwise, it will create a new file. This facilitates the process of following up on the situation of each user.

5. Results

After the application test was finished, different discussion groups were organized with entities and organizations from the social and health sectors. The purpose of these meetings was to assess the usability of the application, its suitability for the rehabilitation of the pathologies to be treated, and the users' reaction to the use of VR as a therapeutic tool. Before starting the exercise, users were provided with an explanation about the VR device, the tasks and the ability needed to work in each of the scenarios.

According to the methods applied and the questions asked during the focus groups with both professionals and users, several interesting features emerged.

In response to the above questions, most users indicated that the glasses were comfortable, except for one user who highlighted their weight as a negative. On the other hand, users and therapists saw it as very positive to not have to use controls. Regarding the design of the scenarios, all users were satisfied with the environments represented and highlighted as a positive the performance of daily life tasks as a means of rehabilitation. Regarding its usefulness in the rehabilitation process, the therapists verified that the activities presented in the virtual scenarios were correctly adjusted to the therapeutic intervention of the pathologies to be trained. In addition, both users and health professionals highlighted the way that the proposed activities made the rehabilitation process motivating and entertaining. Regarding the removal or addition of components to the scenarios, several users suggested an increase in the size of the text of the instructions. Likewise, the therapists recommended the incorporation of instructions in audio format. All users expressed feeling good during the development of the test. Only one of the users reported slight tiredness at the end of the exercise. None of the users felt dizzy, something that frequently occurs when using VR systems.

In the different organized discussion groups, all of the tests carried out were completed, and it was concluded that the developed application can be a powerful tool in rehabilitation sessions. It must be emphasized that these activities represent situations of daily life and that their implementation makes them motivating, entertaining, and safe.

6. Discussion

This project aimed to create VR scenarios as a tool during the rehabilitation process for people with disabilities. In order to do this, an application was created consisting of different environments that simulate everyday life situations configured in a way that allows training in different physical and cognitive abilities. In this section, the main developments and designed virtual scenarios are presented, highlighting their application for improving the rehabilitation process of people with disabilities.

In this article, the benefits of using an HTC VIVE device in combination with the Leap Motion device have been demonstrated. However, other devices of great interest are starting to emerge, such as the Oculus Quest. This VR equipment stands out for its lower cost and for being autonomous (that is, it does not require a computer or external sensors to configure the VR space), as well as for incorporating hand tracking into the device itself. These characteristics make it a very attractive device and invite future research on new more affordable and accessible devices, which could facilitate its implementation in homes and medical centers. We were able to find works [24] in which these devices are beginning to be used, and in which a resulting simplification can be observed in terms of installation and configuration of the VR space. Taking into account the above, it would be of great interest to carry out a comparative study between HTC VIVE and Oculus Quest. This would make it possible to observe in detail the advantages and disadvantages of each device and consequently choose the most appropriate for particular rehabilitation processes.

Leap Motion device is an optical system capable of tracking the movement of a person's hands and fingers, which allows her/him to interact with the different environments directly without the need for controls. The incorporation of this device was an added value in our project, as several work environments under study were intended to work with mobility problems in the hands, which is undoubtedly carried out more effectively by avoiding the use of controls as a means of interaction with the virtual space. In addition, at a general level it allows a greater immersion of the user in any type of environment. The cost of such a device is approximately USD 90, a very low cost considering the benefits it brings to the project.

After analyzing different tools existing in the current market in the Background section, the main differentiating features of the tool proposed in this project are highlighted, along with possible points for improvement.

In the study "Immersive Virtual Reality in older people: a case study" [6], one of the participants suggested improving the ergonomics of the controls. Our proposal would contribute to eliminating the ergonomics problem of the controllers, as the integration of the Leap Motion device avoids the use of the controllers, allowing the use of the hands as a means of interaction with the VR environments. In addition, it would provide greater realism and a wide range of possibilities in terms of interaction in the scenarios. The difference would be remarkable; think of the movements a person can make with the joints of one hand compared to the limitations of interacting with a controller as a medium.

On the other hand, programming our scenario would allow an option to be added to the interface in which we could configure the inclination of the camera in charge of capturing the elements of the scene. In this way, we could adapt this inclination to the user's problems. In the study in question, we would have avoided having to recline the wheelchair of the user with dorsal kyphosis.

In the analysis of the tool proposed by the Polytechnic University of Barcelona [11], we can highlight as very positive that the scenarios were developed by the researchers themselves; this allows better adaptation of the environments to the pathologies to be treated. The need to interact with controls can be a limitation for certain users, as, due to their pathology, they may have problems with the grip or with pressing the different buttons. This inconvenience can be solved using the Leap Motion device.

In a study similar to the previous one [25], in addition to using personalized scenarios, the Leap Motion device was used as a means of interaction. However, unlike our proposal, less realistic scenarios were observed, with an appearance more similar to a game than to real life. The same was the case in the interaction with the elements of the stage, which were not entirely realistic. For example, to open or close a faucet the users were required open or close their hand, instead of having to make the opening movement.

The WalkinVR application [12] provides both users and healthcare professionals with a wide range of possibilities by making all the games in the SteamVR catalog accessible to people with disabilities.

In fact, in the SteamVR catalog there are experiences derived from the application of games for health, with a focus on their impact on motivation, therapeutic results, and the behavior of users. However, these applications fail in the adequacy of the principles of universal design, and there are not any guidelines of accessibility oriented to the developers of the virtual environments [26].

Nevertheless, when using commercial apps as a complementary resource in rehabilitation sessions with people with disability it is possible to find several drawbacks, such as lack of customization, as these are general-purpose applications with few configuration options. Another difficulty is language, as most of the applications are only in English, which can negatively affect the VR experience; for example, non-English speakers may not know the objective of the game. For the most part, commercial applications are complex, their instructions are not easy to understand, and they do not give feedback to the user to guide him or her during the game. [27]. The need to use the controls to manage the elements in the VR environment is an important barrier and handicap for people with disabilities involving movement limitations in their arms and hands.

In a review of the application of Virtual Reality in Complex Medical Rehabilitation, it was concluded that "the use of virtual environments has proven effective for the recovery of impaired motor skills in people with disabilities". Nevertheless, it is necessary to provide personalized scenarios, goals, and tasks in order to achieve the maximum possible improvement in physical and cognitive skills as well as to register the results obtained during the sessions for each user [28].

Therefore, in the present project, a customizable VR application has been created to improve the rehabilitation process of people with disabilities. The virtual scenarios can be adapted to the needs of the user, who can interact with the environment directly with his or her hands thanks to the Leap Motion device, avoiding the use of controls. The visual and auditory feedback offered through the application is another strong point, and is a differential element compared to other commercial applications.

During the execution of each environment, the user receives feedback in order to always know the goal of the game and their progress in it. After finishing the task, a report is generated with data of interest on the specific skill that is being worked on, allowing the therapist to assess the evolution of the user.

During and after the development of the presented scenarios, the researchers were advised by health professionals from rehabilitation centers. Several meetings and trials with real users were carried out in order to configure and improve the scenarios and the inputs and outputs of the application. This collaborative work allowed for the creation and design of interactive solutions in which virtual scenarios meet the preferences, needs, and specifications of both professionals and users. Just this type of implication from all stakeholders suggests that it is possible to develop useful, practical, and relevant scenarios for the process of rehabilitation.

Future Research Directions

A new VR application is currently being developed that will allow even greater customization of the scenarios. This application will have an initial menu through which a professional will be able to select the virtual scenario and configure the available options to adapt them to a specific user. Several of the customizable options will take into account the degree of difficulty of performing the activity, both physically and cognitively. Different scenarios are currently being developed:

- An environment recreating an outdoor orchard: The user has to collect different vegetables indicated in the initial instructions. The scenario will allow training in both physical and cognitive skills. The professional can configure the main features, such as the level of difficulty or the height where the vegetables are located, to adapt the task to the progression of the person.
  - To train physical abilities, the scenario can be oriented to strengthen the thoracolumbar musculature in people who use a wheelchair. One of the tasks implicates that the person must collect tomatoes, located at different heights, and place them into boxes. The height and size of the tomatoes can be adapted by a professional according to the skills to be trained.
  - Concerning the training of cognitive ability, the orchard has tomatoes as well a different types of vegetables. Therefore, the goal can be to complete each box with a different combination of vegetables, indicated in the initial instructions. The number and variety of vegetables can be configured by the professional.

- An indoor supermarket as a virtual scenario: This environment proposes a great variety of options related to the performance of daily life activities, and will be able to present them to the user to practice during the task. The proposed activities that are being developing are:
  - From a shopping list, take the listed products
  - Starting from a cooking recipe, acquire all the necessary ingredients to prepare it
  - Practice paying for the purchase and handling money
  - Purchase of products without exceeding a previously set budget.

These scenarios simulate situations of daily life, although the development of other environments focused on other areas, such as work, leisure, and sports, is contemplated as well.

Our research team is studying the possibility of measuring a greater number of parameters during the execution of the application. These parameters could provide information about how the user feels while performing an activity, as well as more data that can provide more information to the therapist when evaluating a user's progress. Possible parameters include heart rate, sweating, time to achieve each goal in the game, or the number of attempts.

In order to obtain a functional and usable tool, the research team is working collaboratively with NGOs, professionals, and people with disabilities. During work meetings, participants use the virtual scenarios in progress and provide their opinions about different features and improvements to the application. The development of the whole project thus has a perspective of user experience.

7. Conclusions

The present project is an innovative and technological development directed to improve the rehabilitation process of people with disabilities as a complementary resource during their intervention. The virtual scenarios that the application incorporates are designed considering the skills to be trained by the users. Training based on activities of daily living were chosen. In this way the user, in addition to improving their physical and cognitive ability, is able to achieve greater independence in their daily life. Setting clear goals in each VR environment and providing feedback to the user during each task helps to increase their motivation in training. In addition, the high realism of the scenarios developed allows greater immersion, contributing to a better experience in carrying out activities. The Leap Motion device is an added source of value to the project, allowing interaction with the application without the need for controls, which is a clear advantage compared to most of the previous studies analyzed. The virtual scenarios developed in this project have potential benefits:

- People with disabilities and rehabilitation professionals can benefit from the use of VR as a complementary technological tool during the intervention process. Using the same space or room of the hospital or clinical center, VR can simulate a great variety of scenarios, both indoor and outdoor.
- The use of personalized scenarios offers clear advantages over the use of commercial options, allowing activities to be adapted to the abilities and needs of each user.
- The application offers clear advantages to complement the traditional rehabilitation process and intervention for people with disabilities. It allows health professionals to have control of the whole application, configuring the appropriate activity for each user and knowing their progression. He or she can monitor the evolution of each user through the results obtained from his or her motor and cognitive inputs thanks to the registers captured by the application.
- The Leap Motion device is a very useful tool when combined with the VR glasses to create a global and completely immersive experience. This combination allows for dispensing with controllers, as the user can use their hands and fingers to directly interact with the scenarios. These movements lead to improvements in coordination and fine dexterity ability.

In the future, the final goal of the project is to consolidate a solid application with more virtual scenarios simulating activities of daily living and to implement this low-cost tool as a complement to the routine intervention of the rehabilitation centers in our environment

Author Contributions: Conceptualization, M.L.R. and Á.G.G.; methodology, M.L.R. and Á.G.G.; software, M.L.R.; validation, J.P.L. and T.P.G.; formal analysis, Á.G.G.; investigation, M.L.R. and T.P.G.; resources, J.P.L.; data curation, M.L.R.; writing—original draft preparation, M.L.R.; writing—review and editing, Á.G.G., J.P.L. and T.P.G.; visualization, J.P.L.; supervision, Á.G.G.; project administration, Á.G.G.; funding acquisition, J.P.L. and T.P.G. All authors have read and agreed to the published version of the manuscript.

Funding: The APC was funded by the National Program of R + D+i oriented to the Challenges of Society 2019 (coordinated research) Grant number: PID2019-104323RB-C33. Ministry of science and innovation.

Institutional Review Board Statement: Not applicable.

Data Availability Statement: Not applicable.

Acknowledgments: The research was done in the Center CITIC, which is a Research Center accredited by Galician University System (Xunta de Galicia). CITIC is partly supported by "Consellería de Cultura, Educación e Universidades from Xunta de Galicia", which provided 80% of funds through ERDF Funds, ERDF Operational Programme Galicia 2014-2020, and the remaining 20% was provided by "Secretaría Xeral de Universidades [Grant ED431G 2019/01]. Research group thanks to the users from associations of people with disabilities, who, with their participation, helped develop the application and obtain results.

Conflicts of Interest: The authors declare no conflict of interest. The funders had no role in the design of the study; in the collection, analyses, or interpretation of data; in the writing of the manuscript, or in the decision to publish the results.

References

1. Castañares, W. Realidad Virtual, mímesis y simlación. *CIC Cuadernos de Información y Comuniación* **2011**, *16*, 59–81. Available online: https://www.redalyc.org/articulo.oa?id=93521629004 (accessed on 20 December 2021).
2. Pazzaglia, C.; Imbimbo, I.; Tranchita, E.; Minganti, C.; Ricciardi, D.; lo Monaco, R.; Parisi, A.; Padua, L. Comparison of virtual reality rehabilitation and conventional rehabilitation in Parkinson's disease: A randomised controlled trial. *Physiotherapy* **2020**, *106*, 36–42. [CrossRef] [PubMed]
3. Erhardsson, M.; Alt Murphy, M.; Sunnerhagen, K.S. Commercial head-mounted display virtual reality for upper extremity rehabilitation in chronic stroke: A single-case design study. *J. Neuroeng. Rehabil.* **2020**, *17*, 1–14. Available online: https://jneuroengrehab.biomedcentral.com/articles/10.1186/s12984-020-00788-x (accessed on 20 December 2021). [CrossRef] [PubMed]
4. Ahmad, M.A.; Singh, D.K.A.; Nordin, N.A.M.; Nee, K.H.; Ibrahim, N. Virtual Reality Games as an Adjunct in Improving Upper Limb Function and General Health among Stroke Survivors. *Int. J. Environ. Res. Public Health* **2019**, *16*, 5144. Available online: https://www.mdpi.com/1660-4601/16/24/5144/htm (accessed on 20 December 2021). [CrossRef] [PubMed]
5. Singh, D.K.A.; Nor, N.; Rajiman, S.; Yin, C.; Karim, Z.; Ruslan, A.; Kaur, R. Impact of virtual reality games on psychological well-being and upper limb performance in adults with physical disabilities: A pilot study. *Med. J. Malays.* **2017**, *72*, 119–121.
6. Campo-Prieto, P.; Carral Cancela, J.M.; Oliveira, I.M.D.; Rodríguez-Fuentes, G. Realidad Virtual Inmersiva en Personas Mayores: Estudio de Casos (Immersive Virtual Reality in Older People: A Case Study). Retos [Internet]. 2020 [cited 2021]; 39:1001-5. Available online: https://recyt.fecyt.es/index.php/retos/article/view/78195 (accessed on 20 December 2021).
7. Wevr. TheBlue. Version 2018_03_20_theBlu_16. Viveport. 2016. Available online: https://www.viveport.com/1b591122-7ab7-4c27-9d31-cbaf9ef8e1e1 (accessed on 20 December 2021).
8. Lightspeed Studios. NVIDIA VR FunHouse. Version 1.3.1. Steam. 2016. Available online: https://store.steampowered.com/app/468700/NVIDIA_VR_Funhouse/ (accessed on 20 December 2021).
9. Kim, W.S.; Cho, S.; Ku, J.; Kim, Y.; Lee, K.; Hwang, H.-J.; Paik, N.-J. Clinical Application of Virtual Reality for Upper Limb Motor Reha-bilitation in Stroke: Review of Technologies and Clinical Evidence. *J. Clin. Med.* **2020**, *9*, 3369. Available online: https://www.mdpi.com/2077-0383/9/10/3369/htm (accessed on 20 December 2021). [CrossRef] [PubMed]
10. Maggio, M.G.; Latella, D.; Maresca, G.; Sciarrone, F.; Manuli, A.; Naro, A.; De Luca, R.; Calabrò, R.S. Virtual reality and cognitive rehabilitation in people with stroke: An overview. *J. Neurosci. Nurs.* **2019**, *51*, 101–105. Available online: https://journals.lww.com/jnnonline/Fulltext/2019/04000/Virtual_Reality_and_Cognitive_Rehabilitation_in.9.aspx (accessed on 14 January 2022). [CrossRef] [PubMed]
11. ad Virtual para la Rehabilitación de Personas que Han Sufrido un Ictus. Polytechnic University of Barcelona. 2019. Available online: https://cit.upc.edu/es/portfolio-item/rv_rehabilitacion_ictus/ (accessed on 15 January 2022).

12. 2MW. WalkinVR. Version 2.1.2.0. Steam. 2020. Available online: https://www.walkinvrdriver.com/ (accessed on 15 January 2022).
13. Jorge, C.S.C. Gesture-Based Locomotion in Immersive VR Worlds with the Leap Motion Controller. In Proceedings of the 11th International Conference on Interfaces and Human Computer Interaction, Lisbon, Portugal, 21–23 July 2017.
14. De Souza, E.S.; Cardoso, A.; Lamounier, E. A Virtual Environment-Based Training System for a Blind Wheelchair User Through Use of Three-Dimensional Audio Supported by Electroencephalography. *Telemed. e-Health* **2018**, *24*, 614–620. Available online: https://www.liebertpub.com/doi/full/10.1089/tmj.2017.0201 (accessed on 20 December 2021). [CrossRef] [PubMed]
15. Mirzaei, M.; Kán, P.; Kaufmann, H. Effects of Using Vibrotactile Feedback on Sound Localization by Deaf and Hard-of-Hearing People in Virtual Environments. *Electronics* **2021**, *10*, 2794. Available online: https://www.mdpi.com/2079-9292/10/22/2794/htm (accessed on 1 February 2022). [CrossRef]
16. Darc, A.; Santos, P.; dos Lya, A.; Ferrari, M.; Medola, F.O.; Sandnes, F.E. Aesthetics and the perceived stigma of assistive technology for visual impairment. *Disabil. Rehabil. Assist. Technol.* **2022**, *17*, 152–158. [CrossRef]
17. HTC Corporation. HTC VIVE. Available online: https://www.vive.com/eu/product/vive/ (accessed on 20 December 2021).
18. Ultraleap. Leap Motion. Available online: https://www.ultraleap.com/product/leap-motion-controller/ (accessed on 20 December 2021).
19. Unity Technologies. Unity. Version 2019.4.5f1. Unity Technologies. 2019. Available online: https://unity.com/ (accessed on 20 December 2021).
20. Lidon, M. *Unity 3D*, 1st ed.; Marcombo: Barcelona, Spain, 2019.
21. Valve Software. OpenVR. Version 1.14.15. Valve Software. 2020. Available online: https://github.com/ValveSoftware/openvr (accessed on 20 December 2021).
22. Valve Software. SteamVR. Version 1.15. Valve Software. 2020. Available online: https://partner.steamgames.com/doc/features/steamvr/ (accessed on 20 December 2021).
23. Ultraleap. Leap Motion Modules. Version 4.5.1. Ultraleap. 2020. Available online: https://developer.leapmotion.com/unity/ (accessed on 20 December 2021).
24. Paraense, H.; Marques, B.; Amorim, P.; Dias, P.; Santos, B.S. Whac-A-Mole: Exploring Virtual Reality (VR) for Upper-Limb Post-Stroke Physical Rehabilitation based on Participatory Design and Serious Games. In Proceedings of the 2022 IEEE Conference on Virtual Reality and 3D User Interfaces Abstracts and Workshops (VRW), Christchurch, New Zealand, 12–16 March 2022; pp. 716–717. Available online: https://ieeexplore.ieee.org/document/9757547/ (accessed on 1 April 2022).
25. Dias, P.; Silva, R.; Amorim, P.; Laíns, J.; Roque, E.; Pereira, I.S.F.; Pereira, F.; Santos, B.S.; Potel, M. Using Virtual Reality to Increase Motivation in Poststroke Rehabilitation: VR Therapeutic Mini-Games Help in Poststroke Recovery. *IEEE Comput. Graph. Appl.* **2019**, *39*, 64–70. [CrossRef] [PubMed]
26. Ekbia, H.R.; Lee, J.; Wiley, S. Rehab Games as Components of Workflow: A Case Study. *Games Health* **2014**, *3*, 215–226. Available online: https://www.liebertpub.com/doi/10.1089/g4h.2014.0039 (accessed on 20 December 2021). [CrossRef] [PubMed]
27. Miranda-Duro, M.D.C.; Concheiro-Moscoso, P.; Lagares Viqueira, J.; Nieto-Rivero, L.; Canosa Domínguez, N.; García, T.P. Virtual Reality Game Analysis for People with Functional Diversity: An Inclusive Perspective. In Proceedings of the 3rd XoveTIC Conference, A Coruña, Spain, 8–9 October 2020.
28. Volovik, M.G.; Borzikov, V.V.; Kuznetsov, A.N.; Bazarov, D.I.; Polyakova, A.G. Virtual Reality Technology in Complex Medical Rehabilitation of Patients with Disabilities. *Sovrem. Tehnol. V Med.* **2018**, *10*, 173–182. Available online: http://www.stm-journal.ru/en/numbers/2018/4/1492 (accessed on 1 April 2022). [CrossRef]

 *electronics*

MDPI

Article

Preoperative Virtual Reality Surgical Rehearsal of Renal Access during Percutaneous Nephrolithotomy: A Pilot Study

Ben Sainsbury [1,*], Olivia Wilz [2], Jing Ren [1], Mark Green [1], Martin Fergie [3] and Carlos Rossa [4]

1 Faculty of Science, Ontario Tech University, Oshawa, ON L1G 0C5, Canada; jing.ren@ontariotechu.ca (J.R.); mark.green@ontariotechu.ca (M.G.)
2 Faculty of Engineering and Applied Science, Ontario Tech University, Oshawa, ON L1G 0C5, Canada; olivia.wilz@ontariotechu.net
3 Division of Informatics Imaging and Data Science, University of Manchester, Manchester M13 9PL, UK; martin.fergie@manchester.ac.uk
4 Department of Systems and Computer Engineering, Carleton University, Ottawa, ON K1S 5B6, Canada; rossa@sce.carleton.ca
* Correspondence: ben@marionsurgical.com

Abstract: Percutaneous Nephrolithotomy (PCNL) is a procedure used to treat kidney stones. In PCNL, a needle punctures the kidney through an incision in a patient's back and thin tools are threaded through the incision to gain access to kidney stones for removal. Despite being one of the main endoscopic procedures for managing kidney stones, PCNL remains a difficult procedure to learn with a long and steep learning curve. Virtual reality simulation with haptic feedback is emerging as a new method for PCNL training. It offers benefits for both novices and experienced surgeons. In the first case, novices can practice and gain kidney access in a variety of simulation scenarios without offering any risk to patients. In the second case, surgeons can use the simulator for preoperative surgical rehearsal. This paper proposes the first preliminary study of PCNL surgical rehearsal using the Marion Surgical PCNL simulator. Preoperative CT scans of a patient scheduled to undergo PCNL are used in the simulator to create a 3D model of the renal system. An experienced surgeon then planned and practiced the procedure in the simulator before performing the surgery in the operating room. This is the first study involving survival rehearsal using a combination of VR and haptic feedback in PCNL before surgery. Preliminary results confirm that surgical rehearsal using a combination of virtual reality and haptic feedback strongly affects decision making during the procedure.

Keywords: PCNL; simulation; surgical rehearsal; haptic feedback; virtual reality; surgery

Citation: Sainsbury, B.; Wilz, O.; Ren, J.; Green, M.; Fergie, M.; Rossa, C. Preoperative Virtual Reality Surgical Rehearsal of Renal Access during Percutaneous Nephrolithotomy: A Pilot Study. *Electronics* **2022**, *11*, 1562. https://doi.org/10.3390/electronics11101562

Academic Editors: Jorge C. S. Cardoso, André Perrotta, Paula Alexandra Silva and Pedro Martins

Received: 25 March 2022
Accepted: 30 April 2022
Published: 13 May 2022

Publisher's Note: MDPI stays neutral with regard to jurisdictional claims in published maps and institutional affiliations.

1. Introduction

Percutaneous Nephrolithotomy (PCNL) is a minimally invasive procedure for the treatment of nephrolithiasis (commonly known as kidney stones). It involves using a needle to puncture the kidney through a small incision in a patient's back. A sheath is then placed through this entry path, and a nephroscope, shown in Figure 1, is passed through the sheath to gain access to kidney stones. Stones are then fragmented and removed through the nephroscope [1,2].

Even though more than 90% of kidney stones are passed without medical intervention or through the use of non-invasive procedures, PCNL is an integral treatment for more severe cases of large or irregularly shaped kidney stones, or where other treatment options have been unsuccessful [1,2]. Despite decades of clinical prevalence, it is challenging for novice surgeons to receive adequate training and gain experience in the procedure [3]. Such a lack of surgical proficiency may lead to poor treatment outcomes.

Despite being one of the main endoscopic procedures for managing kidney stones, PCNL remains a difficult procedure to learn with a long training period [4]. Traditional simulations such as cadavers are expensive and in short supply [5]. Other available PCNL training resources such as porcine training models require the use of fluoroscopy for tool guidance, which leads to unnecessary radiation exposure to trainees [6,7]. Virtual reality (VR) is emerging as a new method of delivering simulations for training in a variety of surgical procedures. It offers benefits for learners and educators through cost-effective, repeatable, and standardized clinical training on-demand [8,9]. Due to their versatility, simulators are becoming a new standard for effectively training novice surgeons in various surgical procedures such as general surgery [9], intracardiac interventions [10], cataract surgery [11], amongst others [8]. Gradually, this training option is being explored for PCNL as well [12]. Sommer et al. [9] found that surgical simulators improved a novice surgeon's visual-spatial ability. Similarly, a cyber-physical teleoperative rehearsal framework is tested in [13], which found that novices benefit from haptic feedback during surgical training.

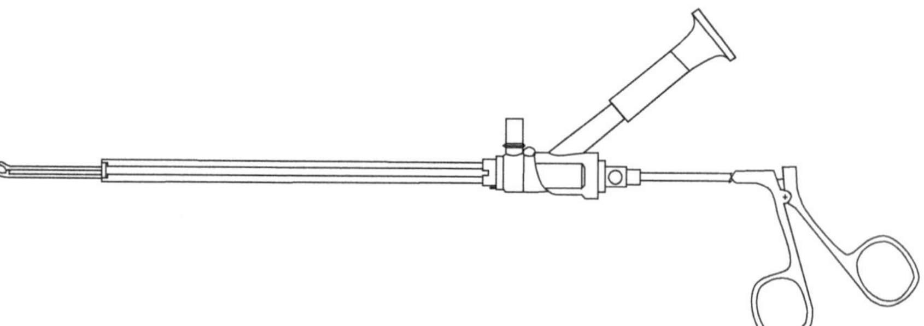

Figure 1. A nephroscope: this tool is inserted into a patient's back during a PCNL procedure. The eyepiece on the top of the nephroscope enables the surgeon to see inside the kidney. Tools are used through the nephroscope to breakup and remove kidney stones.

Once a novice surgeon has been trained, a variety of challenges still exist when performing this procedure in the real world. Arnold et al. [14] emphasized that health care is the only high-risk industry where rehearsals are not yet part of daily work. The development and growth of health care simulations can put an end to this model and provide an opportunity to rehearse high-risk, complex, and rare surgical procedures in a safe environment rather than on an actual patient. Yiasemidou et al. [15] conducted a meta-analysis of studies comparing preoperative rehearsals to standard treatment with two distinct groups of patients and demonstrated that real procedures were performed quicker if preoperative rehearsal took place. However, the immediate clinical outcome was similar for practiced and non-practiced operations. Current evidence suggests that patient-specific preoperative preparation is feasible, safe, and decreases operational time [15–18].

An example of surgical rehearsal is the SNAP VR 360 software (Surgical Theatre, Pepper Pike, OH, USA). It provides a neurosurgeon with a virtual walk-through and preplan of a keyhole surgery [19]. A 3D model used during the walk-through is generated from a patient's computed tomography angiography and magnetic resonance imaging (MRI) [19]. While Surgical Theatre has gained FDA approval for their software to be used in cerebral and spine surgery rehearsal, it only provides a walk-through of the procedure and does not provide integrated real-time tactile feedback during the rehearsal. The implementation of haptic feedback has been proven to be beneficial to surgical rehearsal [13] and can be implemented to simulate surgical complications, including abnormal patient kidney anatomy, such as horseshoe kidneys, malrotated kidney, or duplex kidneys. Additionally, tactile feedback can imitate kidney movements from patient breathing, heart pumping, and general tissue resistance forces. Rehearsing with integrated real-time force feedback allows the surgeon to plan an appropriate path toward the kidney stones while receiving real-time feedback about the tissue displacement, which can ultimately reduce tissue damage during surgery.

Other surgical rehearsal approaches use patient-specific preoperative imaging to create a physical model of the relevant anatomy. The limitations of 3D printed models are that they are static, and thus, they lack the ability to simulate the dynamic conditions of real-world organs that result from pulsations of the heart or lung expansion and contraction. Therefore, incorporating accurate dynamic functionalities into the organ models is a key aspect to achieve more realistic surgical rehearsal [20].

Parkhomenko et al. [16] explored the effect of virtual reality models of a patient's anatomy on preoperative planning for PCNL. Surgeons had the opportunity to interact with a 3D model (constructed from a patient's CT scan) in a VR environment; 10 of the 25 surgeons altered their operative plan based on their interaction with the 3D model. Additionally, surgeons that used the rehearsal model inflicted less blood loss to patients, fewer incisions through the skin, and used fluoroscopy for shorter periods of time, while showing a higher stone clearance rate after the procedure when compared to surgeons that did not perform the rehearsal. The study thus provides significant evidence for the efficacy of virtual reality models based on patient-specific anatomy as beneficial rehearsal tools [16]. In [21], the surgeon could view and interact with a 3D model of a patient's lung displayed next to other operative imaging, allowing them to have a better understanding of the patient's anatomy.

While VR simulations provide a surgeon with a better understanding of patient anatomy, this paper takes the approach one step further and presents a PCNL rehearsal framework that includes 3D model generation from patient data, while including haptic feedback during the rehearsal training using a complete PCNL simulator. A patient agreed to have their preoperative full-body CT scan used in this study. First, a 3D model of the patient's anatomy is constructed based on the preoperative imaging. The surgeon then rehearsed the procedure in the K181 simulator (Marion Surgical, Toronto, ON, Canada); this simulator provides haptic feedback to the user by mimicking tissue resistance forces. The PCNL surgery was then performed on the patient to remove their kidney stones. Questionnaires were provided to the surgeon pre/postoperatively to assess the benefit and quality of the simulated surgical rehearsal. The objective of this study is to assess the impact of both the 3D model and the real-time haptic feedback on the surgery.

A detailed description of the surgical simulator is provided in Section 2 and the process for generating 3D models from 2D imaging is discussed in Section 3. Preoperative and postoperative questionnaires were given to the surgeon to explore the viability of the simulator as surgical rehearsal, and its possible benefits. The nephrolithiasis case details, questionnaires, experimental procedure, surgical outcomes, and questionnaire results are described in Section 4. Finally, concluding remarks and a description of future work are given in Section 5.

2. Marion K181 PCNL Simulator

The Marion K181 PCNL simulator depicted in Figure 2 is a virtual reality PCNL surgical simulator. It provides users with real-time haptic feedback while they control a fluoroscopic arm and a needle for calyceal puncture. The user enters an immersive, 3D virtual operating room using a virtual reality headset from where they gain percutaneous renal access into virtual kidneys rendered from real patient anatomic data obtained from CT scan images [22]. The procedure is practiced/rehearsed in a virtual environment, which eliminates radioactivity exposure for the operator and allows the operator multiple attempts to perform the procedure. This will also enable a surgeon to explore the use of different entry points or directions if they are unsure which would be most appropriate.

While the headset provides the user with an immersive visual environment, a tool connected to a haptic device allows the user to experience real-time haptic feedback. Users control the tool connected to a haptic device while performing the virtual surgery. The haptic device is then able to generate resistive forces, which mimic tissue resistive forces while collecting accurate position data from the tool.

Figure 2. The virtual operating room of the Marion K181 PCNL simulator is shown with a user operating the haptic device, a virtual patient undergoing PCNL, virtual X-ray imaging, and the medical instruments and imaging device, which would be present in the real world operating room. The user is wearing a VR headset with the leap motion attachment for hand tracking. The TV screen behind the user shows a representation of what the user sees in VR, the X-ray view and the Virtual Reality Operating room. The user's hands are holding the needle tool that is attached to the haptic robots that provide real-time force feedback.

The 3D patient models are created by taking anonymized patient venous, delay, non-contrast, and full-body CT scans that are registered and segmented to generate 3D models of the abdominal organs, skin, and bone. These models are decimated and re-meshed into low-polygon versions while maintaining anatomical accuracy [23].

Preliminary Testing of Simulator

At the University of Toronto, a total of 18 participants with varying levels of PCNL experience benchmarked the Marion K181 against other commercially available surgical simulators, such as the PercMentor [6] and the porcine PCNL model by Cook [7]. Study participants concluded that the novel PCNL simulator was comparable to a high-fidelity porcine inanimate model and had adequate content validity evidence to support its use for beginner-level PCNL training. Participants felt it was a valuable teaching tool, equivalent to a high-fidelity porcine model, with the additional advantage of not requiring radiation exposure [7].

In another independent study conducted at the Department of Urology at Boston Medical Center, 20 participants with various levels of PCNL experience evaluated the efficacy of the K181 in the following categories: virtual reality experience, image control, and economy of motion of an immersive virtual reality simulator for percutaneous nephrostomy tract access [24]. This study concluded that the Immersive VR simulator for percutaneous collecting system access is a realistic and unique platform for surgical education and is highly recommended by participants. Almost all (95%) of the participants rated the VR simulation as a realistic experience.

The VR simulator has performed well in previous assessments of its quality and application for teaching novice surgeons. However, this paper explores its applicability to surgical rehearsal. The purpose of this study is to act as a pilot study for using Marion's surgical simulator as a surgical rehearsal tool for PCNL. Specifically, this study aims to determine if a large study with more participants (patients and surgeons) is appropriate, and whether or not a haptic-assisted VR simulator is a suitable surgical rehearsal tool.

Once a patient has agreed to take part in this study, their CT scan data is used to construct a 3D model of the kidney anatomy and surrounding tissue. The surgeon can then rehearse the procedure in the simulator prior to performing the surgery on the actual patient.

3. Generating 3D Model Patient CT Scans

Patients considered for this study were anonymous and provided informed consent for their information to be used in these studies. Once a patient's preoperative imaging was completed, the imaging was used to generate a 3D model for the simulator. The final 3D model consists of a finite element mesh containing all relevant structures from patient imaging.

3.1. Method

A combination of 3D Slicer, Maya, and Blender were used to generate the 3D models from CT scans. The algorithms utilized here are described in more detail by Wu et al. [23], although the general process is described below.

Converting the CT scans to 3D models first requires generating a voxel (3D pixel) representation from the various 2D image segments. Each image segment is stacked with the distance between them corresponding to the depth at which each segment is taken, see Figure 3. Pixel intensities are interpolated between image segments to generate voxels. Once this process has been completed, a basic 3D image of a patient's anatomy exists. However, this 3-dimensional representation lacks clearly defined boundaries between anatomy and tool/tissue interaction and cannot be determined directly from it since this representation does not include specific tissue characteristics, shapes, or boundaries. Thus, it is necessary to create 3D meshes that represent anatomical structures.

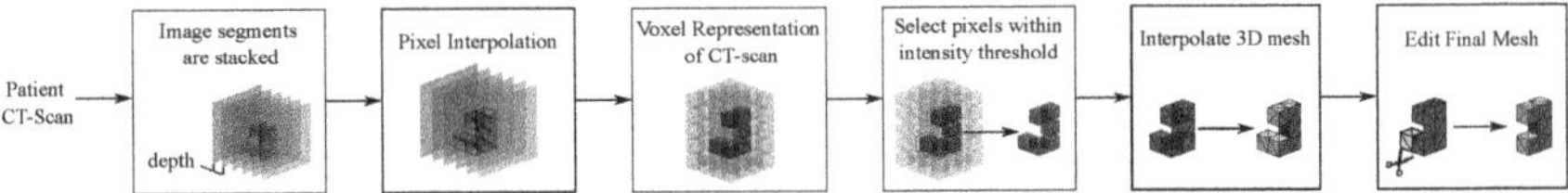

Figure 3. Workflow process of converting CT scan data to 3D models to be displayed as virtual elements within the simulator. Image segments are layered so pixels can be interpolated between images. The pixels are interpolated into voxels (3D pixels), which can then be used to create approximate 3D meshes. The meshes are then completed through a final manual editing process.

These 3D models are constructed by considering voxels within a specified intensity threshold. Voxels within the determined threshold are used to determine the approximate geometric boundaries of an anatomical structure; these boundaries are used to construct a finite element mesh representation of the anatomical structure. The 3D model is then manually edited to create a final smooth, clean, and thoughtfully segmented model. Since the general 3D model is constructed from voxel intensity, some voxels may have been included or excluded incorrectly, leading to uneven mesh surfaces. Further, the editing process can ensure a particular mesh resolution (polygon count) is achieved, in addition to partitioning model components such as vascular components, different tissues, or different structures.

One of the most important components when considering PCNL is the specific size and location of each kidney stone. Incorrectly representing the size of kidney stones can lead to improper planning or practice for the procedure. Accurate 3D models are also integral to generating accurate haptic feedback within the simulator since haptic feedback is based on the mesh models.

3.2. Haptic Feedback Based on 3D Models

Haptic feedback is designed to mimic tissue resistance forces during PCNL. To generate these forces the system simulates how the virtual tool interacts with the tissue.

The simulator tracks the motion of the surgical tool and the user's motion through the virtual reality motion tracking cameras. The x-y-z positional data is then recorded at 100 Hz throughout the 2–10 min simulation. See Figure 4 for the interactions between various components used to generate an immersive simulation. The simulator's physics engine is able to calculate the forces on the tissue, the total length of the path taken by the tool and the surgeon's hand, and the direction of the surgeon's gaze. The system uses three separate components in parallel at different frequencies:

Component 1: The dynamic model of the tool/tissue interaction that calculates tissue deformation and contact forces, and generates a virtual X-ray image. This component provides data for the subsequent two components.

Component 2: The graphical representation of the model displayed in the VR operating room takes the simulation information created in the first component and displays it to the user. The virtual operating room is based on the direction of the user's eye line as well as their actions within the simulator.

Component 3: The haptic controller generates and applies force feedback to approximate real soft tissue interactions based on the virtual patient's tissue model. This component takes the tool/tissue interaction that is determined in the first component to calculate the appropriate haptic forces and apply them through the haptic device.

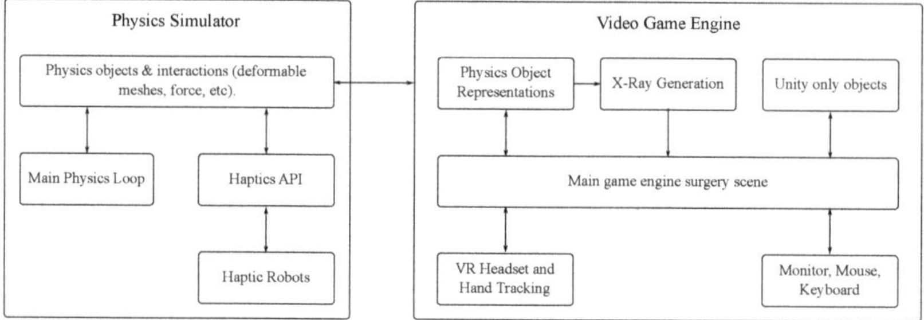

Figure 4. Flow chart of the interactions between the physics simulator, the game engine, and the peripherals to generate an immersive VR experience for users.

4. Results from Surgical Rehearsal

Three anonymous patient's agreed to have their imaging data used within the Marion K181 Simulator for use as a preoperative planning tool. Of these patients, one case has been selected to undergo a contrast-enhanced CT scan. A special contrast material was injected to help highlight the kidney duct system. The contrast material appears white on images, which emphasizes blood vessels, intestines, or other structures that are required to generate the 3D model. A contrast CT scan is necessary to create accurate 3D models of the patient's anatomy. Some CT scan segments from the patient are shown in Figure 5 as well as the 3D model of the patient's calyceal structure. The nephrolithiasis case being considered is one that qualifies for PCNL surgery, although it is a relatively simple case since the patient does not have anatomical abnormalities or a significantly large or severe case of nephrolithiasis. Thus, this case is an excellent way to demonstrate the effectiveness of haptic virtual reality simulation for preoperative planning.

4.1. Experimental Procedure

Once the patient was selected for the surgical rehearsal trial, their CT scans were used to construct 3D models. The 3D model constructed of the patient's calyx structures within the kidney are shown in Figure 5d. This figure also depicts some of the CT scans taken of the patient that were used to construct the 3D model.

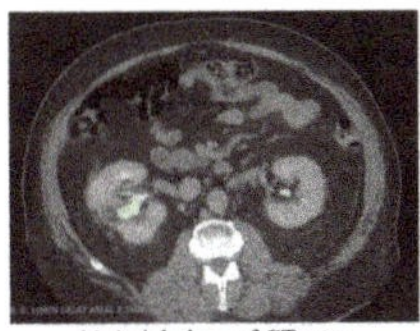 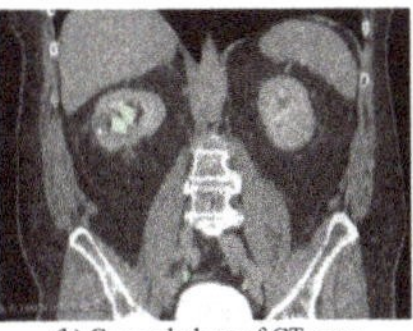 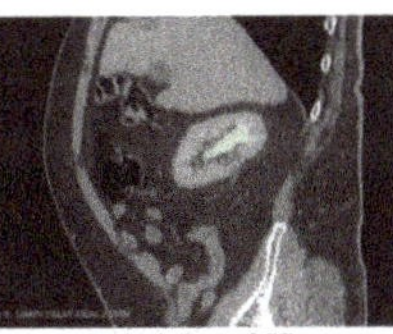 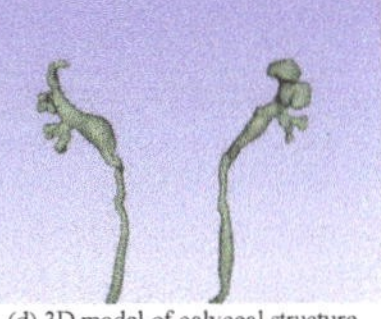

(a) Axial plane of CT-scan   (b) Coronal plane of CT-scan   (c) Sagittal plane of CT-scan   (d) 3D model of calyceal structure

Figure 5. 3D model of the calyxial system within the kidney in (**d**) and overlayed onto the patient's CT scans in (**a–c**), which are the axial, coronal, and sagittal planes of the CT scans, respectively.

The surgeon then uses the simulator to run a virtual walk-through of his procedural plan. Once the surgeon is satisfied with his experience within the simulator, he fills out the preoperative questionnaire. The surgeon then performed the surgery at St. Joseph's Hospital, Hamilton, Ontario, Canada. After the surgery has been completed, the surgeon fills out a postoperative questionnaire. These questionnaires aim to determine how beneficial the surgeon found the simulator for the use of preoperative planning. The pre and postoperative questionnaires contain questions aimed at identifying the surgeon's skill level as well as aspects of the surgery performed. Several assessment questions are given both before and after the surgery to identify if the surgeon's opinion about the rehearsal experience changed as a result of performing the surgery in the real world.

4.2. Results

After completing the surgical rehearsal in the simulator, the surgeon filled out the preoperative questionnaire. The first part of the questionnaire contained three questions aimed at identifying the surgeon's skill level. These questions and their answers are provided in Table A1 in the Appendix A.

The results from part 1 of the questionnaire show that the surgeon is experienced when using a C-arm and has experienced some intraoperative errors as a result of technical errors (see Table A1 of the Appendix A). The second part of the questionnaire is also completed prior to surgery and focuses on the surgeon's experience within the simulator and its realism. The results for the second part of the questionnaire are given in Table A1 in the Appendix A; most of these questions ask the surgeon to rate their experience with the simulator on a scale of 1 to 5, with 1 being poor or very unrealistic and 5 being very realistic. Part three of the preoperative survey concerns the construct validation of the simulator. Construct validation attempts to measure how well the simulator represents and measures the physical phenomena it is attempting to recreate. Finally, the surgeon gives an overall score for their experience with the simulator at the end of the preoperative survey.

The postoperative survey is more in-depth, asking the surgeon to reflect on the overall rehearsal experience with the simulator after performing the real-world surgery. Additionally, several of the same questions were repeated from the preoperative survey to determine whether the surgeon's impression of the simulator was altered after performing the real-world procedure. The postoperative questionnaire results are separated into two tables in the Appendix A with Table A2 focusing on the surgeon's impression of the simulator, while Table A3 contains questions specific to the surgeon's surgical rehearsal experience within the simulator. This surgical rehearsal section specifically focuses on evaluating the simulator in terms of a surgical rehearsal tool. Within the rehearsal portion of the questionnaire, the participant is asked to evaluate how helpful the rehearsal was on a scale of 0 to 10 where 0 indicates that it was not helpful at all, while 10 indicates that it was very helpful.

The limitations of this study are largely due to its size, as only a single case is being considered. Even though the preliminary results indicate that the simulator improves kidney access during PCNL, a larger scale study with several patients and surgeons is required to fully determine the effectiveness of the simulator. Furthermore, quantitative rather than qualitative performance metrics are required to fully evaluate the performance of the simulator in a future study.

4.3. Discussion

The simulator was rated highly in most categories (see Tables A1–A3 in Appendix A) before and after the surgery. The simulation appears to accurately depict the surgery performed in October 2021. It is suggested that the simulation is helpful in decision making on difficult cases to minimize fluoroscopy time (radiation exposure for clinician and patient). The rehearsal can result in less bleeding and can improve the success rate of the surgery.

The ability during the surgical rehearsal to determine the approach to take with the location of the ribs in relation to the targeted calyx in the kidney was rated as excellent and helped the surgeon get real-life access during the actual procedure. Pre-planning saves the surgeon time during the rehearsal phase since one can try different approaches to reach the kidney stones while getting familiar with the patient's anatomy. The haptic feedback provided during the simulation was rated to be helpful to interpret the shape and texture of the skin and kidney using the needle. The force feedback that the surgeon experienced during rehearsal influenced the surgeon's plan for the real surgery. The surgeon rated the construct validation higher in the postoperative survey, noting that the haptics had an influence on decisions in the actual surgery. The surgeon also noted that the simulator can potentially minimize fluoroscopy time and bleeding during the actual surgery, which could, in turn, improve the surgery success rate. This correlates with the surgeon having had time to reflect on the surgical rehearsal and having trained the procedure before going into the surgery.

5. Conclusions

Virtual reality simulators are becoming an essential tool in surgical training. Through virtual reality, novice surgeons can develop their surgical skills without posing any danger to the patient. Expert surgeons, on the other hand, can use a virtual reality simulator to plan a surgical intervention and practice it, before going into the actual surgery. The benefits of surgical rehearsal using virtual reality has been proven in several studies, including craniotomy [25], thoracic surgery [21] and PCNL [16].

Percutaneous renal surgery is a difficult procedure to learn and perform due to challenges with obtaining and/or maintaining percutaneous access [26]. The Marion Virtual Reality PCNL simulator with haptic feedback is a novel tool to allow the surgeon to rehearse and practice the difficult access part of the surgery without harming the patient. This paper describes the first pilot study using a combination of virtual reality and haptic feedback for kidney access rehearsal before PCNL surgery. An experienced surgeon used the simulator before the surgical procedure to plan and practice kidney access on a routine PCNL case before surgery. The survey data collected from the surgeon after the surgical rehearsal on the simulator and after performing the actual surgery on the patient indicates that the simulator improves confidence in the procedure, reduced the time taken by the surgeon to complete kidney access and reduced blood loss. To the best of the author's knowledge, this is the first study combining virtual reality and haptic feedback in PCNL. It shows promising preliminary data for the efficacy of the simulator as a rehearsal tool.

In this paper, a pre-surgical rehearsal was conducted for a single patient case study. Even though the preliminary results indicate that the simulator does help improve kidney access during PCNL, a large-scale study with several patients and surgeons is required to fully determine the effectiveness of the simulator. Furthermore, quantitative rather than qualitative performance metrics are required in a future study. Further studies with a larger sample size of surgeons and residents at various levels of PCNL access experience are required to confirm the findings of this preliminary study. Such studies will run the rehearsals on more difficult cases to determine if the use of the simulated surgical rehearsal improves the outcomes of the actual surgery. Ultimately, a large clinical study to analyze and compare clinical outcomes of surgeries performed with and without the surgical rehearsal platform would confirm the suitability of simulation training in improving surgical outcomes for PCNL.

Author Contributions: B.S., J.R., M.G. and M.F. conceptualized the study and the methodology, B.S. carried out experimental validation, data collection and analyses, B.S. and O.W.; prepared the manuscript, C.R., B.S. and OW. reviewed and edited the manuscript. All authors have read and agreed to the published version of the manuscript.

Funding: This research received no external funding.

Institutional Review Board Statement: This study is a case report. Ethical review and approval were waived.

Informed Consent Statement: Informed consent was obtained from all subjects involved in the study.

Conflicts of Interest: The first author of this paper is affiliated with Marion Surgical.

Appendix A. Surgical Rehearsal Questionnaires and Results

Table A1. Preoperative Assessments: Parts 1 Through 4.

Questionnaire Part 1: User Demographics	
1. *How many PCNL access procedures have you performed in the last year with a C-arm?*	44
2. *Have you experienced any intraoperative errors during PCNL procedures?*	Yes
3. *Was there error due to:*	
Answer: Technical error (i.e., excessive force, tissue injury, etc.)	
Questionnaire Part 2: Face Validation	
How would you rate the virtual reality PCNL simulator with respect to:	
1. *Visual realism*	4/5
2. *Tactile feedback*	4/5
3. *Movement and instruments*	4/5
4. *Anatomical realism*	4/5
5. *How stable were the graphics and your sense of self inside the simulator?*	5/5
6. *Describe your experience/comment on any areas for improvement in realistic representation of the operating environment:*	
Answer: Graphics were very good. Graphics sometimes jumpy.	
Questionnaire Part 3: Construct Validation	
1. *Were you able to interpret the shape and texture of the skin and kidney using the needle tool?*	Yes
2. *Do you feel the tactile information was amplified in the simulation?*	No
3. *If applicable, did motion and force feedback influence your decision?*	No
Questionnaire Part 4: Content Validation	
1. *Do you feel the tasks performed in the simulator reflected the real surgical procedure?*	Yes
2. *Please provide comments of the overall simulation experience (VR) in content accuracy?*	
Answer: Location of rib impacting access was excellent. Help me for real life access.	
Overall Rating	
1. *Rate the experience training with the Marion Surgical PCNL simulator:*	4/5

Table A2. Postoperative Assessments: Parts 1 through 3.

Questionnaire Part 1: Face Validation	
How would you rate the virtual reality PCNL simulator with respect to:	
1. *Visual realism*	4/5
2. *Tactile feedback*	4/5
3. *Movement and instruments*	4/5
4. *Anatomical realism*	5/5
6. *How stable were the graphics and your sense of self inside the simulator?*	4/5
7. *Describe your experience/ comment on any areas for improvement in realistic representation of the operating environment:*	
Answer: The virtual rendering are an excellent simulation of actual patient anatomy. Tactile feedback can always be improved	

Questionnaire Part 2: Construct Validation	
1. *Were you able to interpret the shape and texture of the skin and kidney using the needle tool?*	Yes
2. *Do you feel the tactile information was amplified in the simulation?*	Yes
3. *If applicable, did motion and force feedback influence your decision?*	Yes

Questionnaire Part 3: Content Validation	
1. *Visual simulation is the most important factor in learning surgical motor skills:*	4/5
2. *Do you feel the tasks performed in the simulator reflected real surgical skills?*	Yes
3. *You see the value in VR PNCL tool as a useful tool in Training PCNL skills:*	4/5
4. *You think this VR PCNL simulator is useful for Assessing/Testing PCNL skills:*	4/5
5. *What is the most difficult skill to learn during a full PCNL procedure?*	
Answer: Actual renal access	
6. *Please provide comments of the overall simulation experience (VR) in having an educational role?*	
Answer: Being able to practice targeting of calyx/stone with needle is valuable. Also without having excess radiation exposure or harm to patient.	
7. *Please provide comments of the overall simulation experience (VR) in content accuracy?*	
Answer: It was help to rehearse access. Became aware of rib in the way of calyx of interest.	

Overall Rating	
1. *Rate the experience training with the Marion Surgical PCNL simulator:*	5/5

Table A3. Postoperative Assessments: Rehearsal.

Rehearsal		
	Did the rehearsal help you determine:	
1.	*The location of the stone (with regard to the bulk of the stone)?*	9/10
	(a) *Specifically, where is the bulk of the stone?*	Lower Pole
2.	*The size of the stone?*	10/10
	(a) *What is the size of the largest stone in three dimensions?*	2 cm
	(b) *What is the total volume of the largest stone?*	2 cm^3
3.	*The shape and orientation of each stone-bearing calix?*	10/10
4.	*The optimal calix of entry to perform the PCNL?*	
	(a) *Into which calix (upper, mid, lower, and anterior or posterior) are you planning to place the nephrostomy track?*	Lower Posterior
5.	*How easily do you think you can navigate this patient's pelvic caliceal system from your planned approach with a rigid nephroscope?*	9/10
	When you performed the actual surgery on this patient:	
6.	*How close was the location of the stone relative to the rehearsal (specifically, with regard to the bulk of the stone)?*	9/10
	(a) *Specifically, where is the bulk of the stone:*	Lower Posterior
7.	*The size of the stone?*	10/10
	(a) *What is the size of the largest stone in three dimensions?*	2 cm
	(b) *What is the total volume of the largest stone?*	2 cm^3
8.	*The shape and orientation of each stone-bearing calix?*	9/10
9.	*The optimal calix of entry to perform the PCNL?*	9/10
	(a) *Into which calix (upper, mid, lower, and anterior or posterior) did you place the nephrostomy track?*	Lower Posterior
10.	*How easily were you able to navigate to this patient's pelvic caliceal system from your planned rehearsal approach with a rigid nephroscope?*	9/10

References

1. Aydın, A.; Al-Jabir, A.; Smith, B.; Ahmed, K. Training in Percutaneous Nephrolithotomy. In *Percutaneous Nephrolithotomy*; Zeng, G., Sarica, K., Eds.; Springer: Singapore, 2020; pp. 195–202._21. [CrossRef]
2. Khan, S.R.; Pearle, M.S.; Robertson, W.G.; Gambaro, G.; Canales, B.K.; Doizi, S.; Traxer, O.; Tiselius, H.G. Kidney stones. *Nat. Rev. Dis. Prim.* **2016**, *2*, 1–23.
3. Bird, V.G.; Fallon, B.; Winfield, H.N. Practice patterns in the treatment of large renal stones. *J. Endourol.* **2003**, *17*, 355–363. [CrossRef] [PubMed]
4. Ng, C.F. Training in percutaneous nephrolithotomy: The learning curve and options. *Arab J. Urol.* **2014**, *12*, 54–57. [CrossRef] [PubMed]
5. Bushey, C. Cadaver supply: The last industry to face big changes. *CRAIN's Chic. Bus.* **2016**, *15*.
6. Mishra, S.; Kurien, A.; Ganpule, A.; Muthu, V.; Sabnis, R.; Desai, M. Percutaneous renal access training: Content validation comparison between a live porcine and a virtual reality (VR) simulation model. *BJU Int.* **2010**, *106*, 1753–1756. [CrossRef]
7. Farcas, M.; Reynolds, L.F.; Lee, J.Y. Simulation-Based Percutaneous Renal Access Training: Evaluating a Novel 3D Immersive Virtual Reality Platform. *J. Endourol.* **2021**, *35*, 695–699. [CrossRef]
8. Pottle, J. Virtual reality and the transformation of medical education. *Future Healthc. J.* **2019**, *6*, 181. [CrossRef]
9. Sommer, G.M.; Broschewitz, J.; Huppert, S.; Sommer, C.G.; Jahn, N.; Jansen-Winkeln, B.; Gockel, I.; Hau, H.M. The role of virtual reality simulation in surgical training in the light of COVID-19 pandemic: Visual spatial ability as a predictor for improved surgical performance: A randomized trial. *Medicine* **2021**, *100*, e27844. [CrossRef]
10. Chiang, P.; Zheng, J.; Yu, Y.; Mak, K.H.; Chui, C.K.; Cai, Y. A VR simulator for intracardiac intervention. *IEEE Comput. Graph. Appl.* **2012**, *33*, 44–57. [CrossRef]
11. Staropoli, P.C.; Gregori, N.Z.; Junk, A.K.; Galor, A.; Goldhardt, R.; Goldhagen, B.E.; Shi, W.; Feuer, W. Surgical simulation training reduces intraoperative cataract surgery complications among residents. *Simul. Healthc. J. Soc. Simul. Healthc.* **2018**, *13*, 11. [CrossRef]
12. Badash, I.; Burtt, K.; Solorzano, C.A.; Carey, J.N. Innovations in surgery simulation: A review of past, current and future techniques. *Ann. Transl. Med.* **2016**, *4*, 453. [CrossRef] [PubMed]

13. Wilz, O.; Sainsbury, B.; Rossa, C. Constrained haptic-guided shared control for collaborative human–robot percutaneous nephrolithotomy training. *Mechatronics* **2021**, *75*, 102528. [CrossRef]
14. Arnold, J.; Cashin, M.; Olutoye, O.O. Simulation-based clinical rehearsals as a method for improving patient safety. *JAMA Surg.* **2018**, *153*, 1143–1144. [CrossRef] [PubMed]
15. Yiasemidou, M.; Glassman, D.; Jayne, D.; Miskovic, D. Is patient-specific pre-operative preparation feasible in a clinical environment? A systematic review and meta-analysis. *Comput. Assist. Surg.* **2018**, *23*, 57–68. [CrossRef] [PubMed]
16. Parkhomenko, E.; O'Leary, M.; Safiullah, S.; Walia, S.; Owyong, M.; Lin, C.; James, R.; Okhunov, Z.; Patel, R.M.; Kaler, K.S.; et al. Pilot assessment of immersive virtual reality renal models as an educational and preoperative planning tool for percutaneous nephrolithotomy. *J. Endourol.* **2019**, *33*, 283–288. [CrossRef]
17. Willaert, W.I.; Aggarwal, R.; Van Herzeele, I.; Cheshire, N.J.; Vermassen, F.E. Recent advancements in medical simulation: Patient-specific virtual reality simulation. *World J. Surg.* **2012**, *36*, 1703–1712. [CrossRef]
18. Won, T.B.; Hwang, P.; Lim, J.H.; Cho, S.W.; Paek, S.H.; Losorelli, S.; Vaisbuch, Y.; Chan, S.; Salisbury, K.; Blevins, N.H. *Early Experience with a Patient-Specific Virtual Surgical Simulation for Rehearsal of Endoscopic Skull-Base Surgery*; Wiley Online Library: Hoboken, NJ, USA, 2018; Volume 8, pp. 54–63.
19. Jean, W.C. Virtual reality surgical rehearsal and 2-dimensional operative video of a paramedian supracerebellar infratentorial approach endoscopic resection of pineocytoma: 2-dimensional operative video. *Oper. Neurosurg.* **2021**, *20*, E51–E52. [CrossRef]
20. Qiu, K.; Haghiashtiani, G.; McAlpine, M.C. 3D printed organ models for surgical applications. *Annu. Rev. Anal. Chem.* **2018**, *11*, 287–306. [CrossRef]
21. Guerrera, F.; Nicosia, S.; Costardi, L.; Lyberis, P.; Femia, F.; Filosso, P.L.; Arezzo, A.; Ruffini, E. Proctor-guided virtual reality–enhanced three-dimensional video-assisted thoracic surgery: An excellent tutoring model for lung segmentectomy. *Tumori J.* **2021**, *107*, NP1–NP4. [CrossRef]
22. Sainsbury, B.; Łącki, M.; Shahait, M.; Goldenberg, M.; Baghdadi, A.; Cavuoto, L.; Ren, J.; Green, M.; Lee, J.; Averch, T.D.; et al. Evaluation of a virtual reality percutaneous nephrolithotomy (PCNL) surgical simulator. *Front. Robot. AI* **2020**, *6*, 145. [CrossRef]
23. Wu, C.O.; Sunderland, K.; Filippov, M.; Sainsbury, B.; Fichtinger, G.; Ungi, T. Workflow for creation and evaluation of virtual nephrolithotomy training models. In Proceedings of the Medical Imaging 2020: Image-Guided Procedures, Robotic Interventions, and Modeling, Houston, TX, USA, 15–20 February 2020; International Society for Optics and Photonics: Bellingham, WA, USA, 2020; Volume 11315, p. 1131524.
24. Resad, S.; Parkhomenko, E.; Wang, D.S.; Wason, S.E. *The Utility and Value of Immersive Virtual Reality Simulation for Percutaneous Nephrostomy Tract Access and Surgical Training*; Boston University School of Medicine: Boston, MA, USA, 2019.
25. Montemurro, N.; Condino, S.; Cattari, N.; D'Amato, R.; Ferrari, V.; Cutolo, F. Augmented reality-assisted craniotomy for parasagittal and convexity en plaque meningiomas and custom-made cranio-plasty: A preliminary laboratory report. *Int. J. Environ. Res. Public Health* **2021**, *18*, 9955. [CrossRef] [PubMed]
26. Rais-Bahrami, S.; Friedlander, J.I.; Duty, B.D.; Okeke, Z.; Smith, A.D. Difficulties with access in percutaneous renal surgery. *Ther. Adv. Urol.* **2011**, *3*, 59–68. [CrossRef] [PubMed]

 *electronics*

Article

Digital Taste in Mulsemedia Augmented Reality: Perspective on Developments and Challenges

Angel Swastik Duggal [1], Rajesh Singh [2], Anita Gehlot [2], Mamoon Rashid [3,*], Sultan S. Alshamrani [4] and Ahmed Saeed AlGhamdi [5]

1. School of Electronics and Electrical Engineering, Lovely Professional University, Jalandhar 144001, India; coffeeannon@gmail.com
2. Division of Research & Innovation, Uttaranchal Institute of Technology, Uttaranchal University, Uttarakhand 248007, India; rajeshsingh@uttaranchaluniversity.ac.in (R.S.); anita.ri@uttaranchaluniversity.ac.in (A.G.)
3. Department of Computer Engineering, Faculty of Science and Technology, Vishwakarma University, Pune 411048, India
4. Department of Information Technology, College of Computer and Information Technology, Taif University, P.O. Box 11099, Taif 21944, Saudi Arabia; susamash@tu.edu.sa
5. Department of Computer Engineering, College of Computer and Information Technology, Taif University, P.O. Box. 11099, Taif 21994, Saudi Arabia; asjannah@tu.edu.sa
* Correspondence: mamoon.rashid@vupune.ac.in; Tel.: +91-78-1434-6505

Abstract: Digitalization of human taste has been on the back burners of multi-sensory media until the beginning of the decade, with audio, video, and haptic input/output(I/O) taking over as the major sensory mechanisms. This article reviews the consolidated literature on augmented reality (AR) in the modulation and stimulation of the sensation of taste in humans using low-amplitude electrical signals. Describing multiple factors that combine to produce a single taste, various techniques to stimulate/modulate taste artificially are described. The article explores techniques from prominent research pools with an inclination towards taste modulation. The goal is to seamlessly integrate gustatory augmentation into the commercial market. It highlights core benefits and limitations and proposes feasible extensions to the already established technological architecture for taste stimulation and modulation, namely, from the Internet of Things, artificial intelligence, and machine learning. Past research on taste has had a more software-oriented approach, with a few trends getting exceptions presented as taste modulation hardware. Using modern technological extensions, the medium of taste has the potential to merge with audio and video data streams as a viable multichannel medium for the transfer of sensory information.

Keywords: digital taste; galvanic taste stimulation; taste augmentation; mulsemedia; taste modulation; augmented reality

Citation: Duggal, A.S.; Singh, R.; Gehlot, A.; Rashid, M.; Alshamrani, S.S.; AlGhamdi, A.S. Digital Taste in Mulsemedia Augmented Reality: Perspective on Developments and Challenges. *Electronics* **2022**, *11*, 1315. https://doi.org/10.3390/electronics11091315

Academic Editors: Jorge C. S. Cardoso, André Perrotta, Paula Alexandra Silva and Pedro Martins

Received: 28 February 2022
Accepted: 18 April 2022
Published: 21 April 2022

1. Introduction

The human body possesses the following five physical sensory systems: auditory, optical, olfactory, tactile, and taste. In the current age, there have been many developments to boost the quality of life by enhancing the sensory experience using artificially induced sensory stimuli targeting the senses of sight, sound, and touch. The experience is further made more immersive by combining those sensory modules and creating an integrated deep-dive system. Traditionally, artificial tastes are given to users through a chemical compound, either in a solid or liquid form. Example ingredients for the five basic tastes (sweet, bitter, sour, salty, and umami) are glucose for sweet, citric acid for sour, caffeine/quinine for bitter, sodium chloride for salt, and monosodium glutamate for umami. There have already been quite successful attempts at replicating this taste using electrostimulation, tackling the taste elements both individually [1] and in a collective configuration [2]. The sub-portion of food-texture replication has had work performed previously but was not followed up with more extensions and advancements [3]. Experimental food with indefinitely sustained taste has also been created and can be made commercially viable [4].

There are certain additional components of taste that have not been reached through electrostimulation, such as aftertaste, chilliness, pungency, and throat feel, since it is expected not to work through the same mechanism as taste buds. Further opportunities for exploration can be found upon conducting an in-depth review of the previous literature.

Recent studies conducted on electrically stimulating systems that are developed for inhibiting or enhancing certain gustatory features via ion transfer are described in Section 2 chronologically. The subject of recording the taste of food elements is also briefly explored. Section 3 delves into the recent research into IoT (Internet of Things) with the perspective of extending utility in the domain of taste-oriented Augmented Reality (AR) research. Section 4 deals with the recent works in the domain of artificial intelligence and machine learning that have been implemented into taste recognition tasks using classification algorithms. Section 5 discusses the existing technologies in depth along with possible extensions that can be merged into the existing taste stimulation methods to increase their efficiency. Lastly, the article draws out conclusive recommendations and their potential results based on the analysis of the prior art.

2. Gustatory Taste Stimulation

The earliest records of artificial stimulation of gustatory senses were in 2004, with a food simulator that recreated the biting force as portrayed by every food category. The two-step mechanism had an end-effector equipped with a sleeved pressure sensor to record the biting force and, subsequently, play it back artificially using end effectors with electronically variable force profiles [3]. Four years later, another study experimentally tested the influence of tactile feedback on the sense of taste by placing five swabs equidistantly over the tongue with sucrose and quinine sulfate instantaneously and five seconds post-contact. The experimental study aimed to establish that the tactile sense of the tongue supports the gustatory senses in extension to the taste buds. Three years later, another paper, in an attempt to extend the gustatory palette, introduced a novel hypothesis on whether the taste buds were capable of sensing more extended stimuli. The experimental study was conducted by constructing a combined olfactory and visual AR system (Figure 1) that would display a 6 DoF overlay-visual on top of the food item being consumed while the olfactory module released suggestive flavor-related odors [5–7].

The sense of taste is stimulated artificially in conjunction with the digitalization of haptic, visual, and olfactory feedback to obtain a thorough taste profile. This phenomenon has been explored, and the correlation between taste and smell has been quantified [6,8]. A principal component analysis conducted to distinguish liquid sample compositions yielded 100% results from the apparatus. Subsequently, the study featured the analysis of both high and low vapor pressure solutions, which were smell-biased and taste-biased, respectively, yielding the same 100% output [8]. In 2013, a thorough review of the commercially available taste sensors was conducted, and it was found that the artificial e-tongues from that year were capable of discerning astringency in addition to the five basic tastes.

Furthermore, it was stipulated that the pungency could be quantified in a short time [9]. The year 2016 witnessed a major upheaval in the domain of galvanic taste as the taste of sweetness was stimulated artificially by the "Digital Lollipop" by the same author who presented the tongue-mounted stimulation prototype. This mechanism had a customizable input galvanic signal, and the sweetness was induced using an inverse current mechanism [10]. The study employed electro-stimulatory means to modulate the sense of taste using only a single channel stimulation mechanism, as shown in Figure 2.

The following year, another inversion of the taste sensing device was presented in a form factor of a short color-changing bottle that would respond to the following three of the five basic tastes: sourness, saltiness, and bitterness, and would alter its shade to green, blue, and red for each taste, respectively [11]. The system used a microcontroller to deliver PWM signals to the buccal electrodes while outputting GPIO (General Purpose Input-Output) signals to the RGB LEDs, as shown in Figure 3.

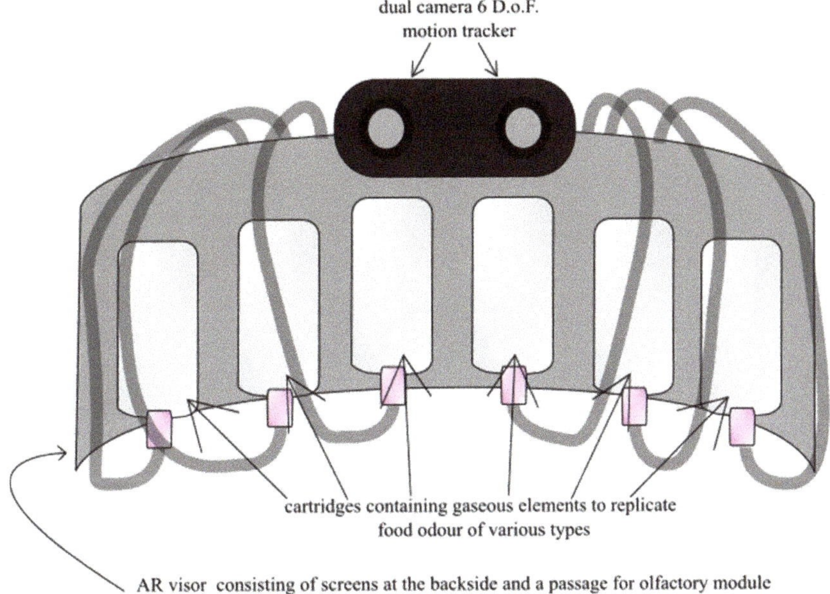

Figure 1. Architecture of the olfactory and gustation sensor module.

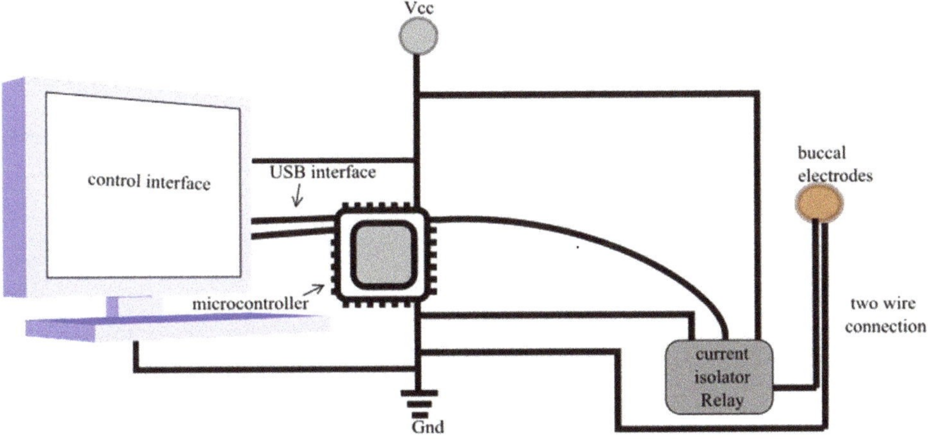

Figure 2. Real-time structure of the Buccal taste augmentation system.

A prototype tongue-mounted module for post-current release taste modulation was developed in 2012 and presented at the 16th international wearable computers symposium as a prototype that would take the form of a digital lollipop in its later stages of development [12,13]. Its taste-modulation output characteristics are depicted in Figure 4.

In a study in 2013, based on previous experimentation by Hettinger, a utensil-based approach was employed to build a salt-taste enhancer by using a cathodal current, and an experimental procedure was conducted to establish the long-term effects of the electrical stimulation and to see if it causes the other tastes to get enhanced as well. The experimental study attempted to create a new "electric taste" with the help of electrically energized utensils such as metal straws and forks powered by small batteries with their circuits closing through the mouth. The process behind the phenomena of taste inhibition via GTS (galvanic taste stimulation) was explored in 2017. It was hypothesized that the inhibited tastes of the five basic tastes from GTS were from the migration of their respective ions, which elicited the tastes [14]. Similarly, the whole process, in terms of physical hardware, from the specification of signal to the product range is sequentially documented and thoroughly detailed [15].

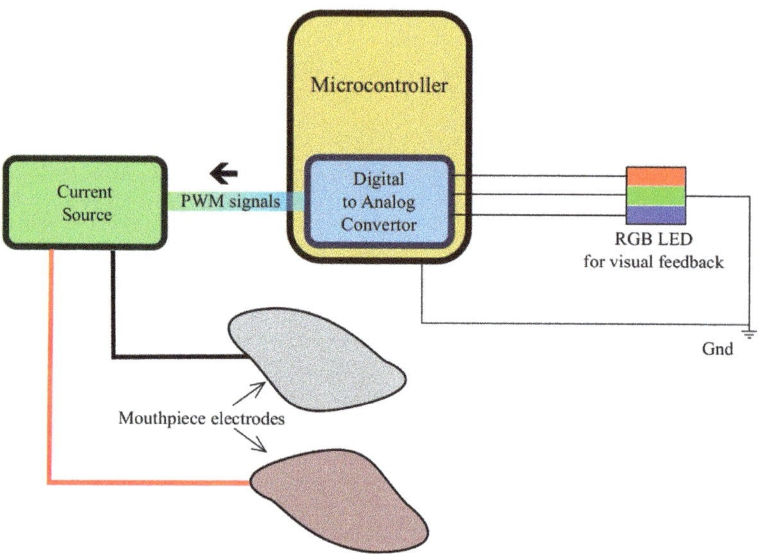

Figure 3. The general block diagram of the taste+ system.

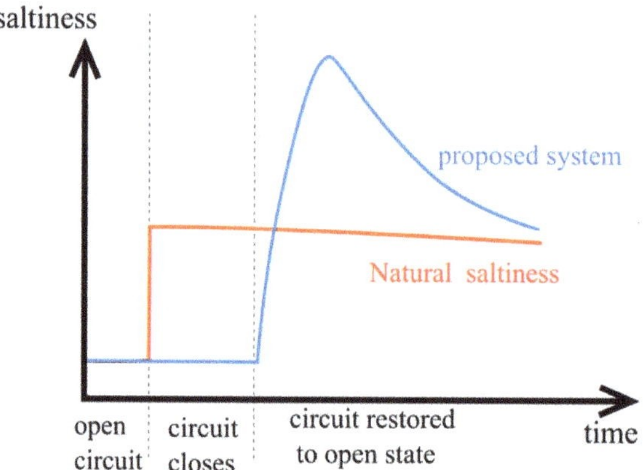

Figure 4. The comparison between salt sensitivity using the natural salt and the electro stimulatory system.

Up until 2017, GTS was performed by the provision of a galvanic charge from inside the buccal cavity. The galvanic jaw stimulation module was presented as an alternative to the earlier GTS, wherein the electrodes had to be thoroughly cleaned and disinfected before reuse. Using electrodes applied externally to the jaw on both sides, the modulation of the taste of brine and whether it was successful in inhibiting the saltiness was tested [16]. Using a collective array of sensory feedback modules, a portable AR system with a rechargeable battery unit was designed with a prime focus on portability and integration with smartphones and tablets [17]. An effort to remove the discomfort of wiring inside the mouth was made by experimentally crafting a galvanic taste-stimulating "gum" that is small enough to fit in the mouth with a piezoelectric element that powered it from biting force [4].

The influence of optical visualization over the gustatory interface has been experimentally examined for correlation in 2019. The proposed system uses an image-to-image translation to modulate the appearance of the food being consumed in real-time by implementing a generative adversarial network (GAN) in an AR setup with a limited taste modulating resultant effect observed upon the completion of the experiment [18]. It was initially termed 'Deep Taste' [19]. Besides using standalone methods for gustatory stimulation, another approach was tried out using the galvanic modules to test whether sending repetitive square waves through the contacts could enhance tastes other than the salty taste that had been observed before [20].

As an extension to taste modulation, the temporal effect of beverages was examined, and the longevity of their aftertaste was increased using electrode stimulation [21]. The throat feel can be replicated by muscle movement in the esophageal periphery. Its artificial replication has been recently tested by external skin deformation using the 'grutio' module [22].

To reproduce preexisting tastes, the initial step would be to record them with a biomimetic device called the 'e-tongue' that tests the concentration of taste-contributing electrolytic ions and outputs and stores the results in digital format [23]. A system that uses ion electrophoresis to emulate taste using six different electrolytes (five to emulate the five basic taste properties and one for control and current regulation) was coupled with taste-measuring instrumentation and a software GUI (graphical user interface) to reproduce the taste of pre-existing substances [24].

Research works in Table 1 provide a general idea of the bleeding edge research conducted upon taste stimulation. Next, we follow up by reviewing the recent literature on IoT from the perspective of its utility in the area of taste modulation. The domain of IoT shows potential in making the galvanic taste modulators portable, wireless, and backed by powerful prepossessing cloud units.

Table 1. A brief overview of the key publications on the development of the gustatory sensor module's hardware.

S. No	Ref. No./ Short Title	Outputs	Potential Extensions
1	[3] Food Simulator	A bite force measurement and replication device with assistive hints for recreating the texture of foods.	The contraption replicated the force of biting with audio and chemical feedback. It could be coupled with an AR overlay headset to replicate the complete experience of consuming virtual food.
2	[10] Digital Lollipop	Application of cathodal current using the body as the closed-circuit conductor for ion transfer. Causes the saltiness to increase upon release of the signal.	The form factor of the output electrodes could be altered to target multiple taste areas simultaneously. More channels could be added to test multiple stimuli.
3	[1] Controlling saltiness without salt	A single-channel bipolar device that is capable of anode/cathode discharge with custom output wave stimulus.	The nature of the conductive electrodes, their respective ion-taste and the toxicity could be experimented with to produce an optimum electrode.
4	[14] Galvanic tongue stimulation inhibits five basic	Externally applied jaw stimulation module for enhancing and inhibiting taste.	The whole system can be made into a compact wearable IoT AR-VR setup.

Table 1. *Cont.*

S. No	Ref. No./ Short Title	Outputs	Potential Extensions
5	[23] Taste sensor: Electronic tongue with lipid membranes	An electronic tongue that measures taste-inducing electrolytic concentration in food and converts it to digital format.	The e-tongue requires a more compact form factor for mobile application. It could be built as a small embedded system with the lipid sensor in a smaller size as was built in 2013 [25].
6	[24] Taste display that reproduces tastes measured by a taste sensor	Software GUI that controls a 5-channel GTS module that is capable of reproducing any taste and calibrating it.	This system could be used in tandem with edge nodes such as AR visors.

3. Internet of Things in Augmented Reality

Upon entering into multimedia as a viable medium, the data on the regulation of the modulator signals requires a suitable architecture built for modern transmission protocols and standardization of input/output modules. Since taste-based modulation became an oblique commercial gimmick after the advent of the internet revolution, it has become customary to conceptualize an IoT-compatible mulsemedia architecture. To fulfill this requirement, a four-layered IoT-architectural concept for mulsemedia data to be transferred through edge nodes was devised for immersive multimedia [26]. The quality of the overall experience of multi-sensory inputs in 360° multimedia was appraised in another study in pursuit of alternative extended approaches toward multimedia enrichment [27].

In addition to electrical taste stimulation, IoT can also be used to deliver AR multi-channel data via the internet [13] to edge nodes that are capable of breaking it down into trigger inputs for multiple AR sensory modules, including taste [17]. Thermal changes are observed to have a considerable impression on the taste buds, causing the user to experience mild sweetness. An edge device was developed for future VR applications with the intent to pursue thermal taste as a viable sensory extension [28]. Tackling the software side, a web server capable of identifying and classifying three distinct taste characteristics (sweet, bitter, and sour) has been presented in a bid to boost taste-enabled I-T AR devices [29].

Besides taste being an output sensation, in the IoT it also serves as an input, with taste-sensing devices employing non-chemical means to ascertain the quality and composition of substances. There have been multiple use-cases wherein the quality of taste has been indirectly monitored by exploiting congruent variables in the process of monitoring food quality. For instance, the magnitude of soluble sugar content in fruit was tested using millimeter wave-permittivity as the congruent variable [30]. Similarly, in milk, the microbial concentration of Lactobacillus has been used to remotely test for quality [31].

Such alternative pathways prove crucial in areas where organic/chemical interference could affect the output negatively by a significant magnitude. Modern food processing focuses on quality through the precise monitoring of the handling and close surveillance of the assembly lines. Minimization of contact with the product has diverted the testing and monitoring procedures to look into photo-analytic testing methodologies such as thermography and photo-spectroscopy testing. In the case of wine, multiple testing venues have been explored, e.g., testing the water quality in plant leaves using infrared thermography [32], and the wine aging level using silicone-wrapped sensor nodes embedded into wine barrels [33]. The data gathered for these use-cases are transferred using an edge node-based IoT architecture. Once the data is obtained, it can be plugged into various algorithms to obtain ML- (machine learning) and AI (artificial intelligence)-based models to build taste recognition and classification systems.

4. Artificial Intelligence and Machine Learning in AR

Machine learning is an effective domain that can seamlessly merge with any other domain and provide much-needed support to enhance its productivity. When applied in AR, it can be used for the optimization of the output provided to the user by the sensors in various ways to ensure a higher degree of immersion. The hardware used in flavor recognition has been implemented in both edible and inedible products and can be implemented in a broader scope.

In an instance of ML being used for testing, the quality of water from natural water bodies was also tested by using datasets containing 54 attributes and 135 instances. The data was processed, and a confusion matrix was created and plugged into various machine learning algorithms, out of which the K-star algorithm performed the best, using only 6 out of the 54 features with an accuracy of 86.67% [34]. Similarly, it has been attempted to predict the contents of a juice using e-nose data to ascertain the elemental content of fruit juice [35].

Coupled with the data output from the e-tongue used to record taste-stimulating ion-electrolyte concentration parameters, an upgraded LDPP (local discriminant preservation projection) model approach was proposed in contrast to the earlier learning algorithms. It was put to the test along with PCA, LDA, LPP (locality-preserving projections), and LPDP (locality-preserving discriminant projections). The machine learning classifiers tested were the SVM (support vector machine), ELM (extreme learning machine), and KELM (kernelized ELM), displaying a maximum accuracy of 98.22%, as depicted in Table 2 [36]. In another similar study, the classifiers used were KNN (K-nearest neighbor), PCA (principal component analysis), NB (naïve Bayes), random forest, EMCC (extended Matthew correlation coefficient), NER (non-error rate), and LDA (linear discriminant analysis), and the maximum accuracy reached was approximately 95% [37].

Table 2. Average performance characteristics of classifiers used in conjunction with subspace projection learning algorithms over multi-beverage identifier e-tongue data [36].

Accuracy in Percentage Classifier Learning Algorithm	SVM	ELM	KELM
PCA	93	93.18	96.48
PCA (Kernelized)	89.49	87.35	91.23
LDA	94.74	94.5	97.35
LPP-S	93.87	94.94	95.61
LPP-H	94.74	94.84	95.61
LPDP-S	96.48	95.48	96.48
LPDP-H	97.35	96.51	97.35
LDPP-S	97.35	96.69	98.22
LDPP-H	98.22	94.56	98.22

S—using a Simple rule affinity matrix. H—using Heat kernel rule affinity matrix.

Using a data set of the physico-chemical parameters of red wine, various machine learning models were pitted against one another to pick out the ideal one for wine classification. Out of the SVM (simple vector machine), ANN (artificial neural network), RR (ridge regression), and GBR (gradient boost regression), the GBR performed the best in the classification of red wine [38].

Another article focusing on the taste characteristics of tea explored the viability of the combined multi-sensory data from e-tongue, e-nose, and camera modules to perceive the taste of tea. It compared three classification algorithms, namely, the 1v1 SVM, VVRKFA (Vector-Valued Regularized Kernel Function Approximation), and ANN, using high and low amplitude pulse and staircase voltammetric datasets with varying sparsity. The results unanimously yielded high accuracy using all three classifiers with low sparsity models [39]. Furthermore, various factors influencing the taste of tea, such as astringency, bitterness, and smell, have been analyzed throughout the course of the decade and have been thoroughly reviewed in a study in 2021 [40]. The study spanned various sensor arrays in combination, such as e-tongue, e-nose, and even computer vision, to perceive and classify tea samples.

Apart from the taste profile, the e-tongue module is also capable of testing the quality of water, which might display certain flavors due to suspended impurities. It used the PCA and the PLS (partial least square regression) methods to quantify the organoleptic analysis data regarding the dissolved impurities [41].

In addition to the taste recognition of edible food products, e-tongue sensors are also used to test the taste profiles of oblique substances such as amino acids. Applying simple linear regression over the data collected from the hydrophobic lipid sensor, the study derived a strong correlation between the sensory score and the values obtained from the lipid sensor [42]. Among several others. These studies followed the trend of taste recognition and classification in AI-ML research about the domain of gustation.

5. Discussion and Recommendations

Upon having explored the domains of IoT and AI-ML for finding applications in taste mulsemedia, a web of intertwined biomedical research pools was uncovered. Once analyzed with the perspective of extending AR utility, numerous potential use cases can be built. Since the connectivity provided by IoT modules brings us closer to the computational processing power of cloud servers, we can employ optimization, filtering, classification, and deep learning models to obtain information that is intrinsically derived from the heaps of data sent in by edge nodes of users' AR helmets.

The introduction of novel hardware has the potential to redirect this trend toward optimization and filtration of taste signals. A network map of IoT coupled communication technology with AI-assisting AR modules of various types is shown in Figure 5.

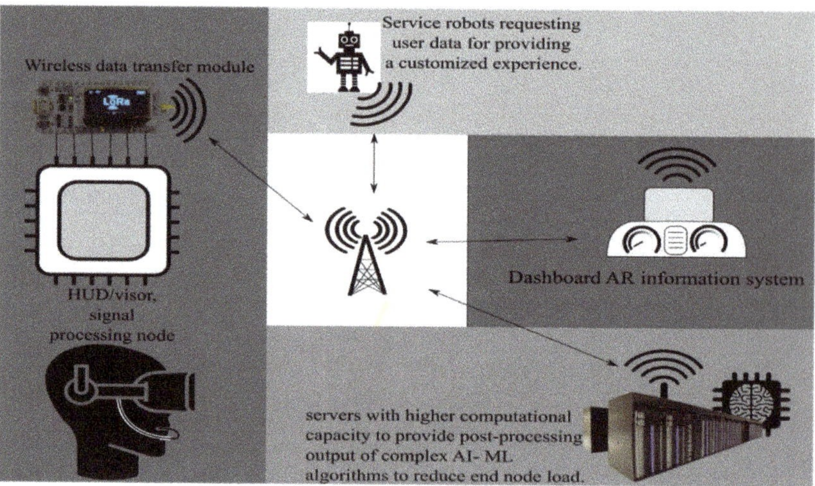

Figure 5. IoT coupled with AI–ML assisting AR modules with all five sensory data inputs.

The introduction of 3Dprinting and spatial scanning has made it possible to scan an individual's buccal cavity and print a custom conductive sleeve to fit the user sans any discomfort. Moreover, the custom buccal sleeve would be much more effective with accurate electrode placement and limited movement and chances of slipping. In addition to the fitting comfort, it is also possible to fit an entire circuit into the 3D printed contraption, thus eliminating the chances of wires protruding out of the mouth, causing excessive salivation. The technology can be merged with a modified form of the food simulator that mimics the texture of the food being simulated along with the bite force required for chewing.

In addition to the dental brace serving as an output module for the gustatory simulation module, it could also function as an acquisition module for obtaining a tongue EEG.

The spectrum of potential taste sensations that can be stimulated through this approach is still limited. Hence, in the early stages of using taste as a stimulation approach in HCI, chemical stimulation has the advantage of covering a broad range of taste experiences over galvanic stimulation. The most recent methodology of taste reproduction still requires external electrolytic solutions, while galvanic stimulation can merely inhibit/enhance an already consumed electrolytic ion cluster by pulling them together to boost taste density instantaneously. The hardware for detection of ion concentration may not be immune to frequent corrosion from reacting with other acidic compounds present in the food.

Limitations of the current prototype of the biting simulator include its cumbersome structure and the metal at the end of the biting end effector [3]. As a large force must be supported, the device must be fabricated in a structure that is capable of sustaining it while also being non-toxic. When the user has finished biting, he/she can feel an unnatural sensation of the thickness of the linkages. The weight of the linkage causes unwanted vibration. The user's teeth contact the flat surface of the linkage. This flat surface degrades food texture. An individual feels the force on an independent tooth while biting into real food. Moreover, the current device applies force only to the teeth. The texture of real food is perceived, in part, by the tongue. However, displaying food texture to the tongue is very difficult.

In the case of the electric gum, for the best approach to solidify its foundations as a viable product, the piezoelectric material has to be analyzed for potential toxicity in a long-term use case, its reaction with saliva examined, and the long-term usage effects studied. The alternative research path would be to explore the potential for generating flavors other than the conventional saltiness from the separated ions. Alternative approaches, such as using real gum containing edible organic batteries using safer electrolyte gels and modulating its taste externally, could be tested out [4].

The apparatus used for the study of the effects of visual feedback on taste can be enhanced by combining similar studies to increase immersion. The delay in the deep taste system can be reduced by using an advanced FFT-GAN-based algorithm over the image data from the webcam. The resolution of the visor can be boosted too with a reduction in computational time [5]. Moreover, it can be merged with auditory and olfactory feedback from the original dish as an experiment on the percentage of effect olfaction has over taste.

Beverage taste modulation using a conductive straw to complete a stimulatory circuit could use an added perk in commercial food packaging in the form of flavor-changing drinks, with the downside of short shelf life due to oxidation of the conductive metal contacts. As a precautionary measure, the straw could be coated in plastic, save for the ends. The design of a sustainable GTS soda can (Figure 6) could be a potentially rewarding research pursuit. It would have great value in the patent sector while being relatively easier as a research option.

Figure 6. Structure for salt-based taste altering soda.

Currently, the optimum process of artificially recording taste is via measurement of taste-contributing ion concentration, and for classification, the ELM algorithm is ranked at the top. The research areas in the software portion of this particular domain are quite saturated, leaving behind either some potential in chemistry-based research, wherein better electrolytes could be sourced for longer-lasting electrode function, or in structural research, wherein the structure of the recording device could be tweaked to be more compact, portable, and multi-functional. The recording medium could feature a storage add-on that could be interfaced in the form of a flash drive, or it could be made wireless using a radio wave-based data transfer protocol, both long- and short-range.

From the perspective of taste being treated as media, it has to establish standardized data-flow protocols, hardware, and media before it can arrive at the same level as audio and video. Once the data from an e-tongue is obtained, its inversions may range from its channels being mapped to audio frequency bands to experience audio through multiple senses to test whether the patterns are as pleasing to taste as industrial approximation mediums. It could be used to plant more memories into the brain with a taste-based stimulus to trigger them, essentially making mankind a bit more "connected".

Being the result of a multivariate experience, texture and memory also play a leading role when recognizing taste sensations. The other senses act as support vectors for the taste sensing system. To replicate the results, an accurate duplicate of the initial environment is required. A GTS system works on low-magnitude currents. Taste buds are localized, but their positions can be altered using galvanic electrostimulation. The initial step in the approach is to experiment with the low voltage levels based on personal threshold values. A gustatory range must be affixed within the experienced band of taste change sensation and is only obtained by testing for it.

The design of the stimulation module in prior use cases was not the primary area of focus in the experimentation. So, it could feature a more ergonomic approach in its structure. The single-channel stimulus can be tested to check for a better-distributed multichannel approach. Taste stimulation can be coupled with various areas of technology to generate more use cases in the commercial sector. The gustatory module can be molded into a portable, wearable module that employs state-of-the-art data transfer channels to integrate seamlessly with the most recent devices.

Wave-shaping techniques can be experimented with to create various taste profiles. Sensory analysis can be performed in real-time while tasting a dish, and the taste can be recorded to allow future replication.

6. Conclusions

The domain of taste electrostimulation holds a lot of promise vis-a-vis AR technology, with possibilities such as integration of the hardware's system into digital multimedia, potential enhancements in memory retention, advancements in gastroenterological research, and many more. The IoT as an extension could provide the required processing prowess to optimize and filter the taste signal. Since the sense of taste usually provides an incentive for appetite, it is indirectly responsible for the amount of nutrients supplied to the body. The currently developed systems require ergonomic structural work to render the technology commercially feasible.

The perception of the utility of GTS systems is still narrowed down to usage in VR systems, rather than a broader radius of inclusivity. For instance, in pursuits such as gardening, farming, or even within the industry, wherein the percentage composition is based on an approach of approximation, the nutrient/threshold levels can be mapped onto different taste channels so that the practitioners can build sensory neural associations with the overall status of the unit. The development of taste stimulation devices is especially critical to the culinary sector, with its prospects extending into online taste sampling, taste copyrighting, establishing a distinguished gustatory range, and taste-based research.

Overall, the entire field of galvanic taste stimulation has tremendous trendsetting potential. This could subsequently tackle major flavor addictions without the user gaining any weight. An alteration of the taste component of nutritious foods to add more flavor could be performed using such modules in daily life. This would boost overall health. The average BMI index could be rigorously pursued using electronic gadgets instead of the tough mechanical weightlifting way. The non-genetic variation of obesity could be potentially eradicated. Moreover, taste stimulation could be used to trigger a state of synesthesia, which could be exploited to boost the power of memory retention. Neural associations could be created to induce certain memories synthetically using a deep dive AR consisting of visual, auditory, olfactory, and gustatory stimulants.

Author Contributions: Conceptualization, A.S.D.; methodology, A.G.; validation, R.S. and M.R.; formal analysis, A.G. and S.S.A.; writing—original draft preparation, A.S.D.; writing—review and editing, M.R. and A.S.A.; supervision, R.S.; funding acquisition, S.S.A. All authors have read and agreed to the published version of the manuscript.

Funding: This research was supported by Taif University Research Supporting Project number (TURSP-2020/215), Taif University, Taif, Saudi Arabia.

Institutional Review Board Statement: Not applicable.

Informed Consent Statement: Not applicable.

Data Availability Statement: Data used in this article will be made available on request to the corresponding author.

Acknowledgments: This study was funded by the Deanship of Scientific Research, Taif University Researchers Supporting Project number (TURSP-2020/215), Taif University, Taif, Saudi Arabia.

Conflicts of Interest: The authors declare no conflict of interest.

References

1.   Nakamura, H.; Miyashita, H. Controlling saltiness without salt: Evaluation of taste change by applying and releasing cathodal current. In Proceedings of the 5th International Workshop on Multimedia for Cooking and Eating Activities, Barcelona, Catalonia, Spain, 21 October 2013. [CrossRef]
2.   Aoyama, K. Galvanic taste stimulation method for virtual reality and augmented reality. In Proceedings of the International Conference on Human-Computer Interaction, Copenhagen, Denmark, 19–24 July 2020; Volume 12184. [CrossRef]
3.   Iwata, H.; Yano, H.; Uemura, T.; Moriya, T. Food Simulator. *IEEE Xplore Comput. Graph. Appl.* **2004**, *24*, 1–4. [CrossRef]
4.   Ooba, N.; Aoyama, K.; Nakamura, H.; Miyashita, H. Unlimited electric gum: A Piezo-based Electric Taste Apparatus Activated by Chewing. In Proceedings of the 31st Annual ACM Symposium on User Interface Software and Technology, Berlin, Germany, 17 October 2018. [CrossRef]
5.   Narumi, T.; Ban, Y.; Kajinami, T.; Tanikawa, T.; Hirose, M. Augmented perception of satiety: Controlling food consumption by changing apparent size of food with augmented reality. In Proceedings of the SIGCHI Conference on Human Factors in Computing Systems, Austin, TX, USA, 5–10 May 2012. [CrossRef]
6.   Narumi, T.; Nishizaka, S.; Kajinami, T.; Tanikawa, T.; Hirose, M. Augmented reality flavors: Gustatory display based on Edible Marker and cross-modal interaction. In Proceedings of the Conference on Human Factors in Computing Systems, Vancouver, BC, Canada, 7–12 May 2011. [CrossRef]
7.   Lim, J.; Green, B.G. Tactile interaction with taste localization: Influence of gustatory quality and intensity. *Chem. Senses* **2008**, *33*, 137–143. [CrossRef] [PubMed]
8.   Cole, M.; Covington, J.A.; Gardner, J.W. Combined electronic nose and tongue for a flavor sensing system. *Sens. Actuators B Chem.* **2011**, *156*, 832–839. [CrossRef]
9.   Tahara, Y.; Toko, K. Electronic tongues—A review. *IEEE Sens. J.* **2013**, *13*, 3001–3011. [CrossRef]
10. Ranasinghe, N.; Do, E.Y.L. Digital Lollipop: Studying Electrical Stimulation on the Human Tongue to Simulate Taste Sensations. *ACM Trans. Multimed. Comput. Commun. Appl.* **2016**, *13*, 1–22. [CrossRef]
11. Ranasinghe, N.; Lee, K.Y.; Suthokumar, G.; Do, E.Y.L. Taste+: Digitally enhancing taste sensations of food and beverages. In Proceedings of the 22nd ACM International Conference on Multimedia, Orlando, FL, USA, 3–7 November 2014. [CrossRef]
12. Ranasinghe, N.; Nakatsu, R.; Nii, H.; Gopalakrishnakone, P. Tongue mounted interface for digitally actuating the sense of taste. In Proceedings of the International Symposium on Wearable Computers, ISWC, Newcastle, UK, 18–22 June 2012. [CrossRef]
13. Ranasinghe, N.; Cheok, A.D.; Nakatsu, R. Taste/IP: The sensation of taste for digital communication. In Proceedings of the 14th ACM International Conference on Multimodal Interaction, Santa Monica, CA, USA, 22–26 October 2012. [CrossRef]
14. Aoyama, K.; Sakurai, K.; Sakurai, S.; Mizukami, M.; Maeda, T.; Ando, H. Galvanic tongue stimulation inhibits five basic tastes induced by aqueous electrolyte solutions. *Front. Psychol.* **2017**, *8*, 2112. [CrossRef]

15. Vi, C.T.; Ablart, D.; Arthur, D.; Obrist, M. Gustatory interface: The challenges of 'how' to stimulate the sense of taste. In Proceedings of the 2nd ACM SIGCHI International Workshop on Multisensory Approaches to Human-Food Interaction, Glasgow, UK, 13 November 2017. [CrossRef]
16. Aoyama, K.; Sakurai, K.; Furukawa, M.; Maeda, T.; Ando, H. New Method for Inducing, Inhibiting, and Enhancing Tastes Using Galvanic Jaw Stimulation. *Trans. Virtual Real. Soc. Jpn.* **2017**, *22*, 137–143.
17. Sardo, J.D.P.; Semião, J.; Monteiro, J.M.; Pereira, J.A.R.; de Freitas, M.A.G.; Esteves, E.; Rodrigues, J.M.F. Portable Device for Touch, Taste and Smell Sensations in Augmented Reality Experiences. In *INCREaSE*; Springer: Cham, Switzerland, 2018; pp. 305–320.
18. Nakano, K.; Kiyokawa, K.; Horita, D.; Yanai, K.; Sakata, N.; Narumi, T. Enchanting your noodles: GAN-based real-time food-to-food translation and its impact on vision-induced gustatory manipulation. In Proceedings of the 26th IEEE Conference on Virtual Reality and 3d User Interfaces, VR 2019, Osaka, Japan, 23–27 March 2019. [CrossRef]
19. Nakano, K.; Horita, D.; Sakata, N.; Kiyokawa, K.; Yanai, K.; Narumi, T. DeepTaste: Augmented reality gustatory manipulation with GAN-based real-time food-to-food translation. In Proceedings of the IEEE International Symposium on Mixed and Augmented Reality, Beijing, China, 14–18 October 2019. [CrossRef]
20. Aoyama, K.; Sakurai, K.; Ando, H.; Maeda, T.; Hara, A. Continuing Enhancement Effect of Repetitive Square Current Stimulation on Five Basic Taste. *Trans. Virtual Real. Soc. Jpn.* **2019**, *24*, 13–21. [CrossRef]
21. Sakurai, K.; Aoyama, K.; Mizukami, M.; Maeda, T.; Ando, H. Saltiness and umami suppression by cathodal electrical stimulation. In Proceedings of the 1st Workshop on Multi-Sensorial Approaches to Human-Food Interaction, Tokyo Japan, 16 November 2016. [CrossRef]
22. Mizoguchi, I.; Sakurai, S.; Hirota, K.; Nojima, T. Grutio: System for Reproducing Swallowing Sensation Using Neck-Skin Movement. *IEEE Access* **2021**, *9*, 105297–105307. [CrossRef]
23. Wu, X.; Tahara, Y.; Yatabe, R.; Toko, K. Taste sensor: Electronic tongue with lipid membranes. *Anal. Sci.* **2020**, *36*, 147–159. [CrossRef]
24. Miyashita, H. Taste display that reproduces tastes measured by a taste sensor. In Proceedings of the 33rd Annual ACM Symposium on User Interface Software and Technology, Virtual Event USA, 20–23 October 2020. [CrossRef]
25. Tahara, Y.; Nakashi, K.; Ji, K.; Ikeda, A.; Toko, K. Development of a portable taste sensor with a lipid/polymer membrane. *Sensors* **2013**, *13*, 1076–1084. [CrossRef] [PubMed]
26. Jalal, L.; Popescu, V.; Murroni, M. IoT architecture for multisensorial media. In Proceedings of the 2017 IEEE URUCON, URUCON 2017, Montevideo, Uruguay, 11 December 2017; pp. 1–4. [CrossRef]
27. Barakabitze, A.A.; Barman, N.; Ahmad, A.; Zadtootaghaj, S.; Sun, L.; Martini, M.G.; Atzori, L. QoE management of multimedia streaming services in future networks: A tutorial and survey. *IEEE Commun. Surv. Tutor.* **2020**, *22*, 526–565. [CrossRef]
28. Karunanayaka, K.; Johari, N.; Hariri, S.; Camelia, H.; Bielawski, K.S.; Cheok, A.D. New thermal taste actuation technology for future multisensory virtual reality and internet. *IEEE Trans. Vis. Comput. Graph.* **2018**, *24*, 1496–1505. [CrossRef]
29. Fritz, F.; Preissner, R.; Banerjee, P. VirtualTaste: A web server for the prediction of organoleptic properties of chemical compounds. *Nucleic Acids Res.* **2021**, *49*, W679–W684. [CrossRef]
30. Yang, Z.; Pathak, P.H.; Sha, M.; Zhu, T.; Gan, J.; Hu, P.; Mohapatra, P. On the feasibility of estimating soluble sugar content using millimeter-wave. In Proceedings of the 2019 Internet of Things Design and Implementation, Montreal, QC, Canada, 15–18 April 2019; pp. 13–24. [CrossRef]
31. Kajal, S.; Yadav, K.S.; Bajaniya, R.S.; Gholap, B.A.; Kadam, R.P. IoT based detection of microbial activity in raw milk by using Intel Galileo Gen. *Int. Res. J. Eng. Technol.* **2017**, *4*, 1.
32. Zia, S.; Spohrer, K.; Merkt, N.; Wenyong, D.; He, X.; Müller, J. Non-invasive water status detection in grapevine (*Vitis vinifera* L.) by thermography. *Int. J. Agric. Biol. Eng.* **2009**, *2*, 46–54. [CrossRef]
33. di Gennaro, S.F.; Matese, A.; Mancin, M.; Primicerio, J.; Palliotti, A. An open-source and low-cost monitoring system for precision enology. *Sensors* **2014**, *14*, 23388–23397. [CrossRef]
34. Muhammad, S.Y.; Makhtar, M.; Rozaimee, A.; Aziz, A.A.; Jamal, A.A. Classification model for water quality using machine learning techniques. *Int. J. Softw. Eng. Its Appl.* **2015**, *9*, 45–52. [CrossRef]
35. Qiu, S.; Wang, J. The prediction of food additives in the fruit juice based on electronic nose with chemometrics. *Food Chem.* **2017**, *230*, 208–214. [CrossRef]
36. Zhang, L.; Wang, X.; Huang, G.B.; Liu, T.; Tan, X. Taste Recognition in E-Tongue Using Local Discriminant Preservation Projection IEEE Trans. *IEEE Trans. Cybern.* **2019**, *49*, 947–960. [CrossRef]
37. Leon-Medina, J.X.; Cardenas-Flechas, L.J.; Tibaduiza, D.A. A data-driven methodology for the classification of different liquids in artificial taste recognition applications with a pulse voltammetric electronic tongue. *Int. J. Distrib. Sens. Netw.* **2019**, *10*, 1550147719881601. [CrossRef]
38. Dahal, K.R.; Dahal, J.N.; Banjade, H.; Gaire, S. Prediction of Wine Quality Using Machine Learning Algorithms. *Open J. Stat.* **2021**, *11*, 278–289. [CrossRef]
39. Saha, P.; Ghorai, S.; Tudu, B.; Bandyopadhyay, R.; Bhattacharyya, N. Tea Quality Prediction by Sparse Modeling of Electronic Tongue Signals. *IEEE Trans. Instrum. Meas.* **2018**, *68*, 3046–3053. [CrossRef]
40. Patil, A.B.; Bachute, M.R.; Kotecha, K. Artificial Perception of the Beverages: An In-Depth Review of the Tea Sample. *IEEE Access* **2021**, *9*, 82761–82785. [CrossRef]

41. Gutiérrez-Capitán, M.; Brull-Fontserè, M.; Jiménez-Jorquera, C. Organoleptic analysis of drinking water using an electronic tongue based on electrochemical microsensors. *Sensors* **2019**, *19*, 1435. [CrossRef]
42. Akitomi, H.; Tahara, Y.; Yasuura, M.; Kobayashi, Y.; Ikezaki, H.; Toko, K. Quantification of tastes of amino acids using taste sensors. *Sens. Actuators B Chem.* **2013**, *179*, 276–281. [CrossRef]

Article

Visual Positioning System Based on 6D Object Pose Estimation Using Mobile Web

Ju-Young Kim [1], In-Seon Kim [1], Dai-Yeol Yun [2], Tae-Won Jung [3], Soon-Chul Kwon [1] and Kye-Dong Jung [4,*]

[1] Department of Smart Convergence, Kwangwoon University, Seoul 01897, Korea; kjyjx@kw.ac.kr (J.-Y.K.); kisidid@kw.ac.kr (I.-S.K.); ksc0226@kw.ac.kr (S.-C.K.)
[2] Institute of Information and Science, Kwangwoon University, Seoul 01897, Korea; hibig10@kw.ac.kr
[3] Department of Immersive Content Convergence, Kwangwoon University, Seoul 01897, Korea; onom@kw.ac.kr
[4] Ingenium College of Liberal Arts, Kwangwoon University, Seoul 01897, Korea
* Correspondence: gdchung@kw.ac.kr; Tel.: +82-2-940-5014

Abstract: Recently, the demand for location-based services using mobile devices in indoor spaces without a global positioning system (GPS) has increased. However, to the best of our knowledge, solutions that are fully applicable to indoor positioning and navigation and ensure real-time mobility on mobile devices, such as global navigation satellite system (GNSS) solutions, cannot achieve remarkable researches in indoor circumstances. Indoor single-shot image positioning using smartphone cameras does not require a dedicated infrastructure and offers the advantages of low price and large potential markets owing to the popularization of smartphones. However, existing methods or systems based on smartphone cameras and image algorithms encounter various limitations when implemented in indoor environments. To address this, we designed an indoor visual positioning system for mobile devices that can locate users in indoor scenes. The proposed method uses a smartphone camera to detect objects through a single image in a web environment and calculates the location of the smartphone to find users in an indoor space. The system is inexpensive because it integrates deep learning and computer vision algorithms and does not require additional infrastructure. We present a novel method of detecting 3D model objects from single-shot RGB data, estimating the 6D pose and position of the camera and correcting errors based on voxels. To this end, the popular convolutional neural network (CNN) is improved by real-time pose estimation to handle the entire 6D pose estimate the location and direction of the camera. The estimated position of the camera is addressed to a voxel to determine a stable user position. Our VPS system provides the user with indoor information in 3D AR model. The voxel address optimization approach with camera 6D position estimation using RGB images in a mobile web environment outperforms real-time performance and accuracy compared to current state-of-the-art methods using RGB depth or point cloud.

Keywords: visual positioning system; convolutional neural network; three-dimensional object pose estimation; voxel; perspective-n-point

Citation: Kim, J.-Y.; Kim, I.-S.; Yun, D.-Y.; Jung, T.-W.; Kwon, S.-C.; Jung, K.-D. Visual Positioning System Based on 6D Object Pose Estimation Using Mobile Web. *Electronics* **2022**, *11*, 865. https://doi.org/10.3390/electronics11060865

Academic Editors: Jorge C. S. Cardoso, André Perrotta, Paula Alexandra Silva and Pedro Martins

Received: 21 January 2022
Accepted: 8 March 2022
Published: 9 March 2022

Publisher's Note: MDPI stays neutral with regard to jurisdictional claims in published maps and institutional affiliations.

1. Introduction

Multi-usage public facilities or large crowded markets without GPS functionality fail to navigation services. Researches on indoor positioning and navigation are developing widely. Recently, machine learning and deep learning methods are applied without sensors for location recognition. However, it is difficult to maintain the quality of location-based AR service without continuous updating the built-in maps as well as constructing indoor maps [1]. Visual positioning system information, which is more innovative than navigation technology obtained using GPS information, resonates with people's lifestyles globally. VPS allows users to use their mobile cameras to visually grasp their surroundings and directions in places where GPS services are difficult, such as indoor spaces [2]. Additionally, these techniques can accurately recognize a location of user through learning only by collecting images from mobile camera. Among recent object pose estimation approaches available for VPS, methods which are counting on depth maps with color images have shown excellent performance [3–5]. However, depth-estimation cameras cannot measure depth outdoors or reflective objects; therefore, this approach is not always reliable. Additionally, depth-estimation cameras consume the battery of an additional mobile device according to the operation of the sensor. Among the indoor positioning methods, though a QRcode method with screenshot have a high accuracy, it also has a problem which the user's position should be determined approximately.

The proposed system is a positioning system based on object pose estimation using images. Our method is used to estimate the position of user at specific indoor locations and provide 3D spatial information in 3D AR. Consequently, this can make user accurately estimate the position and pose of a camera in 3D space using a single-shot deep CNN based web application on a mobile device. After estimating the position of the camera in the 2D box of the object in 3D space. Pose estimation of the camera can stably determine the position through voxel indexing of the voxel database and provide 2D bird-eye view information. In addition, one of the eight vertices of the 3D box of the object is assigned as an anchor point of the 3D AR, and position information of the indoor space is provided to the user with 3D AR model.

The main contributions of this study are as follows.

- We propose an indoor positioning system using a mobile web browser that users can easily access. The mobile client system uses a smartphone camera to acquire images and estimate the pose of the camera in the server system to ensure real-time indoor space.
- We improve a single-shot deep CNN based on 2D object recognition. The pose of the camera calculated using PnP is indexed to the voxel database. A visual positioning system is designed to determine the user location using a spatial voxel address.
- With the help of object pose estimation of single-shot Deep CNN, one object box in the camera pose is used as an anchor point for 3D AR to provide information on a 3D indoor space in 3D AR model.

Unlike previous studies that require user interface applications, our method does not require additional application installation. It is a low cast, fast, and sustainable VPS method with a mobile web browser and can provide users with a variety of location-based AR services.

The remainder of this paper is organized as follows. Section 2 reviews related work. The details of the system and method are described in Section 3. Experiments and evaluation are presented in Section 4, and the conclusions are presented in Section 5.

2. Related Research

2.1. 6D Object Pose Prediction

Recently, machine-learning-based 6D pose-prediction technologies have attracted more attention owing to the increased prevalence deep learning and neural networks. However, 6D pose estimation techniques based on deep learning encounter a unique problem. The accurate estimation of 6D poses of symmetrical objects using conventional deep learning methods is difficult. This is because the shape and the 6D pose of the object do not change on rotation when viewed from a fixed point. However, the corresponding ground truths differ. Zhang and Qi [6] generated the key point-wise features of the point clouds as input features and predicted the keypoint coordinates using a hierarchical neural network involving global point clouds with local information.

PoseCNN estimates the 6D poses of the translation and rotation of an object; 3D translation is performed by determining the center of the image and estimating its distance from the camera, and 3D rotation is performed by regressing to a quaternion representation [7]. This network consists of two stages: in the first stage, feature maps with different resolutions are extracted from the input image. These extracted data are shared across all the tasks performed by the network. In the second stage, the high-dimensional feature maps generated in the first stage are embedded into low-dimensional task-centric features. Subsequently, the network estimates 6D pose by performing the following three tasks: semantic labeling, 3D translation estimation, and 3D rotation regression. Augmented Autoencoder [8] enables the estimation 3D object orientation to facilitate the implicit representation of rotation using auto-encoders; the rotation vector that is most representative of the estimated rotation is obtained from a coded book and assigned to the corresponding estimated rotation.

The state-of-the-art method of 6D object pose estimation using RGB camera input is characterized by the following approaches: (1) detecting the 2D target of the object in the given image, and (2) matching the 2D–3D correspondence using the perspective-n-point (PnP) method for the 6D pose. This type of algorithm can be categorized into keypoint-based and dense 2D–3D correspondence approaches. The 6D pose of the camera using RGB-D image and 3D model and estimated by PnP algorithm is a structure-based localization method of visual positioning system [9].

Keypoint-based method: The pixel-wise voting network (PVNet) [10] regresses pixel-wise unit vectors to determine keypoints, uses these unit vectors to vote for keypoint locations using Random sample consensus (RANSAC) [11], and creates a flexible representation to localize keypoints. HybridPose [12] involves intermediate representation prediction networks and pose regression. The prediction networks take an image as an input and provides the corresponding predicted keypoints, edge vectors, and symmetry correspondences as output. The pose regression consists of two processes, namely initialization and refinement. Initialization solves a linear system problem to obtain an initial pose using the predicted intermediate estimations. HybridPose is robust against occlusion and truncation. BPnP [13] backpropagates the gradients through the PnP solver to update the weights and achieves learning using a solver from a geometric vision problem and an objective function. BB8 [14] is a comprehensive approach that applies a convolutional neural network (CNN) to the detected objects to predict their 3D poses based on 2D projections of the corners of their cuboid 3D bounding boxes. Single-shot deep CNN [15] predicts 2D projections of a cuboid by creating a 3D bounding box around objects using the CNN. The 6D pose is calculated using a PnP algorithm that employs these 2D coordinates and the 3D ground points for the bounding box corners.

DPOD [16] uses an additional refinement network that provides a truncated image of an object and an image patch that must be rendered separately using the predicted pose of the first step and provides the refined pose as output. CDPN [17] untangles the pose to predict rotation and translation separately. For detection, a fast-lightweight detector and fixed-size segmentation are used to determine the exact object region. For translation, estimation is conducted from the detected object region to avoid scale errors. Pix2Pose [18] predicts the 3D coordinates of individual pixels using the truncated area containing the object. In the pose estimation process, image and 2D detection results are inputs. While removing backgrounds and uncertain pixels, the predicted results are used to represent important pixels and adjust bounding boxes. Pixels with valid coordinates and small error predictions are obtained using the PnP algorithm with RANSAC.

2.2. 2D–3D Correspondence

Single-photo resection (SPR) is a basic element in photogrammetry and computer vision. SPR addresses the restoration of earth orientation parameters (EOPs) of a given image/object. The SPR problem is also known as space resection, the perspective 3-points (P3P) problem, or PnP for n-points.

Grunert (1841) introduced the first solution to P3P by applying the cosine law for light emitted from the perspective center to three image points and the corresponding object points from the perspective center. Lepetit et al. [19] reduced the problem to four virtual control points, which is expressed as a weighted sum for n ($n \geq 4$) object points and developed an efficient PnP solution (EPnP). Li et al. [20] introduced a robust PnP (RPnP) solver that utilizes a subset of three points and produces an ($n - 2$) quaternary polynomial. The sum of squares of polynomials and the cost function are used to determine the minimum value via differentiation. A seventh-order polynomial of the differentiation of cost function is solved using the eigenvalue method [21].

The second SPR solution is an iterative method, which is the best approach to achieve high accuracy with minimal or redundant noisy data. However, these iterative methods are slow and approximate the position and orientation of parameter values.

The PnP problem and pose estimation from the projective observation of known points are related to the restoration of 6D poses given the central projection of $n \geq 3$ known 3D points in the calibrated camera. It is extensively used in geometric computer vision systems and determines the camera pose (orientation/position and rotation/translation) from observations of n 3D points.

In the case of a minimum PnP with a finite number of solutions, three observations ($n = 3$) are required in a nondegenerate configuration. This is called the P3P problem. P3P solvers are either directed or triangulated. Direct methods parameterize the pose of the input coefficient using projection invariances. Therefore, feasibility constraints should be applied as a post-processing step on obtaining a solution. The triangulation method triangulates points under pose invariants in the camera coordinate system, considers the distance as an unknown and solves the pose. In this triangulation method, a user can determine the rotation by choosing either a quaternion or $R \in SO\,(3)$. The geometric feasibility constraints, wherein each point is placed in front of the camera, limit the solutions before estimating the pose.

3. System and Methodology

This section outlines the proposed method and details the main modules and important algorithms involved. The proposed system consists of a mobile web and server. After the smartphone takes an image, it predicts a 6D object pose with an image which is transmitted to the server which estimates the pose of the camera, implements the remaining algorithms, and returns the result to. Figure 1 shows that the overall architecture of the proposed method includes three components: (a) acquiring images with a mobile web and single-shot deep CNN, (b) single-shot 6D object pose estimation, and (c) 3D voxel-based VPS.

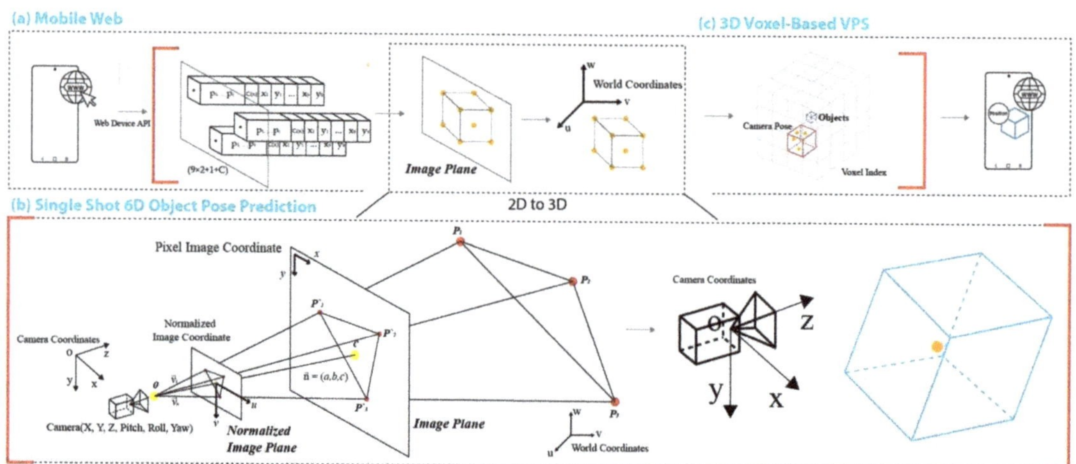

Figure 1. Overview of the proposed visual positioning method (system) (VPS). The process comprises (**a**) pose estimation stages of extended single-shot deep CNN; (**b**) estimate the 6D pose from the correspondences between the 2D and 3D points using a PnP pose estimation method; and (**c**) mobile web with voxel indexing through VPS.

3.1. System Overview

The proposed method is a mobile web implementation mechanism that outsources computing-intensive tasks to cloud servers, allowing web users to gain better location-based services and benefit from the server's stronger computing capabilities. However, additional communication delays and deployment costs are two critical issues that should be simultaneously addressed. The 5G network may achieve a data rate of 1 Gb/s and an end-to-end delay of milliseconds.

Figure 1a shows pose estimation of a single-shot deep CNN 2D object, acquiring an image from a mobile web with a camera. A single-shot deep CNN algorithm uses the acquired image to estimate eight corner points and one central coordinate of the 2D object box in the image. (b) 2D to 3D conversion and camera position estimation: estimates the 3D box and pose of the object with the PnP algorithm of the computer vision with the 3D box and central coordination of the object estimated from the image and the mesh model of the 3D object and finally estimates the pose of the camera. (c) The pose and camera position of the camera are estimated through the displayed voxel index and the mobile web VPS: (a) and (b) processes may be different from ground truth. To reduce this error and estimate more accurate camera location (user location), the estimated location of camera is matched to a voxel index in the voxel database and transmitted to a mobile an updated voxel index.

3.2. 6D Object Pose Estimation

This section focuses on determining an accurate pose estimation method. The proposed method is designed to localize and estimate the orientation and translation of an object accurately without correction. An object pose is expressed as a rigid transformation (RT) from the object to the camera coordinate system, where R and t represent 3D rotation and transformation, respectively.

First, a 6D object pose estimation using RGB image data input is described to obtain rotation information.

If converting a point x_1 into x_2 in a three-dimensional space is represented via a matrix R, a mapping function from a point $X_1 = [x_1 y_1 z_1]^\top$ to $X_2 = [x_2 y_2 z_2]^\top$ is expressed as follows [22].

$$f : \mathbb{R}^3 \rightarrow \mathbb{R}^3 \quad \begin{bmatrix} x_2 \\ y_2 \\ z_2 \end{bmatrix} = R \begin{bmatrix} x_1 \\ y_1 \\ z_1 \end{bmatrix} \tag{1}$$

In this case, the 3×3 matrix R set in which the inverse matrix exists corresponds to the general linear group GL $(3, \mathbb{R})$. Among these R, orthogonal matrices with a determinant of ± 1 are referred to as orthogonal groups. Therefore, there is a relationship between O $(3) \subset$ GL $(3, \mathbb{R}))$. Among these transform matrices, the transformation in which the distance between two pairs of points does not change is called isometries; a matrix with a determinant of +1 is called property isometries. This special orthogonal group is referred to as SO (3). The SO (3) group which is under (SO $(3) \subset$ O $(3))$ can only express pure rotation. Therefore, a 4×4 matrix is considered to express translation as shown in Equation (2); 3D points are extended to homogeneous coordinates. (GL $(4, \mathbb{R}))$.

The complete 6D pose is a three-dimensional orthogonal group, consisting of two parts: 3D rotation R $\in$ SO (3) and 3D transformation t $\in$ R^3, as shown in Equation (3).

$$\begin{bmatrix} X_2 \\ 1 \end{bmatrix} = T \begin{bmatrix} X_1 \\ 1 \end{bmatrix} \tag{2}$$

$$\begin{bmatrix} x_2 \\ y_2 \\ z_2 \\ 1 \end{bmatrix} = \begin{bmatrix} R_{11} & R_{12} & R_{13} & t_x \\ R_{21} & R_{22} & R_{23} & t_y \\ R_{31} & R_{32} & R_{33} & t_z \\ 0 & 0 & 0 & 1 \end{bmatrix} \begin{bmatrix} x_1 \\ y_1 \\ z_1 \\ 1 \end{bmatrix} \tag{3}$$

The 6D pose represents a rigid body transformation from object to camera coordinate system. This entire task has already been resolved in recent tasks in the field of relatively mature 2D object detection, as it includes several sub-tasks, such as detecting objects first in 2D images and processing multiple object categories and instances. In this study, we use the 2D object detection approach and improve it to predict the 6D pose of an object.

The proposed method is capable of end-to-end training that enables 6D pose prediction in real time and predicts the 2D projection of 3D bounding box corners surrounding objects. To regress the 2D boundary box as in the conventional YOLOv3 [23] and predict the projection of the 3D boundary box edge in the image, several additional 2D points are predicted for each object instance in the image. Considering these 2D coordinates and the 3D ground control point at the edge of the boundary box, 6D poses can be algebraically calculated using an efficient PnP algorithm [19].

The 6D pose estimation problem is formulated in terms of predicting the 2D image coordinates of the virtual 3D control point related to the 3D model of the object of interest. When considering 2D coordinate prediction, the 6D pose of the object is calculated using the PnP algorithm. The 3D model of each object is parameterized into nine control points. For these control points, eight corners of a tight 3D boundary box suitable for the 3D model are selected. Additionally, the center of the object's 3D model is used as the ninth point. This parameter designation is common and can be used for all robust 3D objects with arbitrary shapes and topologies.

3.3. 2D–3D Correspondence—3D Position Estimation Utilizing Perspective-n-Point

The camera pose estimation method through 2D point response with n 3D data in computer vision is a fundamental problem. The most common approach to the problem is to estimate six degrees of freedom and five correction parameters (focus distance, pub, aspect ratio, and slope) of the pose. A well-known direct linear transformation (DLT) algorithm is used to set at least six correspondence relationships. However, there are several simplifications to the problem of changing to numerous algorithms that improve the accuracy of DLT. The most common simplification is to assume a known correction parameter, the so-called perspective-n-point problem.

Figure 2 shows that, when there are 3D points (in world coordinates) that match the 2D projection points (in image coordinates) for the object in the image acquired by the camera, the values of the camera's orientation and position are estimated from the object. When a correspondence set between the 3D points $p_i(X_i, Y_i, Z_i)$ expressed in the reference frame of the spatial world coordinate system and the 2D projection $p_i'(u_i, v_i)$ for the image is given, the poses (R and T) for the camera are calculated.

$$s \begin{bmatrix} u \\ v \\ 1 \end{bmatrix} = \begin{bmatrix} f_x & \gamma & c_x \\ 0 & f_y & c_y \\ 0 & 0 & 1 \end{bmatrix} \begin{bmatrix} r_{11} & r_{12} & r_{13} & t_1 \\ r_{21} & r_{22} & r_{23} & t_2 \\ r_{31} & r_{32} & r_{33} & t_3 \end{bmatrix} \begin{bmatrix} x \\ y \\ z \\ 1 \end{bmatrix} \tag{4}$$

3.4. Voxel Index Database Using Camera Pose Optimization

The voxel database uses high-performance 3D sensors to scan indoor spaces. The sizes of the X, Y, and Z axes of the point cloud are calculated using the maximum and minimum values of the scanned point cloud coordinates. Voxel addresses are generated by dividing the calculated X, Y, and Z axes of the indoor space by the predefined voxel size and assigning a voxel address. The voxel address determines the location in the user's space. The voxel database is reconstructed including the real location of the object (3D box central coordinates). The pose of the camera estimated from the image is converted into coordinates of the voxel database. The converted coordinates determine the location of the user using the voxel index.

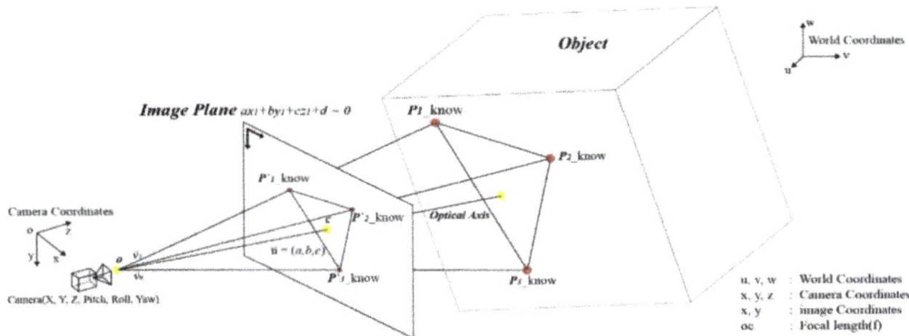

Figure 2. 3D Position Estimation using Perspective-n-Point.

3.5. Voxel Addressing vs. VPS Distance Error

The pose estimation of the estimated object is proportional to the center coordinates x, y, and z of the object and the rotational values of the object pitch, yaw, and roll, and the distance to the origin coordinates (0, 0, 0) of the camera. However, because VPS is made to the address of the voxel, the coordinates of the objects in the voxel database space are the same within the box of voxel labeling. Figure 3 shows that the VPS error rate is on average as much as the center distance of the voxel when the position of the camera and actual camera coordinates are not the same voxel in the voxel database space estimated by PnP of the improved single-shot deep CNN.

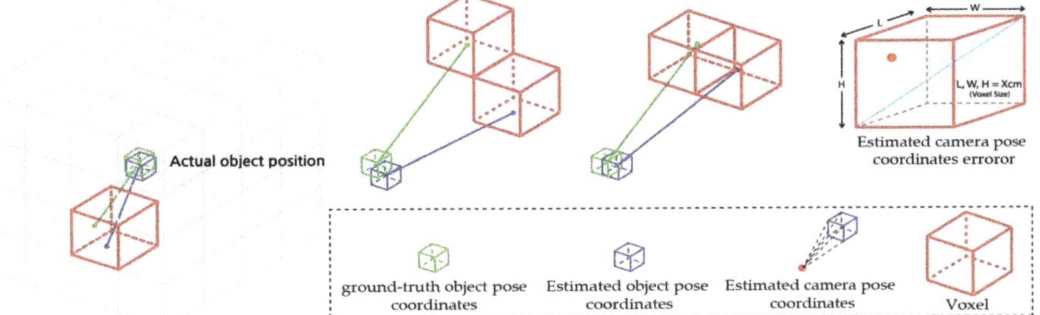

Figure 3. Visual positioning system (VPS) error and measurement.

The actual camera position of the camera corresponding to the object center point of the voxel database of the object selected in the indoor space: $(x_2, y_2, z_2) \in$ Voxel No.x2 and VPS predicted camera position through the pose of the camera estimated by the PnP algorithm: $(x_1, y_1, z_1) \in$ Voxel No.x1. When points belong to the same voxel, the proposed method maps these points to same voxel index. This indexing reduces distance error of estimation of location. Assuming that length, width, and height have same size of α, the actual distance error in the same voxel space does not exceed Equation (5). Therefore, when the voxel number does not match, the distance error of the corresponding voxel is calculated by Equation (6).

$$\left(\text{VPS Distance Error} = \text{Max } \sqrt{3\alpha^2} \right) \in \{\text{Voxel No.X2} = \text{Voxel No.X1}\} \tag{5}$$

$$\left(\text{VPS Distance Error} = \sqrt{(x_2 - x_1)^2 + (y_2 - y_1)^2 + (z_2 - z_1)^2} \right) \in \{\text{Voxel No.X2} \neq \text{Voxel No.X1}\} \tag{6}$$

4. Evaluation

In this section, we compare the CNN of the 6D pose estimation base on RGB on LineMOD [24] with other 6D pose estimation methods for a single individual to measure the performance of the proposed system. It was designed on the premise of indoor use, and night, day, and lighting were not considered. Experiments focusing on real-time execution on the mobile web measured the network speed of cutting-edge methods. We compared three voxel sizes that can stabilize the camera's pose with the proposed method's voxel addressing. The VPS real-time criterion is at least 30 FPS; we describe the experiments we performed as experimental settings and error measurements with implementation details.

4.1. Experimental Setup

4.1.1. System Setup

The system facility conditions used in the experiment are as follows.

Mobile Web: A smartphone Galaxy Note 20 Ultra (SM-N986N) equipped with 108 million pixels and 12 GB RAM and tested in a Web (Chrome Browser) environment with 5G (fifth generation technology standard) mobile communication.

Server: The implementation was written in Python 3.6, using PyTorch for graphics processing unit (GPU) computation. The evaluation details measured the inference times on a desktop using a Linux Ubuntu 16.04 LTS, Ryzen 9 3900X CPU, and RTX 2080 SUPER 8G GPU.

4.1.2. LineMOD Dataset

The LineMod dataset is a popular and widely used benchmark dataset for 6D object pose estimation. It consists of 13 different entities arranged in 13 complex scenes. For each scene, only one object is annotated with a 6D pose; other objects can be viewed simultaneously. There is an example with approximately 1200 annotations per individual.

4.2. Comparison of 6D Pose Estimation Convolutional Neural Network Using RGB

We evaluated the runtime of the 6D pose estimation network for LineMOD datasets that have become the de facto standard benchmarks for 6D pose estimation. Among the latest methods of 6D pose estimation RGB in LineMOD that can be applied to our method, efficient pose [25], RePOSE [26], DPOD [16], HRNet (DSNT + BPnP) [13], HybridPose [12], CDPN [17], PoseCNN + DeepIM [27], E2E6DoF [28], PVNet [10], CullNet [29], SSD-6D [30], keypoint detector localization [31], single-shot deep CNN [15], BB8[14], Pix2Pose [18], and augmented autoencoder [32], which focused not only on accuracy but also on time cost were selected and evaluated. Because the proposed method and voxel index can optimize the user's location accuracy by correcting the VPS error, the experimental evaluation selected the network based on real-time data on the mobile web rather than accuracy. Figure 4 depict the experimental results of the following three networks that were selected for evaluation on the LineMOD dataset considering the runtime: efficient pose, SSD-6D, and single-shot deep CNN. We used the trained model provided in each study. Figure 4 shows that, for each of the 13 classes provided by LineMOD, the efficient pose is $\varphi = 0$. The single trained tape model and SSD-6D used the provided trained bench vise model and trained hole puncher model weight provided by single-shot deep CNN to create boxes for supervised learning and boxes through 6D object pose estimation with 1000 evaluation datasets per class.

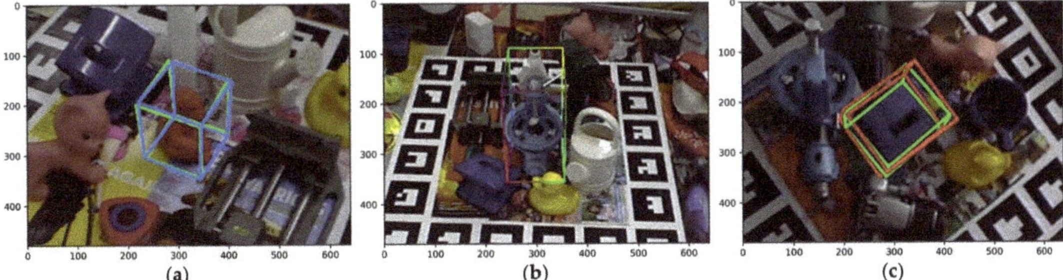

Figure 4. Results of convolutional neural networks applied on the LineMOD dataset 6D for accurate object pose estimation. We show an input RGB image, estimated pose, and ground-truth pose: (**a**) EfficientPose pose estimation; (**b**) SSD-6D pose estimation; (**c**) single-shot deep CNN pose estimation. The 2D matrix markers used in (a), (b) and (c) are only for the learning phase. They are interim results which is utilized to configure datasets.

Table 1 shows the average frame per second (FPS) evaluation table experimented with the learned weight provided by each network using 13 classes of 6D pose estimation networks in the widely used 6D pose estimation benchmark dataset LineMOD using RGB input. A total of 1000 evaluation images were used for each of the 13 classes. The 13 object classes are ape, bench vise, cam, can, cat, driller, duck, eggbox, glue, hole puncher, iron, and lamp. Although there is a slight difference between the runtime speed revealed in each paper and the system environment, similar runtime speeds could be identified overall, as revealed by the author of the network. The average FPS evaluated in the proposed system environment is as follows: Efficient pose 20.50 FPS, SSD-6D 11.74 FPS, and single-shot deep CNN 54.38 FPS were measured. To ensure the best real-time nature of the proposed method through these comparative evaluations, we selected a 3D pose estimation network for single-shot deep CNNs with a runtime rate higher than 50 FPS.

Table 1. Runtime performance comparison between single object pose estimation algorithms. LineMOD dataset is used.

6D Object Pose	Efficient Pose [25] (FPS)	SSD-6D [30] (FPS)	Single-Shot Deep CNN [15] (FPS)
Ape	20.56	11.98	54.36
Bench vise	20.50	11.32	53.99
Cam	20.69	11.45	54.30
Can	20.87	11.87	54.49
Cat	21.01	11.94	54.69
Driller	20.91	11.57	54.53
Duck	19.88	11.74	54.47
Eggbox	19.53	12.43	54.56
Glue	20.29	12.03	55.50
Hole puncher	19.84	11.83	54.16
Iron	21.67	11.78	53.96
Lamp	20.32	11.10	54.04
Phone	20.47	11.59	53.95
Average FPS	20.50	11.74	54.38

Figure 5 shows the overall process of the proposed method. When the mobile web client sends a request to the server with the image and receives the image from the server, it detects the object through the single-shot deep CNN network and converts the ratio of the coordinates of the 3D box on 2D into coordinate values suitable for the picture size. Using the PnP algorithm, converted 2D box coordinates, and the camera internal parameter of the detected object size, the camera pose coordinates relative to the object is obtained. VPS is performed by determining the relative coordinates as voxels in the voxel database. Figure 5 shows the process of responding to the user's camera pose to the client of the mobile wed again and Table 2 summarizes the running time of each process for each step. The operating time of the entire system is 733.1268 ms, which can transmit VPS to the user's mobile web once a second. The voxel indexing step includes the step of drawing the voxel on the server; however, it does not include the time required to send the image to the smartphone and the time taken to load the image.

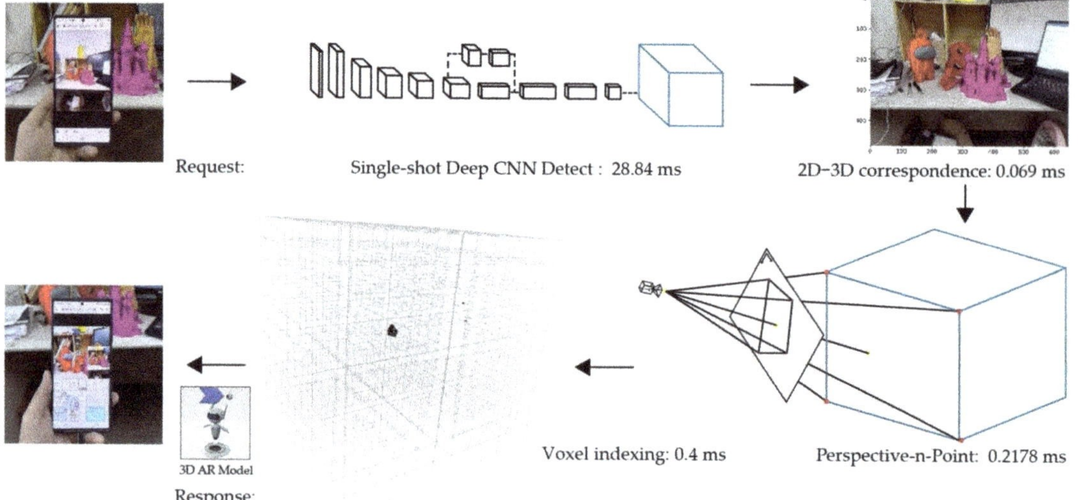

Figure 5. Runtime analysis and comparison of method performing single object pose estimation. LineMOD dataset is used.

Table 2. VPS speed measured by the proposed method system.

Request + Response	Detect	2D–3D Correspondence	Perspective-n-Point	Voxel Indexing	Total
700 ms	28.84 ms	0.069 ms	0.2178 ms	0.4 ms	733.1268 ms

4.3. VPS Results of Voxel Index

The pose estimation error of the improved single-shot deep CNN is proportional to the x, y, and z coordinates of the object center and the rotational pitch, yaw, and roll values of the object, and it is proportional to the origin coordinates (0, 0, 0) of the camera. However, because the VPS targets the address of the voxel, the coordinates of the camera in space remain unchanged within the indexed voxel box. Our method consists of a network module and an algorithm module, and it is computed using our equation in the algorithm, and the measurement uncertainty in our system is proportional to the estimations of network-specific. The measurement uncertainty estimated by the network is corrected using our method, by positioning through voxels. The improved single-shot deep CNN has an average error of VPS in the database space estimated by VPS when the position of the camera coordinates and the actual camera coordinates are not the same voxel. The actual camera position of an object selected in an indoor space, $(x_2, y_2, z_2) \in$ Voxel No.X2 and VPS Predicated camera position through pose estimation of an extended single-shot deep $(x_1, y_1, z_1) \in$ Voxel No.X1 are in the range of Equation (5). The actual distance error in the same voxel space does not exceed that obtained via Equation (5). However, when the voxel numbers do not match, the distance error is determined via Equation (6). Table 3 shows the VPS distance error of the extended single-shot deep CNN pose estimation obtained using Equation (6). The voxel size of the voxel database is tested for the 20, 50, and 100 sizes, and the position may be localized within the accuracy of the sub meter level with respect to 80% or more at a voxel size of 50 cm. Table 3 shows that more than 95% of the 100 cm voxel size is successfully identified in the ground truth position.

Table 3. Distance errors for the ground truth and estimated camera poses.

Voxel	Ape	Bench Vise	Cam	Can	Cat	Driller	Duck	Eggbox	Glue	Hole Puncher	Iron	Lamp	Phone	Average
Distance	10 cm	5 cm	7 cm	5 cm	8 cm	7 cm	9 cm	9 cm	8 cm	8 cm	11 cm	14 cm	11 cm	8.61 cm

Based on the object box center point of the indoor space, voxels of 2 m in width and height were divided into 1 m units along the x, y, and z axes, and addresses are formed in the divided voxel database space. Table 4 shows that 55.5% of the total voxels can be classified into the same voxel address when the voxel is divided into 20 cm. Table 5 shows that 81.7% is indexed to the same voxel address when divided by 50 cm, and Table 6 shows that 95.2% is indexed within the same voxel address when divided by 1 m.

Table 4. Distance error comparison for voxel size (20 cm).

Voxel Index Error (20 cm)	Ape	Bench Vise	Cam	Can	Cat	Driller	Duck	Eggbox	Glue	Hole Puncher	Iron	Lamp	Phone	Average
1 Voxel (%)	34.8	27.3	35.5	27	36.5	32.9	36.4	36.3	31.8	35.8	37.9	33.6	38.4	34.2
2 Voxel (%)	10.2	4.7	6.9	4.1	10	8.6	10	10.8	8.4	7.6	13.5	9.9	13.4	9.1
3 Voxel (%)	2.2	0.6	1.0	0.1	1.3	1.1	1.2	1.1	1.1	0.8	2.5	1.0	1.8	1.2
4 Voxel + (%)	0.3	0.2	0.0	0.0	0.1	0.0	0.2	0.1	0.3	0.3	0.3	1.4	0.3	0.3

Table 5. Distance error comparison for voxel size (50 cm).

Voxel Index Error (50 cm)	Ape	Bench Vise	Cam	Can	Cat	Driller	Duck	Eggbox	Glue	Hole Puncher	Iron	Lamp	Phone	Average
1 Voxel (%)	19.4	11.7	15.4	12.5	19.7	14.1	16.6	19	17.4	16.8	23.5	14.7	21	17.1
2 Voxel (%)	2.1	0.6	0.7	0.3	1.6	0.8	1	1.9	1	0.9	2.5	1.3	1.4	1.2
3 Voxel (%)	0.1	0.0	0.0	0.0	0.1	0.0	0.9	0.0	0.0	0.0	0.1	0.4	0.0	0.1
4 Voxel + (%)	0.0	0.0	0.0	0.0	0.0	0.0	0.0	0.0	0.0	0.0	0.0	0.6	0.1	0.1

Table 6. Distance error comparison for voxel size (100 cm).

Voxel Index Error (100 cm)	Ape	Bench Vise	Cam	Can	Cat	Driller	Duck	Eggbox	Glue	Hole Puncher	Iron	Lamp	Phone	Average
1 Voxel (%)	4.6	3.7	4.7	3.9	6.0	4.5	5.0	4.5	4.0	5.6	5.6	4.8	5.2	4.8
2 Voxel (%)	0.0	0.0	0.1	0.0	0.0	0.0	0.2	0.0	0.1	0.1	0.0	0.4	0.0	0.1
3 Voxel (%)	0.0	0.0	0.0	0.0	0.0	0.0	0.0	0.0	0.0	0.0	0.0	0.0	0.0	0.0
4 Voxel + (%)	0.0	0.0	0.0	0.0	0.0	0.0	0.0	0.0	0.0	0.0	0.0	0.0	0.0	0.0

As shown in Table 7, the 6D pose estimation accuracy excluding the eggbox and glue classes of EfficientPose in our experiment is 5.68% higher than the estimated accuracy of

Single-shot Deep CNN; however, the time rate achieved by EfficientPose is 20.50 FPS, as shown in Table 7, and the error rate is higher in a specific class. Figure 6 shows an example of the LineMOD dataset. Figure 6a displays the input RGB image and the ground truth pose in red with the estimated pose of the extended single-shot Deep CNN in blue, and Figure 6b displays of a voxel database with a voxel labeling index.

Table 7. Distance error comparison of 6D pose estimation via the EfficientPose network according to VPS Voxel Size.

Voxel Index Error (50 cm)	Ape	Bench Vise	Cam	Can	Cat	Driller	Duck	Eggbox	Glue	Hole Puncher	Iron	Lamp	Phone	Average
1 Voxel (%)	13.3	13.2	10.0	14.0	12.8	11.3	12.3	3.2	7.7	11.5	9.4	10.8	10.8	10.8
2 Voxel (%)	0.6	0.6	0.7	0.9	1.0	0.4	1.2	15.3	12.4	0.3	0.4	0.3	0.4	2.7
3 Voxel (%)	0.0	0.1	0.1	0.3	0.1	0.0	0.1	5.8	4.6	0.0	0.0	0.0	0.0	0.9
4 Voxel + (%)	0.1	0.0	0.0	0.1	0.1	0.0	0.0	61.7	34.3	0.0	0.0	0.0	0.0	7.4

(a)

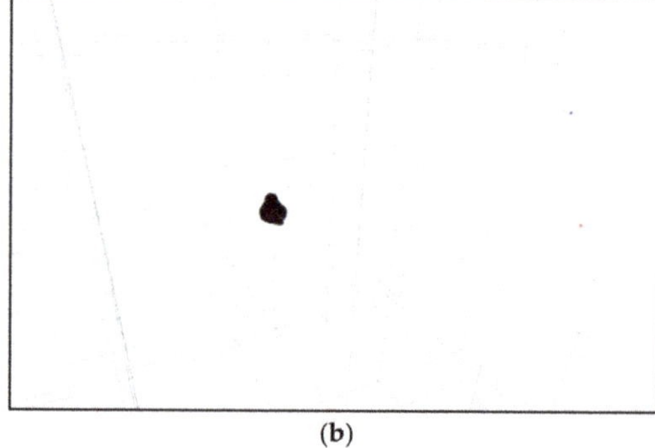

(b)

Figure 6. Results on the LineMOD dataset: (**a**) the input RGB images, poses estimated with the extended single-shot deep CNN in blue, and ground truth poses in red; (**b**) visual positioning system in voxel database with voxel labeling index.

5. Conclusions

In this study, we introduced a system that determines a user's location using a highly scalable, end-to-end 6D object posture estimation approach based on the state-of-the-art 2D object detection architecture of the single-shot deep CNN. We improved the architecture in an intuitive and efficient manner to perform 6D object pose estimation of multiple objects and instances and 2D object detection while maintaining the advantages of the underlying network and keeping additional computational costs low. Based on the object, a positioning system in a large indoor space using a smartphone camera was proposed. The system used a web on smartphones to detect specific objects indoors and calculated a user's location. The system integrated deep learning and computer vision algorithms and proposed the VPS that could determine the position of an object and pose estimated through deep learning by matching the position and pose of the object in space with a predefined. It is a visual positioning system that used a voxel address that can determine a user's location by learning images acquired by a camera on the mobile web through deep learning, estimating the pose of an object, and matching the camera pose in a predefined voxel indexing space. The proposed method organized a database with voxel addresses to determine a location of user. This shows that the proposed method can efficiently lead to high location accuracy and direction estimation in a well-known space. The proposed system uses web-based images of mobile devices that users can easily access when GPS is insufficient, and is a deep learning-based visual positioning which uses fixed specific location to provide 3D AR contents to users. The proposed method is particularly suitable for scenarios that ensure real-time performance.

Author Contributions: Conceptualization, J.-Y.K. and I.-S.K.; Methodology, J.-Y.K., I.-S.K., and T.-W.J.; Software, J.-Y.K. and T.-W.J.; Investigation, J.-Y.K., I.-S.K., D.-Y.Y. and T.-W.J.; Writing—Original Draft Preparation, J.-Y.K. and I.-S.K.; Writing—Review and Editing, K.-D.J., S.-C.K., D.-Y.Y. and T.-W.J.; Supervision, K.-D.J.; Project Administration, K.-D.J. All authors have read and agreed to the published version of the manuscript.

Funding: This research received no external funding.

Data Availability Statement: The code and dataset will be made available on request to the first author's email with appropriate justification. The public site for each dataset was as follows: 6D pose estimation benchmark dataset LineMOD using RGB: https://bop.felk.cvut.cz/datasets/ (accessed on 20 Janaury 2022).

Acknowledgments: The present research has been conducted by the Research Grant of Kwangwoon University in 2021.

Conflicts of Interest: The authors declare no conflict of interest.

References

1. Huang, H.; Garther, G. A survey of mobile indoor navigation systems. In *Central and Eastern Europe*; Section III: Multimedia Cartography; Springer: Berlin/Heidelberg, Germany, 2009; pp. 305–319.
2. Zhang, X.; Wang, L.; Su, Y. Visual place recognition: A survey from deep learning perspective. *Pattern Recognit.* **2021**, *113*, 107760. [CrossRef]
3. Brachmann, E.; Krull, A.; Michel, F.; Gumhold, S.; Shotton, J.; Rother, C. Learning 6D Object Pose Estimation Using 3D Object Coordinates. In Proceedings of the European Conference on Computer Vision(ECCV), Zurich, Switzerland, 6–12 September 2014; pp. 536–551.
4. Choi, C.; Christensen, H.I. RGB-D Object Pose Estimation in Unstructured Environments. *Robot. Auton. Syst.* **2016**, *75*, 595–613. [CrossRef]
5. Kehl, W.; Milletari, F.; Tombari, F.; Ilic, S.; Navab, N. Deep Learning of Local RGB-D Patches for 3D Object Detection and 6D Pose Estimation. In Proceedings of the European Conference on Computer Vision(ECCV), Amsterdam, The Netherlands, 11–14 October 2016; pp. 205–220.
6. Zhang, W.; Qi, C. Pose Estimation by Key Points Registration in Point Cloud. In Proceedings of the 2019 3rd International Symposium on Autonomous Systems (ISAS), Shanghai, China, 29–31 May 2019; pp. 65–68. [CrossRef]
7. Xiang, Y.; Schmidt, T.; Narayanan, V.; Fox, D. Posecnn: A convolutional neural network for 6d object pose estimation in cluttered scenes. *arXiv* **2017**, arXiv:1711.00199.
8. Sundermeyer, M.; Marton, Z.C.; Durner, M.; Triebel, R. Augmented Autoencoders: Implicit 3D Orientation Learning for 6D Object Detection. *Int. J. Comput. Vis.* **2020**, *128*, 714–729. [CrossRef]
9. Bai, X.; Huang, M.; Prasad, N.R.; Mihovska, A.D. A survey of image-based indoor localization using deep learning. In Proceedings of the IEEE Conference on 2019 22nd International Symposium on Wireless Personal Multimedia Communications (WPMC), Lisbon, Portugal, 24–27 November 2019; pp. 1–6.
10. Peng, S.; Liu, Y.; Huang, Q.; Zhou, X.; Bao, H. Pvnet: Pixel-wise voting network for 6dof pose estimation. In Proceedings of the IEEE/CVF Conference on Computer Vision and Pattern Recognition, Long Beach, CA, USA, 15–20 June 2019; pp. 4556–4565.
11. Fischler, M.A.; Bolles, R.C. Random sample consensus: A paradigm for model fitting with applications to image analysis and automated cartography. *Commun. ACM* **1981**, *24*, 381–395. [CrossRef]
12. Chen, S.; Song, J.; Huang, Q. Hybridpose: 6d object pose estimation under hybrid representations. In Proceedings of the IEEE Conference on Computer Vision and Pattern Recognition (CVPR), Seattle, WA, USA, 13–19 June 2020; pp. 428–437.
13. Chen, B.; Parra, Á.; Cao, J.; Li, N.; Chin, T.J. End-to-end learnable geometric vision by backpropagating PnP optimization. In Proceedings of the IEEE/CVF Conference on Computer Vision and Pattern Recognition, Seattle, WA, USA, 13–19 June 2020; pp. 8097–8106.
14. Mahdi, R.; Vincent, L. BB8: A scalable, accurate, robust to partial occlusion method for predicting the 3D poses of challenging objects without using depth. In Proceedings of the IEEE International Conference on Computer Vision (ICCV), Venice, Italy, 22–29 October 2017; pp. 3848–3856.
15. Tekin, B.; Sinha, S.N.; Fua, P. Real-time seamless single shot 6d object pose prediction. In Proceedings of the IEEE Conference on Computer Vision and Pattern Recognition (CVPR), Salt Lake City, UT, USA, 18–23 June 2018; pp. 292–301.
16. Zakharov, S.; Ivan, S.; Slobodan, I. Dpod: 6d pose object detector and refiner. In Proceedings of the IEEE/CVF International Conference on Computer Vision, Seoul, Korea, 27 October–2 November 2019; pp. 1941–1950.

17. Li, Z.; Gu, W.; Xiangyang, J. Cdpn: Coordinates-based disentangled pose network for real-time rgb-based 6-dof object pose estimation. In Proceedings of the IEEE/CVF International Conference on Computer Vision, Seoul, Korea, 27 October–2 November 2019; pp. 7677–7686.

18. Kiru, P.; Timothy, P.; Markus, V. Pix2pose: Pixel-wise coordinate regression of objects for 6d pose estimation. In Proceedings of the 2019 IEEE/CVF International Conference on Computer Vision (ICCV), Seoul, Korea, 27 October–2 November 2019; pp. 7667–7676.

19. Lepetit, V.; Moreno-Noguer, F.; Fua, P. EPnP: An Accurate O(n) Solution to the PnP Problem. *Int. J. Comput. Vis.* **2008**, *81*, 155–166. [CrossRef]

20. Li, S.; Xu, C.; Xie, M. A Robust O(n) Solution to the Perspective-n-Point Problem. *IEEE Trans. Pattern Anal. Mach. Intell.* **2012**, *34*, 1444–1450. [CrossRef] [PubMed]

21. Press, W.; Teukolsky, S.; Vetterling, W.; Flannery, B. *Numerical Recipes: The Art of Scientific Computing*; Cambridge University Press: Cambridge, UK, 1989; Volume 1.

22. Blanco, J.L. A Tutorial on se (3) Transformation Parameterizations and on-Manifold Optimization. Available online: https://citeseerx.ist.psu.edu/viewdoc/download?doi=10.1.1.468.5407&rep=rep1&type=pdf (accessed on 20 January 2022).

23. Redmon, J.; Farhadi, A. YOLOv3: An Incremental Improvement. *arXiv* **2018**, arXiv:1804.02767.

24. Hinterstoisser, S.; Holzer, S.; Cagniart, C.; Ilic, S.; Konolige, K.; Navab, N.; Lepetit, V. Multimodal templates for real-time detection of texture-less objects in heavily cluttered scenes. In Proceedings of the 2011 International Conference on Computer Vision, ICCV'11, Barcelona, Spain, 6–13 November 2011; pp. 858–865.

25. Bukschat, Y.; Vetter, M. EfficientPose: An efficient, accurate and scalable end-to-end 6D multi object pose estimation approach. *arXiv* **2020**, arXiv:2011.04307.

26. Shun, I.; Xingyu, L.; Rawal, K.; Rio, Y.; Kris, M.K. RePOSE: Fast 6D Object Pose Refinement via Deep Texture Rendering. *arXiv* **2021**, arXiv:2104.00633.

27. Yi, L.; Gu, W.; Xiangyang, J.; Xiang, Y.; Fox, D. DeepIM: Deep Iterative Matching for 6D Pose Estimation. *arXiv* **2018**, arXiv:1804.00175.

28. Gupta, A.; Medhi, J.; Chattopadhyay, A.; Gupta, V. End-to-End Differentiable 6DoF Object Pose Estimation with Local and Global Constraints. *arXiv* **2020**, arXiv:2011.11078.

29. Gupta, K.; Lars, P.; Richard, H. Cullnet: Calibrated and pose aware confidence scores for object pose estimation. In Proceedings of the IEEE/CVF International Conference on Computer Vision Workshops, Seoul, Korea, 27–28 October 2019; pp. 2758–2766.

30. Wadim, K.; Fabian, M.; Federico, T.; Slobodan, I.; Navab, N. SSD-6D: Making rgb-based 3D detection and 6D pose estimation great again. In Proceedings of the IEEE Conference on Computer Vision and Pattern Recognition (CVPR), Venice, Italy, 22–29 October 2017; pp. 1530–1538.

31. Zhao, Z.; Peng, G.; Wang, H.; Fang, H.S.; Li, C.; Lu, C. Estimating 6D pose from localizing designated surface keypoints. *arXiv* **2018**, arXiv:1812.01387.

32. Sundermeyer, M.; Marton, Z.C.; Durner, M.; Brucker, M.; Triebel, R. Implicit 3D Orientation Learning for 6D Object Detection from RGB Images. *arXiv* **2019**, arXiv:1902.01275.

Article

Effects of Using Vibrotactile Feedback on Sound Localization by Deaf and Hard-of-Hearing People in Virtual Environments

Mohammadreza Mirzaei [1,*], Peter Kán [1,2] and Hannes Kaufmann [1]

[1] Institute of Visual Computing and Human-Centered Technology, Vienna University of Technology (TU Wien), 1040 Wien, Austria; peterkan@peterkan.com (P.K.); hannes.kaufmann@tuwien.ac.at (H.K.)
[2] Department of Computer Science, Aarhus University, 8200 Aarhus, Denmark
* Correspondence: mohammad.mirzaei@tuwien.ac.at

Abstract: Sound source localization is important for spatial awareness and immersive Virtual Reality (VR) experiences. Deaf and Hard-of-Hearing (DHH) persons have limitations in completing sound-related VR tasks efficiently because they perceive audio information differently. This paper presents and evaluates a special haptic VR suit that helps DHH persons efficiently complete sound-related VR tasks. Our proposed VR suit receives sound information from the VR environment wirelessly and indicates the direction of the sound source to the DHH user by using vibrotactile feedback. Our study suggests that using different setups of the VR suit can significantly improve VR task completion times compared to not using a VR suit. Additionally, the results of mounting haptic devices on different positions of users' bodies indicate that DHH users can complete a VR task significantly faster when two vibro-motors are mounted on their arms and ears compared to their thighs. Our quantitative and qualitative analysis demonstrates that DHH persons prefer using the system without the VR suit and prefer mounting vibro-motors in their ears. In an additional study, we did not find a significant difference in task completion time when using four vibro-motors with the VR suit compared to using only two vibro-motors in users' ears without the VR suit.

Keywords: virtual reality; haptic feedback; tactile sensation; sound source localization; deaf and hard-of-hearing

Citation: Mirzaei, M.; Kán, P.; Kaufmann, H. Effects of Using Vibrotactile Feedback on Sound Localization by Deaf and Hard-of-Hearing People in Virtual Environments. *Electronics* **2021**, *10*, 2794. https://doi.org/10.3390/electronics10222794

Academic Editor: Jorge C. S. Cardoso

Received: 17 October 2021
Accepted: 11 November 2021
Published: 15 November 2021

Publisher's Note: MDPI stays neutral with regard to jurisdictional claims in published maps and institutional affiliations.

1. Introduction

Deafness and hearing loss are issues that affect millions of people around the world (World Federation of the Deaf (Available Online: http://wfdeaf.org/our-work/ (accessed on 5 October 2021))). These issues are manifested in different intensities related to different causes and can affect different aspects of Deaf and Hard-of-Hearing (DHH) persons' social life [1]. Recent advances in technology and medicine, such as Cochlear Implants (CIs), help these groups of people to use and enjoy technology more than before [2], but CI technology has some disadvantages. It is costly and not practical for all DHH persons [3]. In addition, it requires invasive surgery, and the surgery's success rate depends on the person's age [3]. Therefore, using alternative technologies to help DHH persons is beneficial.

One of the low-cost technologies to help DHH persons is inexpensive vibrotactile devices that can transmit information as vibration signals through the DHH person's body. Previous studies have shown that using haptic sensors in wearable devices such as suits, belts, bracelets, shoes, gloves, and chairs can help DHH persons perceive information from their environments [4–6]. DHH persons detect sounds differently, so they have limitations in completing sound-related tasks, especially in Virtual Reality (VR). However, DHH persons can perceive sound information using their other senses, such as tactile sensation [7]. They can sense vibrations and feel sounds in the same part of their brain that hearing persons use to hear sounds [8].

Previous research has shown the usability of vibrotactile systems for hearing and DHH persons as they complete different tasks related to navigation [9] and sound awareness [10].

In VR, haptic feedback is usually used in special VR suits or other wearable devices, and it improves the immersive VR experience. A few versions of vibrotactile-based VR suits are available on the market, such as "TactSuit" (Available Online: https://www.bhaptics.com/ (accessed on 5 October 2021)) and "TeslaSuit" (Available Online: https://teslasuit.io/ (accessed on 5 October 2021)), but they have not been tested on DHH users. Haptic VR suits can help DHH persons to perform sound-related VR tasks, but more comprehensive studies are needed about the effects of using different setups of haptic VR suits for DHH users. We also need to know if the number of haptic devices positively affects the completion of sound-related VR tasks among DHH users. This paper investigates the capabilities and effects of using different setups of haptic VR suits among DHH users. Our main hypotheses in this study are as follows:

Hypothesis 1 (H1). *DHH persons can complete sound-related VR tasks faster using different VR suit setups compared to not using the VR suit.*

Hypothesis 2 (H2). *Increasing the number of haptic devices on a VR suit does not significantly affect the performance of sound source localization in VR among DHH persons.*

A special VR suit with four adjustable vibro-motors was designed for this study to analyze different aspects of using haptic VR suit setups for DHH users. We intended to find the optimal number of haptic devices (vibro-motors) necessary for DHH persons to perform sound source localization in VR. In addition, we determined the best positions for mounting vibro-motors on DHH persons' bodies for the completion of sound-related VR tasks by analyzing the results of questionnaires about discomfort scores and the desire to use different haptic setups of our proposed VR suit. In summary, the main contributions of our study are as follows:

1. The effects of using haptic VR suit setups among DHH users on the completion of sound-related VR tasks are analyzed.
2. The optimal number of haptic devices necessary for DHH persons to perform sound source localization in VR is identified.
3. The best positions for mounting haptic devices on DHH persons' bodies for completing sound-related VR tasks are defined.

The rest of the paper is organized as follows. In Section 2, related work on the use of wearable vibrotactile feedback systems in VR is presented. In Section 3, we explain the study design and methodologies of our approaches. In Section 4, the experimental results of different setups of our proposed haptic VR suit are presented. In Section 5, the results of a complementary study related to the optimal number of haptic devices are presented. In Section 6, we discuss the effects, limitations, and future work of our proposed VR haptic device, and finally, we conclude the paper in Section 7.

2. Related Work

Vibrotactile feedback has been widely used in many previous studies for navigation or spatial awareness and to show different application scenarios for delivering information using vibrations through the skin. Hashizume et al. [9] developed a special wearable haptic suit called "LIVEJACKET" that can improve the quality of music experience among users when listening to digital media by using vibrotactile feedback [9]. They did not test the "LIVEJACKET" on DHH persons, but there are some other studies, such as Petry et al. [5] and Shibasaki et al. [11], that used a similar approach to improve the music experience among DHH persons.

Some other researchers have tried to deliver haptic cues from virtual environments to users' bodies. Lindeman et al. [12] implemented a system that can deliver vibrotactile stimuli to the user's whole body from virtual environments. Their proposed system could improve the immersive VR experience and the feedback time in critical situations in a VR environment. Kaul et al. [13] proposed a system called "HapticHead" that can utilize multiple vibrotactile actuators around the head for intuitive haptic guidance through moving tactile cues, and it was able to effectively guide users towards virtual and real targets in 3D environments. Peng et al. [14] proposed a system called "WalkingVibe" that uses vibrotactile feedback for walking experiences in VR to reduce VR sickness among users. Vibrotactile devices are also used in some specific gloves. G. Sziebig et al. [15] and Regenbrecht et al. [16] presented vibrotactile gloves with vibration motors to provide sensory feedback from a virtual environment to the user's hands.

A few other researchers have proposed systems that help DHH persons with sound awareness, such as Saba et al. [6], Jain et al. [10], and Mirzaei et al. [17]. Saba et al. proposed a wearable interaction system called "Hey yaa". This system allows DHH persons to call each other using sensory-motor communication through vibration. In a qualitative study, Jain et al. [10] showed the importance of sound awareness and vibrotactile wearable devices among DHH persons. In addition, Mirzaei et al. [17] proposed a wearable system for DHH users called "EarVR" that can be mounted on VR Head-Mounted Displays (HMDs) and can locate sound sources in VR environments in real time. Their results suggest that "EarVR" can help DHH persons to complete sound-related VR tasks and can also encourage DHH users to use VR technology more than before [17].

Almost all of these studies show the positive effects of using vibrotactile feedback systems in VR or the real world. However, to the best of our knowledge, none of them investigated the effects of using different setups of haptic VR suits by mounting vibration devices on different body positions of DHH persons for completing sound-related VR tasks.

3. User Study

For our study, we designed a special VR suit with four vibro-motors to demonstrate the four main directions of incoming sounds (front, back, left, and right) for deaf users (Figure 1). This VR suit can deliver vibrotactile cues from a VR environment to DHH persons' bodies. The vibro-motors are controlled wirelessly using an Arduino (Available Online: https://www.arduino.cc/ (accessed on 5 October 2021)) processing unit with a Bluetooth module.

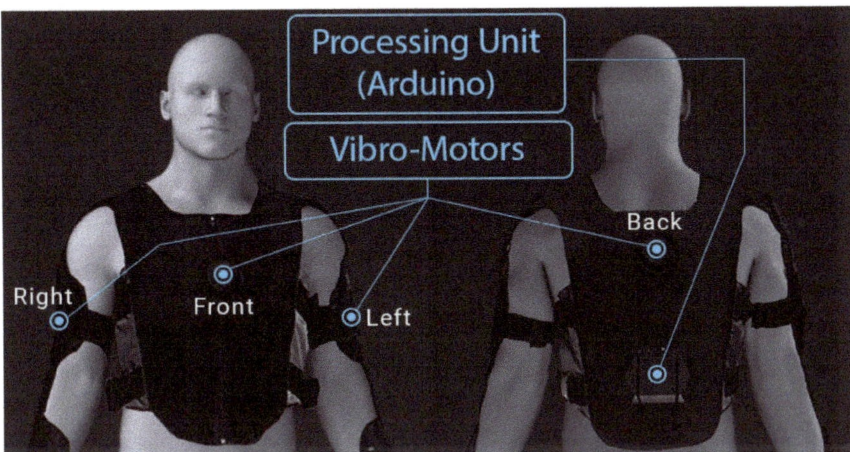

Figure 1. The concept design of our proposed VR suit.

We conducted a test to investigate the effects of mounting vibro-motors on different sections of DHH persons' bodies, such as thighs, arms, and ears. At the end of the test, we asked participants to fill out questionnaires about their preferred position for mounting the vibro-motors and the discomfort score of different setups of our proposed VR suit.

3.1. Hardware Design

An Arduino Micro Pro with an HC-06 Bluetooth module controls coin vibro-motors (10–14 mm) wirelessly from the host computer that is running the VR application. We assembled the processing unit in a mountable package on the back of the VR suit (Figure 2). The whole system is powered by a customized rechargeable Lithium-ION (Li-ON) battery with a capacity of 8000 mAh and a voltage of 7.4 V. The vibro-motors have an operating voltage range of 3 V to 4 V at 40–80 mA with a frequency of 150–205 Hz.

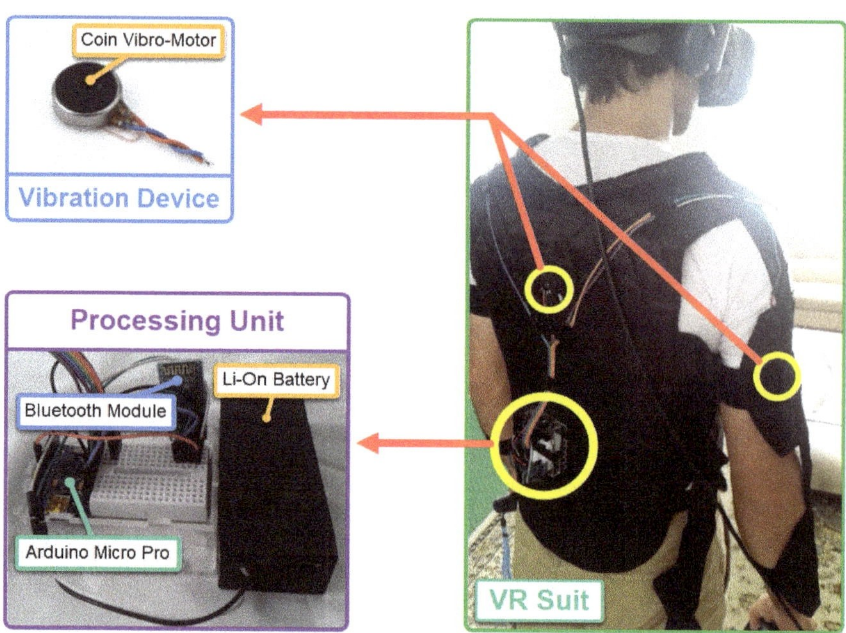

Figure 2. The prototype version of our proposed VR suit.

The flat surfaces of the vibro-motors were very close to the user's body so that the user could feel the vibrations very well. Previous studies, such as Rupert [18] and Toney et al. [19], have reported that a major problem is difficulty in maintaining good contact between vibro-motors and the users' bodies. They suggest that haptic devices (active motors) be optimally fit in their positions with an appropriate degree of pressure to ensure the perception of haptic feedback. Therefore, we designed our VR suit with special Velcro tapes to help to maintain the vibro-motors in their fixed positions on the suit during the tests. The Velcro tapes also allowed us to easily change the positions of vibro-motors on the users' bodies. We fitted the VR suit for each participant before the main test and asked them to wear thin clothes for the main experiment to ensure that they felt the vibrations from all four vibro-motors on the VR suit.

3.2. Software Design

We designed a simple VR task using the Unreal Engine 4 (UE4) (Available Online: https://www.unrealengine.com/en-US/ (accessed on 5 October 2021)) game engine with an Arduino plugin to communicate with the Arduino processing unit. Arduino Integrated Development Environment (IDE) was used to develop the code for the Arduino Micro Pro. In the VR task, the player is spawned at the center of an enclosed VR room and can only rotate around. We also added a "FRONT" label to one of the four walls in the VR room to show the front direction in the VR environment (Figure 3a). We used this label as an index corresponding to the vibro-motor mounted on the front of the VR suit.

The player had to start the task by standing in front of the wall labeled "FRONT". This procedure let us know the exact positions of the sound sources in the VR environment with respect to the user's orientation so that we could send the proper signals to correct vibro-motors on the VR suit. The player was able to start the VR task by pressing the grip button on the VR controller after standing in the mentioned position and select sound sources (speakers) by pressing the trigger button on the VR controller with the help of a ray-cast laser pointer (Figure 3b). We designed the task so that every time it was started, only one sound source (speaker) would appear randomly in one of the four main positions in the VR environment (front, back, left, and right).

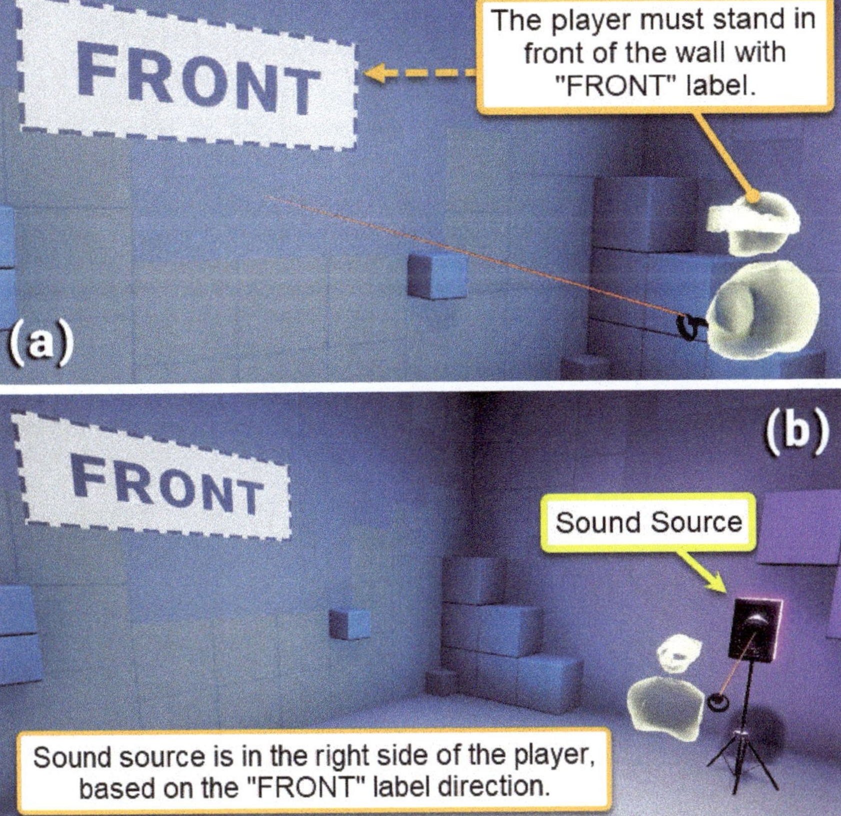

Figure 3. The VR environment, (**a**) The "FRONT" label position, (**b**) Sound source position based on the "FRONT" label direction.

3.3. Main Experiment

For the main experiment, we prepared three different setups of our proposed haptic VR suit. We used two fixed vibro-motors to represent the front and back directions but changed the positions of the other two vibro-motors in different setups as follows:

1. Setup 1: Two vibro-motors mounted on the front and back and two on the left and right sides of the thighs (left side of the left thigh and right side of the right thigh), Figure 4a;
2. Setup 2: Two vibro-motors mounted on the front and back and two on the left and right sides of the arms (left side of the left arm and right side of the right arm), Figure 4b;

3. Setup 3: Two vibro-motors mounted on the front and back and two on the left and right ears, Figure 4c.

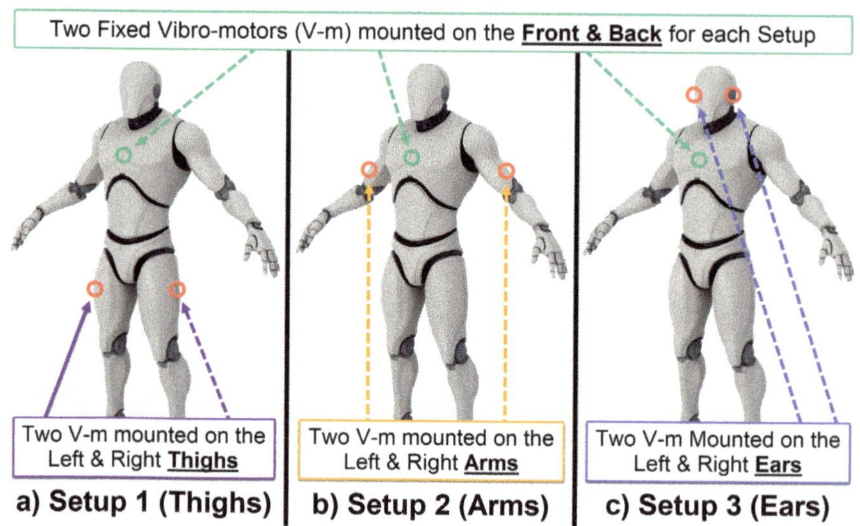

Figure 4. Suggested setups of the VR suit.

We called these setups "Setup 1 (thighs)", "Setup 2 (arms)", and "Setup 3 (ears)" respectively. We also considered an additional setup without using the haptic suit or any other assisting tool. In this condition, deaf users had to search and find sound sources in the VR environment only by using vision without the help of a haptic suit by rotating around to find the rendered 3D model of the sound source (speaker). We called this additional condition "Setup 0 (no suit)" and compared the results of using the VR suit in Setups 1 (thighs), 2 (arms), and 3 (ears) with this new condition.

3.4. Procedure

We recruited 20 DHH participants from the deaf community with no hearing in either ear (12 men, 8 women; ages: 20–40, $\bar{X} = 30.4$). We briefly introduced the goal of our experiment to the participants and described the equipment and the environment of the test. Because none of the participants had tested a haptic suit before, we asked each of them to wear our proposed haptic suit with different setups before the main experiment and asked them about their opinion in an open-ended question. Thus, the participants became familiar with different setups of the haptic suit and could withdraw from the experiment if they felt uncomfortable with any of the haptic suit setups. In addition, we carefully stated all VR safety warnings in our consent form, so all participants were completely aware of them and signed the form before the main experiment. None of the participants withdrew from the experiment, and all of them participated in our study.

For "Setup 3 (ears)", we asked the participants to answer a question about their preferred position on the head for mounting the vibro-motors. All 20 participants preferred mounting the vibro-motors inside the ears compared to the temples or behind the ears. We attached soft and flexible plastics to the ears' vibro-motors to prevent unpleasant feelings when DHH users put the vibro-motors inside their ears. Many participants mentioned in their comments that putting vibro-motors inside their ears felt like wearing headphones, similar to persons without hearing problems.

For the main experiment, we asked each participant to play each condition (setup) for 10 rounds. The player was able to start the VR task by pressing the grip button on the VR controller. After starting the VR task, one speaker appeared randomly in one of the four main positions in the VR environment (front, back, left, and right) based on the player's starting position.

Every time a speaker appeared in the VR environment, a vibro-motor related to the speaker's position would vibrate (controlled by the Arduino) so that the player could find the correct speaker position and select it. This meant that the vibro-motor on the player's chest started when the sound was in front of the user, the one on the back started when it was behind the user, and the left and right vibro-motors started based on the respective sound location. After selecting the speaker, it would disappear, and the player had to face the wall labeled "FRONT" to start the next round. This process continued until the player selected 10 speakers (completing 10 rounds).

For each setup, the completion time of every round was saved for the player. Then, the average task completion time of the setup was calculated for that player. In the end, the overall average task completion times of all players were calculated for the setups. A Friedman test with significance level of $\alpha = 0.05$ and a post hoc Wilcoxon signed-rank test with Bonferroni correction resulting in a significance level set at $\alpha < 0.008$ were used to assess statistical significance between average task completion times of the setups.

We asked participants to answer a question about the discomfort score for each setup, with the possibility of changing their answers at the end of the experiment (after completing all of the setups). To determine the preferred setup, we used a questionnaire at the end of the experiment with a 5-point Likert-scale question ("1 = most negative" and "5 = most positive") about which setup is more desirable to use, and to identify the discomfort level, we collected discomfort scores ranging from 1 to 10 (lower value = more comfortable) and calculated the preference score and discomfort level by averaging the scores of all participants for each setup at the end of the experiment (the final results). We analyzed the data for each setup based on the participants' average task completion time and their responses to our questionnaires about discomfort and preferred setup in our study.

We asked the participants to complete each setup in one day, starting with Setup 0. Starting the experiment with "Setup 0 (no suit)" and then "Setup 1 (thighs)" was important for our further study because using the sense of touch in the lower body as a cue for audio direction was very unusual among DHH participants, and they wanted to test it first (after Setup 0). Therefore, we fixed the order of setups to "Setup 0 (no suit)" on day 1, "Setup 1 (thighs)" on day 2, "Setup 2 (arms)" on day 3, and "Setup 3 (ears)" on day 4.

4. Results

The average task completion time of each setup is shown in Figure 5. Deaf users completed the VR task with average task completion times of 27.05 s for "Setup 0 (no suit)", 23.39 s for "Setup 1 (thighs)", 21.7 s for "Setup 2 (arms)", and 21.41 s for "Setup 3 (ears)".

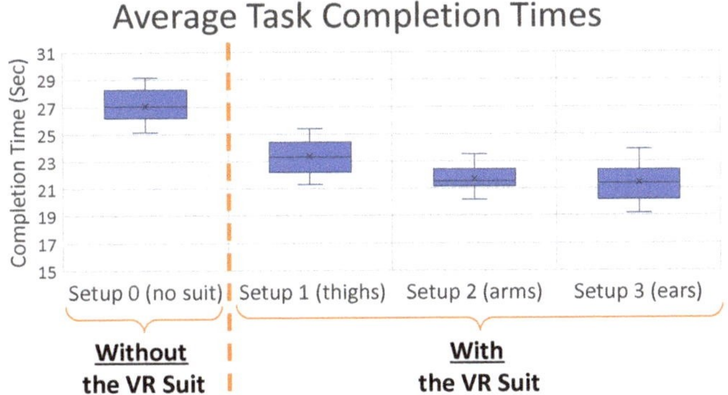

Figure 5. Average task completion times of each setup.

The Friedman test revealed a statistically significant effect of using different setups on the average task completion times ($\chi^2 = 49.020, p < 0.001$). Figure 6 shows the Wilcoxon test results for different setups. The Wilcoxon test revealed a statistically significant main effect of "Setups 1 (thighs), 2 (arms), and 3 (ears)" vs. "Setup 0 (no suit)", as well as "Setup 2 (arms)" and "Setup 3 (ears)" vs. "Setup 1 (thighs)", but no significant main effect was found for "Setup 2 (arms)" vs. "Setup 3 (ears)". These results indicate that the position of vibro-motors on the VR suit affects the task completion times of DHH users.

VR Suit Setups	Setup 1 (Thighs)	Setup 2 (Arms)	Setup 3 (Ears)
Setup 0 (No VR Suit)	$Z = -3.920$ $p < 0.001$	$Z = -3.920$ $p < 0.001$	$Z = -3.920$ $p < 0.001$
Setup 1 (Thighs)	----	$Z = -3.509$ $p < 0.001$	$Z = -3.808$ $p < 0.001$
Setup 2 (Arms)	----	----	$Z = -1.456$ $p = 0.145$

Figure 6. Wilcoxon test results on differences in average task completion times between each pair of setups.

The results of comparing Setups 1, 2, and 3 vs. Setup 0 indicate that using a haptic VR suit with different setups positively affects task completion times of DHH users in VR. Therefore, hypothesis H1 is supported. In addition, the results show that the arms and ears are preferred to the thighs for mounting the vibro-motors. Some deaf participants mentioned in their comments that feeling the sense of touch in their upper body, such as arms and ears, was more familiar to them as a warning sign to focus attention in a specific direction compared to the sense of touch in their lower body (thighs). We assume that this is the main reason why mounting vibro-motors on the upper body was more effective for completing sound localization tasks in VR among DHH users. Future studies are required to understand why the arms and the ears are better than the thighs for DHH persons.

Figure 7 shows the average responses to question about discomfort scores of Setups 0 (no suit), 1 (thighs), 2 (arms), and 3 (ears). The Friedman test revealed significant differences between different setups for the average discomfort level ($\chi^2 = 58.898, p < 0.001$).

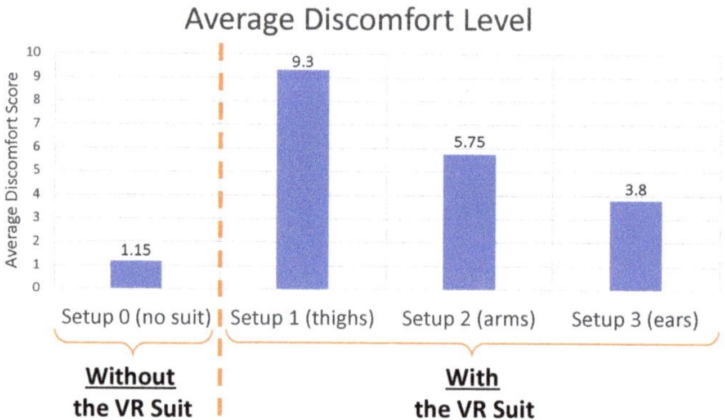

Figure 7. Average discomfort level of each setup.

Figure 8 shows the Wilcoxon test results of the average discomfort level for different setups. The Wilcoxon test (with $\alpha < 0.008$) revealed a statistically significant main effect of "Setup 0 (no suit)" vs. "Setups 1 (thighs), 2 (arms), and 3 (ears)", "Setup 2 (arms)" and "Setup 3 (ears)" vs. "Setup 1 (thighs)", and "Setup 3 (ears)" vs. "Setup 2 (arms)".

These results indicate that participants rated "Setup 0 (no suit)" significantly better (more comfortable) than all of the other setups. The results of comparing setups of the VR suit show that "Setup 3 (ears)" was rated as significantly more comfortable than Setups 1 and 2.

VR Suit Setups	Setup 1 (Thighs)	Setup 2 (Arms)	Setup 3 (Ears)
Setup 0 (No VR Suit)	$Z = -4.005$ $p < 0.001$	$Z = -3.939$ $p < 0.001$	$Z = -3.747$ $p < 0.001$
Setup 1 (Thighs)	----	$Z = -3.762$ $p < 0.001$	$Z = -3.947$ $p < 0.001$
Setup 2 (Arms)	----	----	$Z = -4.089$ $p < 0.001$

Figure 8. Wilcoxon test results on average discomfort level between each pair of setups.

Figure 9 shows the frequencies of responses to a question about the desire to use the setups. The Friedman test revealed significant differences between different setups for the desire to use the setup ($\chi^2 = 54.279, p < 0.001$).

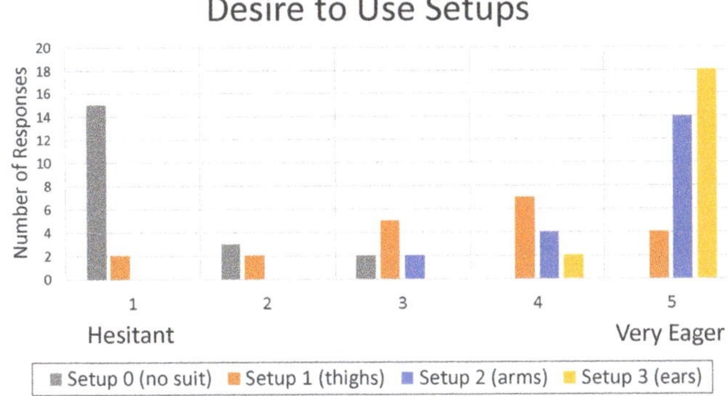

Figure 9. Desire to use the setups.

Figure 10 shows the Wilcoxon test results for the desire to use different setups. The Wilcoxon test (with $\alpha < 0.008$) revealed a statistically significant main effect of "Setups 1 (thighs), 2 (arms), and 3 (ears)" vs. "Setup 0 (no suit)", "Setup 2 (arms)" and "Setup 3 (ears)" vs. "Setup 1 (thighs)", and "Setup 3 (ears)" vs. "Setup 2 (arms)".

VR Suit Setups	Setup 1 (Thighs)	Setup 2 (Arms)	Setup 3 (Ears)
Setup 0 (No VR Suit)	$Z = -4.005$ $p < 1.113$	$Z = -4.970$ $p < 1.113$	$Z = --.176$ $p < 1.113$
Setup 1 (Thighs)	〰〰	$Z = -4.711$ $p < 1.113$	$Z = -4.756$ $p < 1.113$
Setup 2 (Arms)	〰〰	〰〰	$Z = -8.600$ $p = 1.11-$

Figure 10. Wilcoxon test results on desire to use the setups.

These results indicate that participants preferred to use at least one setup with the VR suit compared to "Setup 0 (no suit)". Many deaf participants commented to us that although setups with the VR suit (Setups 1, 2, and 3) felt a little uncomfortable because of the VR suit, these setups were very useful compared to "Setup 0 (no suit)" in VR and helped them to complete the VR task much more quickly and easily. In addition, the results of comparing setups with the VR suit show that participants preferred "Setup 3 (ears)" more than other setups.

5. Experiment with the Number of Vibro-Motors

Almost all of the participants (18 out of 20) commented that they prefer to use an assistive system in VR without a VR suit. Therefore, we decided to conduct another experiment without the VR suit and with only two vibro-motors mounted in deaf users' ears. We used soft plastic covers for each vibro-motor to minimize the unpleasant feeling of mounting them inside the users' ears (Figure 11). All participants commented that putting vibro-motors inside their ears felt like using regular in-ear headphones without any unpleasant sensation.

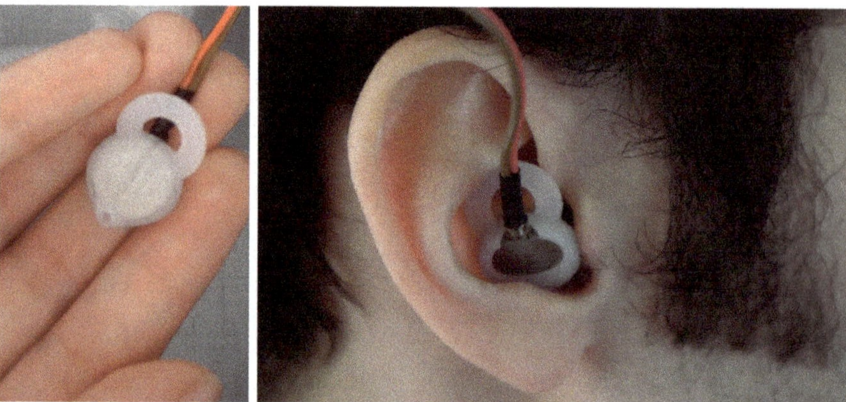

Figure 11. Mounting vibro-motors with soft covers inside users' ears.

This new condition is very similar to "Setup 3 (ears)", and we used the same VR task as in the previous experiment. The only difference was the removal of the VR suit and the two vibro-motors on the front and back. For this new condition, when the sound source was in the front, none of the vibro-motors vibrated because the player could see the sound source immediately, and when the sound source was in the back, both of the vibro-motors in the left and right ears vibrated at the same time. We intended to determine if the human brain can handle the new situation for sound source localization in VR and compared the results with the results of using the VR suit.

We called this new condition "Setup 4 (only ears)", and we tested both "Setup 3 (ears)" from the main study and this new "Setup 4 (only ears)" on a new group of participants comprising 10 DHH persons from the deaf community (7 men, 3 women; ages: 25–35, $\bar{X} = 28.3$, with no hearing in either ear) with the same procedure as that explained in Section 3.4. Each participant completed both the "Setup 3 (ears)" and "Setup 4 (only ears)" tests on one day, in random order. After finishing both tests, we asked the participants to complete a questionnaire about the discomfort score and the desire to use each setup. Then, we calculated the average task completion time, average discomfort level, and the desire to use each setup and compared the results. Figure 12 shows the average task completion time of "Setup 4 (only ears)—without the VR suit" compared to "Setup 3 (ears)—with the VR suit". Participants completed the VR task with an average task completion time of 21.058 s for "Setup 3 (ears)" and 21.389 seconds for "Setup 4 (ears only)".

We applied a Wilcoxon signed-rank test with $\alpha = 0.05$ on the average task completion times of setups. The Wilcoxon test suggested that these average task completion times were close, with no significant difference ($Z = -0.714, p = 0.475$). This result indicates that using the VR suit with four vibro-motors does not significantly affect the average task completion time of deaf users in comparison to using only two vibro-motors on the ears. Therefore, hypothesis H2 is supported.

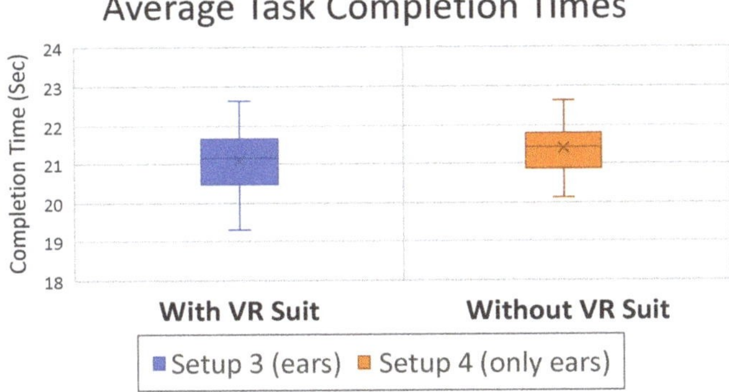

Figure 12. Average task completion times of "Setup 3 (ears)" and "Setup 4 (only ears)".

Figure 13 shows the average responses to questions about discomfort scores and the desire to use "Setup 3 (ears)" and "Setup 4 (only ears)". Comparison of the average discomfort level of "Setup 4 (only ears)—without the VR suit" with that of "Setup 3 (ears)—with the VR suit" using the Wilcoxon test showed significant differences between these two setups ($Z = -2.809, p = 0.005$); participants rated "Setup 4 (ears only)" as more comfortable than "Setup 3 (ears)".

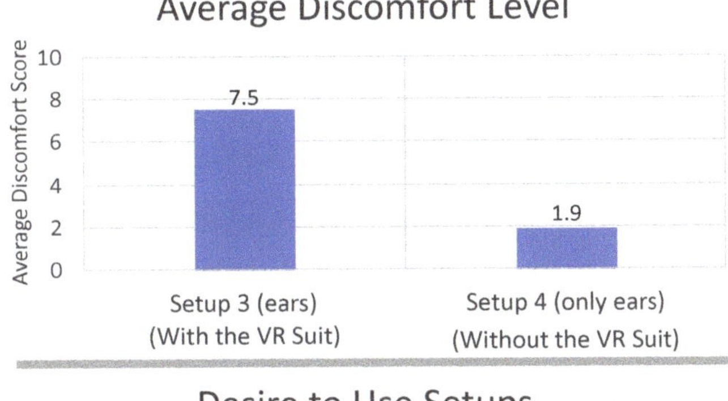

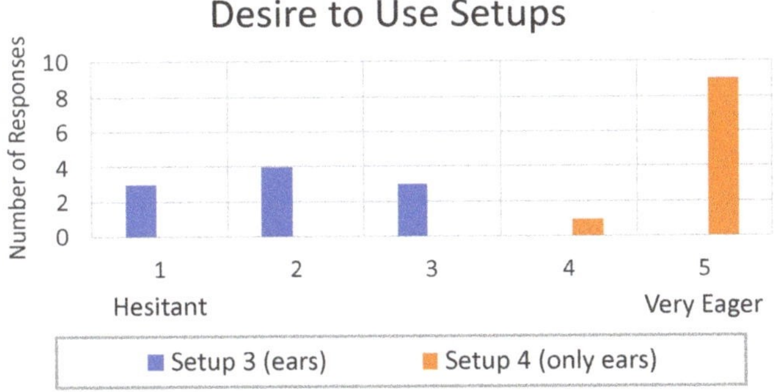

Figure 13. Results of average discomfort level and desire to use "Setup 3 (ears)" and "Setup 4 (only ears)".

In addition, the Wilcoxon test for the results on the desire to use the setups showed significant differences between these two setups ($Z = -2.831, p = 0.005$). DHH participants preferred to use "Setup 4 (only ears)—without the VR suit" more than "Setup 3 (ears)—with the VR suit" for completing sound-related VR tasks. This is an exciting result because it will help us to develop small portable VR assistants for deaf persons without using VR suits in the future.

6. Discussion and Future Work

The results of our first experiment suggest that DHH persons complete sound-related VR tasks significantly faster with our proposed haptic VR suit than without haptic feedback. In addition, the results for the discomfort level and the desire to use different setups show that DHH persons prefer mounting the vibro-motors on their upper body sections, such as arms and ears, compared to lower body sections, such as thighs, when using the VR suit. However, the results of our second experiment suggest that DHH persons prefer mounting vibro-motors on their ears when not using a VR suit. According to these results, both of our hypotheses in this study are supported, but more neurological studies are required to understand why DHH persons react differently to tactile sensations from different parts of their bodies.

In this study, the experiment was entirely new to DHH participants, so we limited the number of vibro-motors to four on a custom VR suit and limited the directions of incoming sounds to the front, back, left, and right. In addition, we only tested our suggested setups of the VR suit on the same group of participants on different days and in a fixed order. Although the VR environment randomly changed in each test for each participant, the fixed order of the setups in our first experiment may have had some learning effects on the participants. Therefore, it will be essential to randomize the order of conditions across participants in our future work.

Our study results can ultimately lead to a more efficient design and save resources and costs while providing a more enjoyable VR experience for DHH persons. In summary, our study suggests the following guidelines for VR developers who intend to design haptic devices for DHH users with a focus on sound source localization in VR:

1. When using a VR suit, DHH persons prefer mounting vibro-motors on the upper body.
2. Front and back vibro-motors (haptic devices) are not mandatory for sound source localization in VR.
3. Haptic VR suits are not preferred by DHH persons for sound source localization in VR, and they prefer compact and portable haptic devices.
4. DHH persons prefer mounting a vibro-motor inside each of their ears for sound source localization in VR.

7. Conclusions

In this study, we investigated different setups of a haptic VR suit for DHH persons that helps them to complete sound-related VR tasks. Our experimental results suggest that sound source localization in VR is not significantly faster with four haptic devices compared to two. Therefore, the efficiency of completing sound-related VR tasks is not related to the number of haptic devices on the suit. However, further studies are required to analyze more complex sound source localization scenarios in VR for DHH persons (e.g., diagonal directions for incoming sounds).

Furthermore, our complementary study shows that DHH persons prefer mounting vibro-motors in their ears when not using a VR suit. We consider this an exciting result because it can help VR developers to develop compact, cheap, and portable haptic devices for DHH users for purposes such as sound source localization in VR.

Our results suggest that using haptic devices in VR can help deaf persons to complete sound-related VR tasks faster. Using haptic devices can also encourage DHH persons to use VR technology more than before. Further studies are required to test and analyze the effects of haptic devices on sound source localization VR tasks and VR enjoyment among DHH persons. We hope to inspire VR developers to design and develop wearable assistive haptic technologies for deaf persons to help them to use and enjoy VR technology as much as persons without hearing problems.

Author Contributions: The authors in this research contributed to this article as follows: Conceptualization, M.M.; data curation, M.M.; formal analysis, M.M., P.K. and H.K.; funding acquisition, M.M.; investigation, M.M.; methodology, M.M., P.K. and H.K.; project administration, M.M.; resources, M.M.; software, M.M.; supervision, H.K.; validation, M.M., P.K. and H.K.; visualization, M.M.; writing—original draft, M.M.; writing—review and editing, M.M., P.K. and H.K. All authors have read and agreed to the published version of the manuscript.

Funding: TU Wien Open Access publication fund. This research received no external funding.

Informed Consent Statement: Signed consent forms were obtained from all subjects involved in the study.

Data Availability Statement: The data presented in this study are available on request from the corresponding author. The data are not publicly available due to privacy request from the DHH participants.

Acknowledgments: The authors acknowledge TU Wien Bibliothek for financial support through its Open Access Funding Program, and they also wish to thank all DHH participants who volunteered for this study.

Conflicts of Interest: The authors declare no conflict of interest.

Abbreviations

The following abbreviations are used in this manuscript:

3D	3-Dimensional
CI	Cochlear Implant
DHH	Deaf and Hard-of-Hearing
IDE	Integrated Development Environment
Li-ON	Lithium-ION
VR	Virtual Reality

References

1. Araujo, F.A.; Brasil, F.L.; Santos, A.C.L.; Batista Junior, L.S.; Dutra, S.P.F.; Batista, C.E.C.F. Auris System: Providing Vibrotactile Feedback for Hearing Impaired Population. *BioMed Res. Int.* **2017**, *2017*, 2181380.
2. Maiorana-Basas, M.; Pagliaro, C. Technology use among adults who are deaf and hard of hearing: A national survey. *J. Deaf. Stud. Deaf. Educ.* **2014**, *19*, 400–410. [CrossRef] [PubMed]
3. Spencer, P.E.; Marschark, M.; Spencer, L.J. Cochlear implants: Advances, issues, and implications. In *The Oxford Handbook of Deaf Studies, Language, and Education*; Marschark, M., Spencer, P.E., Eds.; Publishing House: Washington, DC, USA, 2011; pp. 452–470.
4. Karam, M.; Branje, C.; Nespoli, G.; Thompson, N.; Russo, F.A.; Fels, D.I. The emoti-chair: an interactive tactile music exhibit. In Proceedings of the CHI '10 Extended Abstracts on Human Factors in Computing Systems (CHI EA '10), Atlanta, GA, USA, 10–15 April 2010; Association for Computing Machinery: New York, NY, USA, 2016; pp. 3069–3074.
5. Petry, B.; Illandara, T.; Nanayakkara, S. MuSS-bits: Sensor-display blocks for deaf people to explore musical sounds. In Proceedings of the 28th Australian Conference on Computer-Human Interaction (OzCHI '16), Launceston, Australia, 29 November–2 December 2016; Association for Computing Machinery: New York, NY, USA, 2016; pp. 72–80.
6. Saba, M.P.; Filippo, D.; Pereira, F.R.; de Souza, P.L.P. Hey yaa: A Haptic Warning Wearable to Support Deaf People Communication. In *Collaboration and Technology. CRIWG 2011. Lecture Notes in Computer Science*; Vivacqua, A.S., Gutwin, C., Borges, M.R.S., Eds.; Springer: Berlin/Heidelberg, Germany, 2011; pp. 7–17.
7. Levänen, S.; Hamdorf, D. Feeling vibrations: Enhanced tactile sensitivity in congenitally deaf humans. *Neurosci. Lett.* **2001**, *301*, 75–77. [CrossRef]
8. Schmitz, A.; Holloway, C.; Cho, Y. Hearing through Vibrations: Perception of Musical Emotions by Profoundly Deaf People. *arXiv* **2020**, arXiv:abs/2012.13265.
9. Hashizume, S.; Sakamoto, S.; Suzuki, K.; Ochiai, Y. LIVEJACKET: Wearable Music Experience Device with Multiple Speakers. In *Distributed, Ambient and Pervasive Interactions: Understanding Humans. DAPI 2018. Lecture Notes in Computer Science*; Streitz, N., Konomi, S., Eds.; Springer: Cham, Switzerland, 2018; pp. 359–371.

10. Jain, D.; Lin, A.; Guttman, R.; Amalachandran, M.; Zeng, A.; Findlater, L.; Froehlich, J. Exploring Sound Awareness in the Home for People who are Deaf or Hard of Hearing. In Proceedings of the 2019 CHI Conference on Human Factors in Computing Systems (CHI '19), Glasgow, UK, 4–9 May 2019; Association for Computing Machinery: New York, NY, USA, 2019; pp. 1–13.

11. Shibasaki, M.; Kamiyama, Y.; Minamizawa, K. Designing a Haptic Feedback System for Hearing-Impaired to Experience Tap Dance. In Proceedings of the 29th Annual Symposium on User Interface Software and Technology (UIST '16 Adjunct), Tokyo, Japan, 16–19 October 2016; Association for Computing Machinery: New York, NY, USA, 2016; pp. 97–99.

12. Lindeman, R.W.; Page, R.; Yanagida, Y.; Sibert, J.L. Towards full-body haptic feedback: the design and deployment of a spatialized vibrotactile feedback system. In Proceedings of the ACM symposium on Virtual reality software and technology (VRST '04), Hong Kong, China, 10–12 November 2004; Association for Computing Machinery: New York, NY, USA, 2004; pp. 146–149.

13. Kaul, O.B.; Rohs, M. HapticHead: A Spherical Vibrotactile Grid around the Head for 3D Guidance in Virtual and Augmented Reality. In Proceedings of the 2017 CHI Conference on Human Factors in Computing Systems (CHI '17), Denver, CO, USA, 6–11 May 2017; Association for Computing Machinery: New York, NY, USA, 2017; pp. 3729–3740.

14. Peng, Y.; Yu, C.; Liu, S.; Wang, C.; Taele, P.; Yu, N.; Chen, M.Y. WalkingVibe: Reducing Virtual Reality Sickness and Improving Realism while Walking in VR using Unobtrusive Head-mounted Vibrotactile Feedback. In Proceedings of the 2020 CHI Conference on Human Factors in Computing Systems (CHI '20), Honolulu, HI, USA, 25–30 April 2020; Association for Computing Machinery: New York, NY, USA, 2020; pp. 1–12.

15. Sziebig, G.; Solvang, B.; Kiss, C.; Korondi, P. Vibro-tactile feedback for VR systems. In Proceedings of the 2009 2nd Conference on Human System Interactions, Catania, Italy, 21–23 May 2009; pp. 406–410.

16. Regenbrecht, H.; Hauber, J.; Schoenfelder, R.; Maegerlein, A. Virtual reality aided assembly with directional vibro-tactile feedback. In Proceedings of the 3rd International Conference on Computer Graphics and Interactive Techniques in Australasia and South East Asia (GRAPHITE '05), Dunedin, New Zealand, 29 November–2 December 2005; Association for Computing Machinery: New York, NY, USA, 2005; pp. 381–387.

17. Mirzaei, M.; Kán, P.; Kaufmann, H. EarVR: Using Ear Haptics in Virtual Reality for Deaf and Hard-of-Hearing People. *IEEE Trans. Vis. Comput. Graph.* **2020**, *26*, 2084–2093. [CrossRef] [PubMed]

18. Rupert, A.H. An instrumentation solution for reducing spatial disorientation mishaps. *IEEE Eng. Med. Biol. Mag.* **2000**, *19*, 71–80. [CrossRef] [PubMed]

19. Toney, A.; Dunne, L.; Thomas, B.H.; Ashdown, S.P. A shoulder pad insert vibrotactile display. In Proceedings of the 7th IEEE International Symposium on Wearable Computers, White Plains, NY, USA, 21–23 October 2003; pp. 35–44.

MDPI

St. Alban-Anlage 66

4052 Basel

Switzerland

Tel. +41 61 683 77 34

Fax +41 61 302 89 18

www.mdpi.com

Electronics Editorial Office

E-mail: electronics@mdpi.com

www.mdpi.com/journal/electronics